FROM
BLACK LAND
TO
FIFTH SUN

ALSO BY BRIAN FAGAN

Eyewitness to Discovery (editor)
Oxford Companion to Archaeology (editor)
Time Detectives
Kingdoms of Jade, Kingdoms of Gold
Ancient North America
The Journey from Eden
The Great Journey
The Adventure of Archaeology
The Aztecs
Clash of Cultures
Return to Babylon
Elusive Treasure
The Rape of the Nile

FROM BLACK LAND TO FIFTH SUN

The Science of Sacred Sites

BRIAN FAGAN

HELIX BOOKS

ADDISON-WESLEY
Reading, Massachusetts

Library of Congress Cataloging-in-Publication Data

Fagan, Brian M.

From black land to fifth sun : the science of sacred sites / Brian Fagan.

p. cm. — (Helix books)

Includes bibliographical references and index.

ISBN 0-201-95991-7

1. Religion. 2. Archaeology. I. Title.

BL250.F34 1998 97-42843

291.3'5—dc21 CIP

Addison-Wesley is an imprint of Addison Wesley Longman, Inc.

Jacket design by Suzanne Heiser
Text design by Karen Savary
Set in 11.5-point Bembo by GAC/Shepard Poorman

1 2 3 4 5 6 7 8 9—MA—0201009998
First printing, May 1998

Find Helix Books on the World Wide Web at
http://www.aw.com/gb/

CONTENTS

PREFACE

From Black Land to Fifth Sun describes how archaeologists use modern science to study ancient cosmologies and religious beliefs. All archaeologists analyze the material remains of human behavior in the past, using such durable finds as pottery, stone tools, building foundations, and food remains. In recent years, some scholars have turned from the material to the intangible. They ask a question that was unthinkable even twenty years ago: What can archaeology tell us about the relationships between ancient cultures and their world as they perceived it? The answer comes from the cutting edge of a still-young discipline that treads a fine line between science and the free-for-all world of imagination and pseudoscience.

Archaeology is, fundamentally, a science that relies heavily on logical observation, reasoning, and plain old-fashioned common sense. Without such common sense, the sophisticated scientific methods described at intervals in these pages would be meaningless. Archaeology is not, however, an exact science replete with irrefutable proofs. We archaeologists study people in all their bewildering diversity and sometimes maddening perversity. Inevitably, we also describe, and interpret, the past through our own cultural biases and perspectives, which means that our conclusions are only mere—although hopefully close—approximations of reality. Imagine asking a Cro-Magnon family of 15,000 years ago to describe and interpret twentieth-century France with the aid of a spark plug, a handful of cow bones, a computer keyboard, a chess piece, and the concrete foundations of a shopping mall, and you will understand the complex and challenging task that archaeologists today are undertaking—albeit in the opposite direction. By combining archaeological and scientific techniques, they are achieving remarkable success in using surviving artifacts as a mirror to reflect the intangible.

Our story begins in the late Ice Age of 15,000 years ago and ends with the Spanish conquest of Aztec civilization in A.D. 1521. Each chapter in this book examines a sacred place from the past and uses multidisciplinary science, and sometimes historical sources, to move beyond architecture and artifacts and into the spiritual world. Rather than provide a fully referenced, and distracting, narrative, I have compiled a Guide to Further Reading, which gives basic references for readers wishing to delve into the technical literature in more detail. A list of works referenced in the text also appears at the end of this book. All radiocarbon dates later than 6000 B.C. have been tree-ring calibrated; all measurements are in metric, the common scientific convention.

From Black Land to Fifth Sun was a lengthy developmental process, involving many years of traveling, site and museum visits, and discussions with colleagues in all parts of the world. I am grateful to all those friends and fellow archaeologists who have answered my questions, tolerated my presence in the field, and shown me around their sites, or, at times, entire ancient landscapes. There are so many that I cannot possibly mention them all by name. I hope they will take this acknowledgment as my heartfelt thanks. In particular, Professor Richard Leventhal of the University of California, Los Angeles, and his field colleagues were wonderful hosts at Xunantunich, Belize, and deserve a special word of thanks.

Pat Leddy undertook the laborious task of editing the rough draft manuscript, and Jack Scott drew the maps and text figures. From the beginning, my agent, Victoria Pryor, has encouraged my continuing on what has proven to be an unusually arduous literary journey. My thanks also to Amanda Cook, Jeff Robbins, and Lynne Reed of Helix Books for their many kindnesses and suggestions. Copy editor Alicia Jones achieved editorial miracles with a complex manuscript. I am grateful to the Department of Special Collections of the Davidson Library, University of California, Santa Barbara, for their assistance with some illustrations.

FROM
BLACK LAND
TO
FIFTH SUN

chapter one

THE ARCHAEOLOGY OF THE INTANGIBLE

All right, I will take you to the world of legend.
You know that time, that place well
where animals talked and walked as men,
untamed, unchanged, real people still . . .

Hilda Austin, a Nieʔkepmx elder from the Sailshan tribe of British Columbia, Canada (quoted from Swan 1994, 320)

Deep in the Central African savanna, the drums mourned the Ila chief all night. One kilometer away, at our excavation camp near the Kafue River, we slept uneasily. I tossed and turned to the insistent drumming, musing on an elder who had died full of years, with hundreds of head of cattle to his name. Only two weeks before, we had drunk beer and talked, as men do, of women and cattle, hunting, spears, and rifles. He had gazed with aged eyes across the open vista of the Kafue Flats, which extended from our camp out to the far horizon, remembering the buffalo hunts of his youth, and carefree days of driving cattle to higher ground when the river flood came. We had talked of rain and the high grasses perfect for grazing, of the right way to burn off a field to fertilize the soil. His stories brought the landscape alive—tales of lechwe antelope slain with spears, of stalking warthog in thick reeds. As he took his leave that evening, the chief said: "I have taught my sons. They live in the city, but they will come home to the cattle." I was never to see him alive again.

As the sun rose in a hazy sky illuminating a grey dawn, heavy with the smell of wood smoke from burning fields, the drums fell silent. The chief's five sons had returned home to the cattle and buried their father before noon. They were indeed city men, dressed in nice clothes and possessed of a Western veneer. Nevertheless, they danced on the grave and, as the fine dust rose, sent his spirit to the bush, thus returning their father to the land that had nurtured him and generations of Ila farmers and cattle herders before him. In a few months the chief would become an honored ancestor, one of those who had gone before—a guardian of the land.

I saw the dancing from a distance and walked across to the ancient village mound to get a better view. As I looked back I noted the deep pit of my excavation that lay dark in the brightness of midday. The stratified layers of ash and charcoal exposed village after village, extending down 3 meters from 300 years ago to a founding date more than 1,000 years in the past. Untold generations of farmers had lived and died in these settlements. The late chief and his Ila people had deep roots in the Kafue Flats.

A new moon rose that evening, a thin sliver set against a tapestry of brilliant stars. The night sky had been our companion since the excavation began. I had watched the moon wax and wane, and tracked constellations moving night by night, hour by hour, across the heavens. Some years before, the Southern Cross high overhead had guided me to distant Pacific islands. An old friend from long nights at sea, it was as familiar as the soft hills of my English homeland. I thought of the deep ancestral Ila roots in the land; of endless cycles of planting, growth, and harvest; of the verities of procreation, fertility, birth, life, and death—all measured by the passage of sun, moon, and stars in the skies. My mind turned to the Roman poet Lucretius: "Thus the sum of things is ever being renewed, and mortals live dependent one upon another. Some races increase, others diminish, and in a short space the generations of living creatures are changed and like runners hand on the torch of life" (Bailey 1947, 75).

The chief's sons had come back to his cattle. They had danced on his grave. The torch of life had passed to new hands, but the material and spiritual worlds remained unchanged as long as the people stayed on the land.

My friend the Ila chief was a famous storyteller, a master of clipped sentences, subtle emphasis, arm gestures, and nuances of facial expression. He told me of a hunt long ago, how he had stalked an impala antelope in tall grass, muzzle loader in hand. His sharply detailed descriptions and dramatic delivery brought the landscape alive—the smells and sounds of dry leaves crackling underfoot, loud as a gunshot in the windless air; the hunter freezing as the suspicious impala looked up; agonizing minutes standing absolutely still, then slowly moving forward crablike toward shelter behind a tree. Silence, stealth, unblinking concentration, moving only when the antelope grazed. A gust of wind, musket being raised ever so slowly, the antelope only a few yards away. Crack! Missed! The impala jumping high and the thump of the startled animal's hooves ringing out.

Throughout his long life, the chief had existed comfortably in the midst of a living landscape, defined for him by his ancestors and by a lifetime of experiences. He shared these events, utilizing his vast skills as a storyteller. The riches of his historical experience vanished with him, but his legacy as an ancestor endured.

The Ila chief's tale reminded me that human history takes many forms—set down in documents, passed from generation to generation by word of mouth, sometimes enduring only in material form—which is where archaeologists come in.

I am an archaeologist, a scientist who studies ancient human behavior and long-dead societies from surviving, material remains of the past. My "canvas" covers more than two-and-a-half million years, from the very beginnings of humanity to the recently vanished cultures of pre-Columbian America, and classical Greece and Rome—and even encompasses the technological achievements of the Industrial Revolution. I converse with material "voices" from the past—with stone artifacts and pot sherds, ruined buildings, broken animal bones, and minute seeds—durable legacies of long-forgotten human behavior, made and used by ancient cultures. My finds tell tales such as those of the Ila chief, but their narratives are usually incomplete, describing for the most part the business of day-to-day living. The real storytellers are long silent. The myths and chants that were an integral part of their vanished world have died with them, but the artifacts offer hints into the lives of people deeply in tune with their environments, who believed that their cosmos was filled with ancestors and spirit beings. Their existence was one in which the material and spiritual worlds came together in mythic world orders that defined the essence of human life in ways quite unlike our own. Ever since I saw the chief's sons dance on his grave, I have grappled with a fundamental question: How can archaeologists use science to recover symbolic worlds of the past, and the mythic and ritual settings that defined them? How can we bridge the gap between the tangible and intangible, to move from the material to the spiritual?

A COMMON FRAMEWORK

We are *Homo sapiens sapiens,* capable of subtlety, and of passing on knowledge and ideas through the medium of language. We possess consciousness, self-awareness, and the ability to foresee events. We can express ourselves and show emotions. Studies of mitochondrial DNA have traced the roots of modern humans back to tropical Africa between 100,000 and 200,000 years ago (see information in box, page 5). Archaeology tells us that *Homo sapiens sapiens* had settled in western Asia at least 90,000 years ago, and in western Europe (replacing earlier Neanderthal populations), 35,000 years before present.

Sometime during this ancient diaspora, these anatomically modern people developed a unique capacity for symbolic and spiritual thought; for defining the boundaries of existence, and the relationship between the individual, the group, and the cosmos. We do not know when these capabilities first developed, but late Ice Age cave art tells us that humans melded the living and spiritual worlds at least 30,000 years ago. By 10,000 years ago, when the first farming societies appeared in western Asia, human cosmology shared several common elements or beliefs, which form the framework of our scientific story.

The first common principle was that the world of living humans formed part of a multilayered cosmos, which included the supernatural otherworlds of the heavens and an underworld sandwiching the human plane. Gods, goddesses, spirit beings, and

MITOCHONDRIAL DNA AND MODERN HUMAN ORIGINS

Molecular biology has played a significant role in dating earlier human evolution and is now yielding important clues as to the origins of *Homo sapiens*. Researchers have utilized mitochondrial deoxyribonucleic acid (mtDNA) in their studies, because it accumulates mutations much faster than nuclear DNA, and is, therefore, a useful tool for calibrating mutation rates. Mitochondrial DNA is inherited only through the maternal line; it does not mix and become diluted with paternal DNA. Thus, it provides a potentially reliable link with ancestral populations. The geneticists studied the rate of mutation in mtDNA samples to determine the lineage of each sample. They determined the differences on the basis of the amount of diversity in the mtDNA, which was then used to calculate mutation rates, and thus establish the approximate age of the population. When genetic researchers analyzed the mtDNA of 147 women from Africa, Asia, Europe, Australia, and New Guinea, they found that the differences among the samples were very small. Therefore, they argued that the 5 populations were all of comparatively recent origin. There were some differences, sufficient to separate the samples into 2 groups—a set of African individuals and another comprising individuals from all groups. The biologists concluded that all modern humans derive from a 200,000-year-old African population. From this population, smaller groups migrated to, and inhabited, the remaining areas of the Old World, with little or no interbreeding with existing, more archaic human groups.

A storm of criticism has descended on this so-called "African Eve" hypothesis, most of it directed against the calculations of the rate of genetic mutation. The methodology is very new and evolving rapidly. Nevertheless, with mitochondrial data now available from some five thousand modern individuals, there is evidence that Africans display more diverse types of mtDNA than other present-day populations elsewhere in the world, which suggests that they had more time to develop such mutations.

Recent research has now moved beyond mtDNA to the use of normal (nuclear) DNA to confirm the original conclusions. An even larger database of normal (nuclear) DNA (also nuclear DNA of blood groups and enzymes), displays a hierarchy of clusters. There was a primary split between Africans and non-Africans, then a later one between Eurasians and southeast Asians. This implies that modern humans originated in Africa, then dispersed from there to split again in Asia. It is thought that the ancestral population lived in Africa between 200,000 and 100,000 years ago. Therefore, assuming a constant rate of genetic diversification, all human variation could have arisen in the past 150,000 years or so.

---

ancestors inhabited these supernatural layers of the cosmos. This universe often began as a dark sea of primordial waters, or, as Genesis puts it, a world "without form."

A second shared belief is that a vertical axis served as a support for the bowl of heaven, and linked the various cosmic layers. Many times this vertical axis was depicted symbolically as a tree. Mircea Eliade, one of the greatest religious historians of the twentieth century, stressed the importance of this *axis mundi* (axis of the world), which joined the living and spiritual worlds at a mythic center—a sacred place. Supernatural power filled the otherworlds and approached the surface of the earth at sacred locations such as caves, prominent mountain peaks, and permanent springs.

Such sacred places—natural or man-made—and the mythic landscapes associated with them, played a vital role in all societies. Eliade called them instruments of orthogenetic transformation: settings for the rituals that ensured the continuity of cultural traditions; a place where the word of the gods rang out in familiar chants passed from one generation to the next. Sacred mountains such as the Hindu Mount Meru, the Greeks' Olympus, and the Lakota Indians' Black Hills, or places such as Condor Cave in southern California's San Rafael Wilderness, famous for its rock paintings, were interchangeable points—cosmic axes where spirit beings could travel over into the natural world.

The Egyptian pharaohs erected pyramids as sacred mountains linking the domain of the sun to the realm of earth. Mayan lords built great ceremonial centers as symbolic representations of their world of sacred mountains, caves, trees, and lakes. To demolish a sacred place was to destroy the essence of human existence itself. Spanish conquistador Hernán Cortés razed the Aztec capital, Tenochtitlán in the Valley of Mexico, knowing its temples and plazas reproduced in stone and stucco a cherished and all-encompassing supernatural world.

The third common belief of human cosmology is that the material and spiritual worlds formed a continuum, with no boundary between them. An "external" landscape on earth was also an "internal" landscape of the mind, or "landscape of memory," where colors, jagged peaks, streams, groves of trees, cardinal directions, and other phenomena had spiritual associations and their places in local mythology. Usually, ancestors—those who had gone before—were the intermediaries between the living and the supernatural worlds. They looked after the welfare of the living and were guardians of the land.

A fourth element that was shared by past societies was that individuals with unusual supernatural powers, either shamans or spirit mediums, had the ability to pass effortlessly in an altered state of consciousness between the material and spiritual realms, to fly free in the supernatural world through ritual and performance.* Such "men and women of power" had direct and personal links to the supernatural world. During solitary quests they experienced visions of dots, lines, spirit animals, gods, and even ancestors. When they returned from their spiritual journeys, they painted images of their vision quests, the spirits and supernatural events they had witnessed. From these dream journeys, shamans acquired the wisdom to keep their world in balance with the sacred, and the power to influence events in the natural world. They were able to cure others, or to become sorcerers who could cause disease; they could also bring the rains, and set off factional strife and even wars.

Lastly, it was commonly believed that human life was governed by the cycles of the seasons—by seasons of planting, growth, and

*The word "shaman" comes from the Siberian word "saman," meaning someone with unusual spiritual powers.

harvest, identified by movements of the heavenly bodies. Notions of fertility, procreation, life, and death lay at the core of such a cyclical human existence. Myth and ritual played an important part in defining this world order, and allowed the material and spiritual worlds to pass one into the other as a single constellation of belief. Through poetry, music, dance, and evocative surroundings, a deep sense of a sacred order emerged.

The intangible assumes many forms, but these commonalities, observed by anthropologists and religious historians in many human societies throughout the world, provide a viable framework for scientific investigation of ancient sacred places—the settings for mythic performance (figure 1.1).

Architecture provided the setting, a powerful form of nonverbal communication that kept the messages of ritual, and of the mythic world, in people's minds. The performers reenacted the idealized roles of gods and ancestors in shrines, in plazas, and atop pyramids. Their scripts changed as the years passed; the choreographed dances twisted and turned according to ever-shifting ritual formulae.

The bright colors have since faded, hieroglyphs have eroded, and the incense has long been extinguished. No banners fly over temples; the dancers, priests, and narrators have vanished on the wind. Only the ruined and weathered settings now provide clues to the evanescent worlds of gods and ancestors. Silent burial mounds, earthworks, pyramids, and temples—once set in natural and cultural landscapes—are archaeologists' documents; the raw material for reconstructing the intangibles of the past, discourses between the living and spiritual worlds.

The Victorian archaeologist Sir Austen Henry Layard, famous for his spectacular excavations at biblical Nineveh and author of an archaeological bestseller, *Nineveh and Its Remains*, once described archaeologists as seeking what ancient societies had left behind, "as children gather up the coloured shells on deserted sands" (1849, 213). We excavate and collect artifacts, food remains, and other material relics of the past. From them, we build mental pictures of ancient societies formed from the fragments of an incomplete archaeological record. Inevitably, our reconstructions fall short of past reality. The stone circles at Avebury and Stonehenge and the courts of Knossos

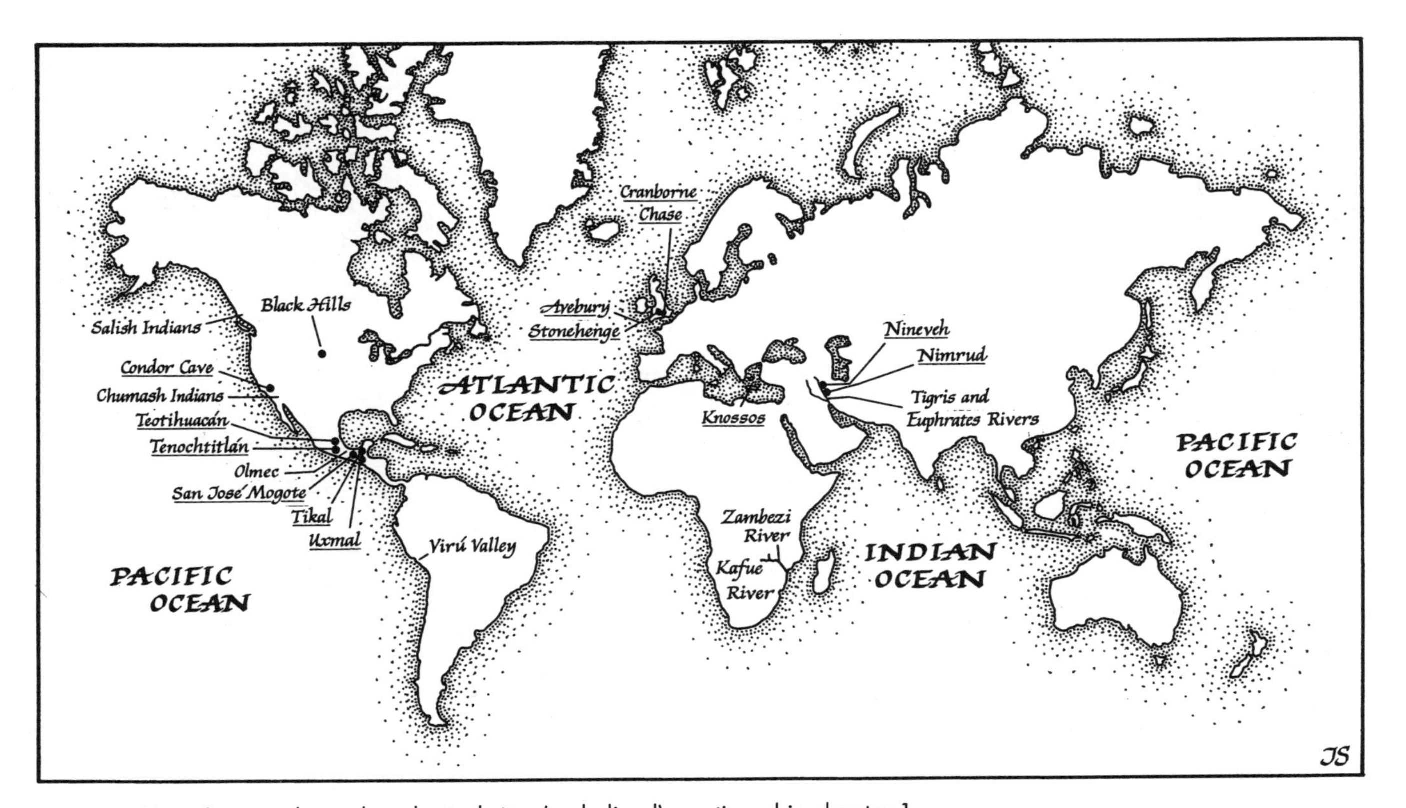

Fig. 1.1. Map showing the archaeological sites (underlined) mentioned in chapter 1.

and the Mayan pyramids are now empty stages, mere skeletal remnants of sacred places that once swirled with charismatic activity. They are Austen Henry Layard's "coloured shells"—the places where reassuring chants to the gods filled the air, where incense drifted toward the heavens, where eloquent orations and equally measured silences resonated—the transparent things through which the cosmos and the past shine. Fortunately, modern science sometimes parts the opaque curtain of the past, and leads us into vanished intangible worlds.

THE ARCHAEOLOGY OF MIND

> And the crippled Smith brought all his art to bear
> on a dancing circle, broad as the circle Daedalus
> once laid out on Cnossos' spacious fields
> for Ariadne the girl with lustrous hair.
> Here young boys and girls, beauties courted
> with costly gifts of oxen, danced and danced,
> linking their arms, gripping each others' wrists . . .
> And now they would run in rings on their skilled feet,
> nimbly, quick as a crouching potter spins his wheel,
> palming it smoothly, giving it practice twirls
> to see it run, and now they would run in rows,
> in rows crisscrossing rows—rapturous dancing.
>
> (Homer, *Iliad* book XVIII, as translated by Fagles 1990, 487)

The boys and girls in Homer's stanza seem to come alive, as if we were there ourselves to see them trace the intricate grapevine steps still performed today in circle dances. Dancing has always played a vital role in performance, at occasions both secular and ritual. So it should not surprise us that British archaeologist Peter Warren located 3 circular platforms of carefully constructed masonry in the most recent levels of the Palace of Knossos, dating to about 1400 B.C. (see chapter 10). The largest is about 8.38 meters in diameter, a solid, low platform that Warren believes was a dancing floor just like that in the *Iliad*. Could it be that Warren unearthed the very dancing floor remembered by Homer centuries later? The possibility is an intriguing one. How can we bridge the chasm between the material and intangible with only art, artifacts, architecture, and food remains to guide us? One way is to study folk dance, and look for any surviving past cultural customs in present-day rural folklore, an art that is, unfortunately, vanishing rapidly in the face of industrial civilization.

A generation ago, many archaeologists threw up their hands in despair when confronted with ancient religion. One anonymous cynic described religion as the "last resort of troubled excavators." Any unexplained artifact or structure ended up in a vaguely defined category called "ritual." About a generation ago, a small group of archaeologists challenged their colleagues to move beyond artifacts and food remains. They asked: Why should we interpret the past in terms of purely ecological, technological, and other material factors? Some of the best intellects in archaeology today are grappling with a scientific methodology for studying human consciousness, especially religion and belief. Such a methodology is critical, but there is a fine line between rigorous science and what archaeologists Kent Flannery and Joyce Marcus call "a kind of bungee jump into the Land of Fantasy" (1993, 261). The emerging "archaeology of mind" is a marriage of cultural systems theory, settlement archaeology, environmental reconstruction, contextual archaeology, and the decipherment of written records.

Cultural Systems

Since the 1960s, a new generation of archaeological thinking, developed by Lewis Binford of the University of Michigan and others, has emerged, in which human cultures are thought of as "cultural systems." Systems theory, developed in the 1950s, is a body of theoretical constructs that are used to search for general relationships in the empirical world. In systems theory, a unit is seen as being a subsystem of a much larger, more complex system. The subsystems within the larger system interact with and affect one another. Thus, to understand a system as a whole, the smaller units (subsystems), and the interactions among them, must be studied. The systems theory approach to archaeology envisages cultural systems as interacting with other systems, such as the natural environment.

Every cultural system has many interacting subsystems—technology, subsistence, art, religious beliefs, and so on—all of them reacting constantly to internal and external stimuli. Binford and his followers call cultural systems "open systems" that interchange both energy and information with their environments. Populations, societies, and cultures as a whole influence human beings who are born, live, and die within an open system. In other words, a society endures much longer than the lifetimes of its individual members.

Binford and others moved away from artifacts and artifact classification to broader studies of relationships between human cultures and their environments; of linkages between the different components of ancient societies, including religious beliefs. The systems approach assumes that human consciousness, cosmology, and religious beliefs touched all aspects of existence and cultural activity. Therefore, human intangibles became as much a part of people's adaptations to their environments as artifacts or architecture.

Settlement Archaeology

The systems approach embraced another fundamental change in archaeology: shifting focus away from excavation of single sites and on to the study of entire regions and changing ancient landscapes—settlement archaeology. Settlement archaeology began in Europe during the 1930s, but gained new impetus in the Americas after World War II. The great Harvard archaeologist Gordon Willey pioneered regional surveying with a remarkable study of the coastal Virú Valley, Peru, in 1948, where he identified entire hierarchies of human settlements, from large ceremonial centers to tiny hamlets—ancient settlement patterns that shifted dramatically across the landscape over a period of more than 3,000 years. In the 1950s, another archaeologist, René Millon, developed a long-term mapping project that surveyed the entire city of Teotihuacán in Mexico. The multiyear survey revealed a sprawling, well-planned metropolis with over 200,000 inhabitants (for more information, see box on p. 350).

These early surveys triggered more comprehensive lanscape surveys in both the Old and New Worlds. Pennsylvania State University archaeologists under William Sanders and Jeffrey Parsons spent the late 1960s and 1970s surveying the archaeological sites of the Valley of Mexico, from the earliest times up to the Spanish Conquest of A.D. 1521. Like Willey, they developed detailed site hierarchies, that were due to changes that occurred over time as different ceremonial centers and cities came into prominence and collapsed into obscurity. They documented the massive population shifts that resulted from the growth of Aztec Tenochtitlán, as thousands of people moved closer to the great capital and its swamp gardens and markets. By studying entire landscapes and changing settlement patterns over many centuries, archaeologists can reconstruct the ways in which peoples such as the Aztecs laid out their sites with reference to prominent landmarks such as caves or mountains, which had astronomical or symbolic significance.

The regional surveys of a quarter century have shown the folly of divorcing archaeological sites from their natural settings—the homelands in which they had once flourished. In recent years, archaeologists engaged in landscape survey have turned to Geographic Information Systems (GIS), computer-aided systems that are used for the collection, storage, and analysis of all kinds of spatial data, including archaeological finds. GIS is best thought of as a computer database with mapping capabilities. It incorporates computer-aided mapping, computerized databases, and statistical packages, and has the ability to generate new information based on existing data. GIS data comes from digitizing maps and from remote sensing devices such as satellites, as well as manual entry of other information of all kinds.

Italian archaeologists have used GIS technology to capture and interpret life in Roman Pompeii, which was destroyed by an eruption of the volcano Vesuvius on August 24, A.D. 79. Their database consists of information collected since 1862. These researchers use an IBM computer to digitize archaeological maps and local terrain, and to link visual representations of artifacts integrally to both detailed descriptions of each find from the city and the locations where they are found. These artifact images are linked in turn to the maps to provide detailed insights into individual houses and rooms. The enormous Pompeii database with its 50 gigabytes of detailed information about the city can be used to study such topics as the relationship between lifestyle and distribution of wealth, or to correlate fresco motifs on house walls found at one end of the site to another. GIS is used to understand the myriad relationships that tie works of art, buildings, and individual artifacts to an entire culture and community, some of which are not readily perceived by the human mind. It even permits archaeologists to create computer simulations of possible settlement patterns in elaborate forms of "what if" scenarios. GIS has enormous, and still largely unrealized, potential for the study of sacred places and ancient landscapes.

Environmental Reconstruction

Although human societies may be a part of the earth's natural history, they leave distinctive imprints on their environments. Stone Age foragers left a light impression. They were predators among many carnivores; perhaps more ruthless than others, but nevertheless as much a part of the animal kingdom as antelope or lions, even though they used tools to adapt to their natural surroundings. Farmers had a much greater impact. They drastically modified their environments within a few generations of adopting food production: settling down in sedentary villages, clearing natural vegetation, planting crops, and then moving on or rotating their fields when the soil became exhausted of its nutrients. Environmental change became even more extreme as human populations grew. Cultivators adopted more intensive agricultural methods such as irrigation or swamp farming, and cattle, sheep, and other domesticated herds caused deforestation and overgrazing. These landscapes also became part of the local mythic geography, of the symbolic world. There is a close relationship between the way people perceived and exploited their environments and their world view, which makes environmental reconstruction a vital part of the study of the intangible.

Powerful scientific tools from many disciplines help archaeologists study environmental change. For example, palynologists can identify ancient pollen grains from samples taken from bogs, marshes, and even ancient land surfaces under earthworks. They compare their pollen diagrams with identifications of seeds and actual plant remains recovered from the same levels. At times, this botanical information intersects with data from soil chemistry, identifications of tiny mollusca, and rodent bones with such precision that one can draw conclusions regarding the placement of burial mounds or other sacred monuments in specific types of environment, such as the fringes of cultivated land (see chapter 7). Beetles and other insects with distinctive habitats, such as refuse heaps, can tell the archaeologist whether an abandoned structure was used as a residence or for some other purpose. For example, a British beetle, *Ptinus fur,* thrives in buildings, food debris, and old straw. Thus, it serves as a useful marker for interpreting well-preserved, but enigmatic wooden structures.

Contextual Archaeology

Artifacts, architecture, and art set in precise contexts of time and space are the ultimate foundations of the archaeology of mind. A synthesis of data recovered from a network of objects, buildings, and other finds recorded in context can reveal much about sacred places. We can admire a plastered human skull from ancient Jericho or a finely painted vessel from a Mayan grave displayed in a museum as a work of art. Such artifacts are, however, meaningless when divorced from their original archaeological context in the ground.

The law of superposition came to archaeology from stratigraphic geology, the assumption that the layers of the earth, or an archaeological site, were laid down in succession, the lowest layers being the earliest. Careful stratigraphic observations provide one element of archaeological context.

The law of association deals with spatial relationships, with the associations among individual features, artifacts, and other finds within a single layer. In its original context, the Mayan painted pot displayed in our hypothetical museum may have lain in a royal grave, *in association with,* and thus contemporary with, the skeleton in the same sepulchre. Associative context is vital to the study of shrines and other sacred places, where the relationships between closely placed artifacts and small rooms, even benches and niches, tell a revealing story.

Careful observation of archaeological context depends on meticulous excavation, backed by careful research design. Unfortunately, many of the world's most spectacular sacred artifacts lie in private collections and public museums, ripped from their historical contexts by looters, or found before the days of rigorous excavation methods. However, with careful digging, even the most prosaic objects can reveal unsuspected ritual associations.

Exotic Artifacts. Exotic artifacts with known ritual associations can provide telling information about new ritual activities. San José Mogote was a large village in Mexico's Valley of Oaxaca in 1300 B.C. Originally, the village of thatched houses with about 150 inhabitants shared 1 larger, lime-plastered public building. During the next century, San José Mogote grew rapidly into a community of 80 to 120 households (400 to 600 people) living in rectangular houses with clay floors, plastered and white-washed walls, and thatched roofs. Public buildings built on adobe and earth platforms have been discovered. We know they were ceremonial structures because fragments of conch-shell trumpets and turtle-shell drums were excavated from them. Clay figurines of masked, costumed dancers were also found in San José Mogote's ceremonial buildings, as well as stingray spines, that were used in personal bloodletting ceremonies by all Mesoamericans for many centuries.[†]

Conch shells and stingray spines reached the highlands from the Gulf of Mexico. The appearance of similar ritual artifacts in other communities like San José Mogote is no coincidence. They arrived at a time when common art styles had come into widespread use throughout Mesoamerica.

Exotic objects have always played important parts in ritual activities, and are sometimes of great symbolic importance themselves. Such ornaments became prized heirlooms of great prestige, passed from individual to individual over many generations. This symbolic exchange of prestigious ornaments validated a huge prosaic commerce in more commonplace items by strengthening individual relationships between powerful leaders and entrepreneurs who controlled trading relationships. High-technology science provides valuable tools for studying trade and ritual exchanges. Spectrographic analysis and other methods identify trace elements in artifacts such as

[†]Conventionally, archaelogists refer to "Mesoamerica" as that region of Central America in which civilizations arose. By the same token, the Andean region is the equivalent area of South America, much of it centered on Peru.

mirrors made of volcanic glass, metal artifacts, and glass beads. This information can then be used to recognize ancient trade routes and connections between major ceremonial centers.

Architecture and Iconography. In societies where information passed orally from one generation to the next, architecture and art communicated powerful messages. As villages became towns and towns became cities and palaces, organized religion replaced the more informal ceremonial rituals of village life, and became an instrument of political power and chiefly authority. As we shall see in chapter 10, Minoan rulers on Bronze Age Crete used their sprawling palaces as settings for formal ceremonies that reinforced their spiritual and secular authority. They adorned the walls of their palaces with friezes of goddesses and male gods and symbols of bulls. Their religious artifacts formed a distinctive nonverbal vocabulary: horns of consecration, libation vessels, and sacrificial altars. Archaeologists study them as if reading picture books without captions. Sets of religious beliefs are revealed through associations—from the placement of sacred artifacts and altars in actual shrines and wall-painting depictions, from ornaments and regalia unearthed in palace rooms, and from meticulous dissection of bone fragments and ritual objects in-situ where they were once abandoned.

The classic iconographies of Mesoamerica and the ancient Andes survived for many centuries. Distinctive artistic motifs on public buildings, statues, clay vessels, and even textiles help scientists decipher the complex meanings of religious symbolism as old as civilization itself. The Olmec people of lowland Mexico used a complex symbolism of forest animals to express an enduring ideology as early as the second millennium B.C. Olmec rulers were the first in Mesoamerica to portray their dominance in carvings, paintings, and sculptures, through grotesque depictions of half-humans and half-felines. Birds, caiman, serpents, and spiders also appear on Olmec works of art, but jaguars dominate.

Fortunately, archaeologists can draw on a large repository of anthropological information from modern native American forest societies to achieve some understanding of Olmec ideology. The jaguar, which prefers to hunt for its prey in watery environments, personifies rain and fertility. The Olmec homeland with its many swamps and waterways was an environment where farmers relied on river levees and canals to control water supplies, and abounded in jaguars. In time, village leaders acquired close supernatural links to the jaguar as their means for controlling rain and floods.

Like their illustrious successors, Olmec lords were shaman-rulers, who ruled by virtue of their ability to cross effortlessly from the living to the spiritual world. Olmec artists depicted the hallucinogenic visions of these shamans: a half-human, half-feline figure with snarling mouth that may have been the Olmec rain god, and other composite creatures that melded the bird, the jaguar, and the serpent, perhaps giving birth to such mosaic deities as the mythic figure Quetzalcoatl ("Feathered Serpent"), a god associated with the wind and air, who was such a prominent part of much later Mesoamerican belief.

Ethnographic Analogy. Ethnographic analogy—the comparison between living and ancient peoples—is a long-established, if often controversial, scientific tool. Rich archives of oral tradition and vital ethnographic information still survive among the many rapidly vanishing South American groups, and have helped greatly in interpreting archaeological finds and ancient art traditions.

A century ago, Victorian archaeologists thought nothing of comparing living Eskimo cultures with entire late Ice Age societies in France, although they were separated by more than 15,000 years. Today, scientists are more cautious, using analogy at many carefully controlled levels. They make comparisons of artifacts such as arrowheads over large chronological distances, probably with reasonable credibility. More commonly, analogy employs the "direct historical method": working back from recent, well-documented societies step-by-step into earlier times. This approach has worked well with Pueblo Indian societies in the American Southwest; with historic Shona peoples in Zimbabwe, Central Africa; and with historic sites such as American Colonial villages. Ethnographic analogies become more controversial when researchers attempt to make less obvious comparisons; for example, taking data on nineteenth-century shamanism in southern Africa, and applying it to Ice Age cave art. Even in this instance, however, the approach has some validity at a general level, for there are broad, and easily demonstrable, similarities in altered states of consciousness and shamanism between ancient and modern societies in many parts of the world. Yet another example involves the revering of ancestors and the layered cosmos. Ethnographic analogy utilizes invaluable archives of information about the spiritual worlds of many living societies, to make intelligent observations about much older cultures. I rely heavily on generalized ethnographic analogy in this book, not because there are precise parallels between specific ancient and living societies, but because analogy at this level gives us *broad* insights into how people conceived of their world as it appears in the material archaeological record.

The Decipherment of Written Records

The most successful applications of scientific methodology to the intangible come from instances where contemporary written records provide amplification of archaeological records. The decipherment of the scripts of Egypt, Mesopotamia, and Mayan civilization rank among the greatest scientific triumphs of the past two centuries. With the cuneiform tablets and papyri of these cultures at our sides, we are able to experience the voices of the past: their thoughts, their prayers, and their beliefs. These writings speak with authority, but require cautious interpretation, coming, as they do, from societies where communal rather than individual thought and action were the norm. Text-aided archaeology requires meticulous critical skills and a healthy awareness of the limitations of ancient documents. For instance, Egyptian hieroglyphs recite the standard litanies of worship and adoration of kings. They flatter and glorify, presenting an ideal world peopled by a masterful, all-powerful pharaoh. However, behind the sycophantic phrases and ritual formulae lie general patterns of religious belief and cosmology that provide a framework for broader scientific interpretation. Utilizing other writings from the same time period, we can reconstruct the general significance of the pharaohs' earliest pyramids, and unravel the complexities of Egypt's first unification 5,000 years ago.

The decipherment of ancient Mayan writings has transformed our perceptions of their civilization. Earlier researchers had assumed that the Maya were peaceful astronomers and priests who were obsessed with the heavenly bodies and the measurement of time. In dramatic contrast, deciphered glyphs reveal a warlike, bloodthirsty society of ambitious, competitive lords vying for power and prestige by every means possible. Mayan inscriptions lay out political histories, set down royal genealogies, and preserve vast amounts of astronomical and cosmological information. The latest multidisciplinary Mayan research shows the full potential of archaeology and glyphs. By using Mayan inscriptions, researchers are beginning to unravel not only the intricacies of Mayan cosmology and religious beliefs, but also the significance of the orientation of individual buildings at major ceremonial centers. A husband and wife team, Harvey and Victoria Bricker, have used astronomical glyphs from an inscription over the central doorway of the classic Mayan Palace of the Governor at Uxmal in Mexico's Yucatan to show how the ruler Lord Chac ordered the carving over his palace door of a zodiac band that replicated the sky at the time Venus rose at its southerly extreme in A.D. 910. The glyphs depict the lord in his role as Skybearer supporting the heavens, enthroned on a building oriented toward a Venus extreme, about 28 degrees south of west (242 degrees). Such research demands a daunting array of archaeological, astronomical, and epigraphic skills.

The voices of ancient people also come down to us in small bits and pieces, set down by anthropologists, explorers, missionaries, and native peoples themselves. The more complete chronicles have priceless value. One such document, the sixteenth-century Franciscan missionary Bernardino de Sahagun's *General History of the Things of New Spain* (Anderson and Dibble, vols 1–12, 1954–1969) almost singlehandedly preserved the lore of Aztec civilization for posterity. Navigating such accounts, however, requires extraordinary care. The researcher has to develop an awareness of the subtle nuances of native tongues from cultures where all knowledge was transmitted orally. For example, Aztec authority came from an ability to speak in public—from oration, song, and chant—words that created a vision of the world. Therefore, in order to document Aztec culture, de Sahagun had to record mnemonic accounts of early Aztec civilization that were drilled into the children of nobility in state schools, setting them down in writing for the first time. (The Aztecs employed codices, a type of picture script, as aide-mémoire for formal recitations, but lacked a formal script like that of the Maya.)

Spanish and other colonial documents provide useful information on early Andean and Mesoamerican societies, but require careful analysis and interpretation, since they were written by people with their own political, religious, and racist biases. However, under careful scrutiny, archives such as the Jesuit *Requirements* (official reports of missionary activity) can be found to contain important ethnographic data on populations and village distributions, and sometimes discussions of indigenous ceremonies and rituals.

Sometimes, controlled use of historical documents and archaeology produces extraordinary results. When archaeologist Olga Linares studied a cemetery of high-ranking people in central Panama, she drew on sixteenth-century Spanish accounts of local chiefdoms engaged in constant warfare and raiding, and on detailed information of local animal species. The graves contained flamboyantly decorated polychrome vessels—open pots designed to be seen from above—where mourners could glimpse the animal motifs painted upon them. The pots were considered so valuable that sometimes they were exhumed from one grave and placed into another. Historical accounts mentioned that the highly competitive chiefs vied with each other for leadership and prestige. They tattooed and painted their bodies with badges of rank and bravery. Each group of warriors wore different symbols associating them with their leader. They went to their graves with helmets, weapons, and painted pottery.

As she excavated the cemetery, Linares studied the graves with sixteenth-century Spanish documents that described the culture and modern information on animal species by her side. Both sources amplified the archaeological data a great deal. She observed that the art styles rarely depicted plants. But many animal species commemorated aggression and bravery. Dangerous beasts like crocodiles, large felines, sharks, stingrays, scorpions, and poisonous snakes often appeared on funerary vessels. The dead lay with shark's teeth and stingray spines adorning their bodies. The Panamanians never commemorated prey animals or innocuous species like monkeys, but instead used carefully selected animals to communicate the qualities most admired in chiefs and warriors.

The archaeology of mind will never be an easy undertaking. Pursuing the intangible requires large data sets, excellent preservation, and sophisticated theoretical models. The pursuit is often frustrating, sometimes unexpected, and relies heavily on a broad array of scientific methods from dozens of scientific disciplines, among them botany, nuclear physics, and zoology. Australian historian Inga Clendinnen, herself the author of a notable book on the Aztecs, calls us "Ahabs pursuing our great white whale." She adds: "We will never catch him . . . it is our limitations of thought, of understandings, of imagination we test as we quarter these strange waters. And then we think we see a darkening in the deeper water, a sudden surge, the roll of a fluke—and then the heart-lifting glimpse of the great white shape, its whiteness throwing back its own particular light, there on the glimmering horizon" (Clendinnen 1991, 275). So it is with her guiding light for encouragement that we embark on such a scientific journey, beginning in the late Ice Age world of 15,000 years ago, and ending with the glittering Aztec capital, Tenochtitlán, in A.D. 1521.

chapter two

DARK CAVES, OBSCURE VISIONS

Caves are dark, mysterious places, where daylight does not exist. Some of the oldest sacred caves lie in the Dordogne region of France, where deep cliffs streaked with dark lines and furrows overlook placid rivers and lush water meadows (figure 2.1). For thousands of years, people have entered caves to feel the power of the earth and gain access to the enigmatic spiritual world. In pitch-black chambers, they confront the unknown and experience revelation as hallucination.

In the past, the cave functioned as a primordial sacred place, with its passages leading to the dark unknown—to the very bowels of the earth. Caves also served as the entrance to the spiritual world, the gateway through which shamans journeyed and flew through the domains of the otherworld. The art of being a shaman involved controlling altered states of consciousness within society. In the flickering darkness, with rhythmic drumming and bone flutes and other instruments imitating animal and bird sounds, the "multimedia effect" may have been mesmerizing, as human chants resonated off the cave walls.

Shamans in search of visions (and novices determined to experience them for the first time) separated themselves from others and penetrated deep into the silent recesses of caves. Here, they controlled the range of images by reducing the personal element and focusing the nervous system in the direction of specific mental imagery, seeking altered states of consciousness that provided their visions. After a time, the quester's visions may have projected themselves on

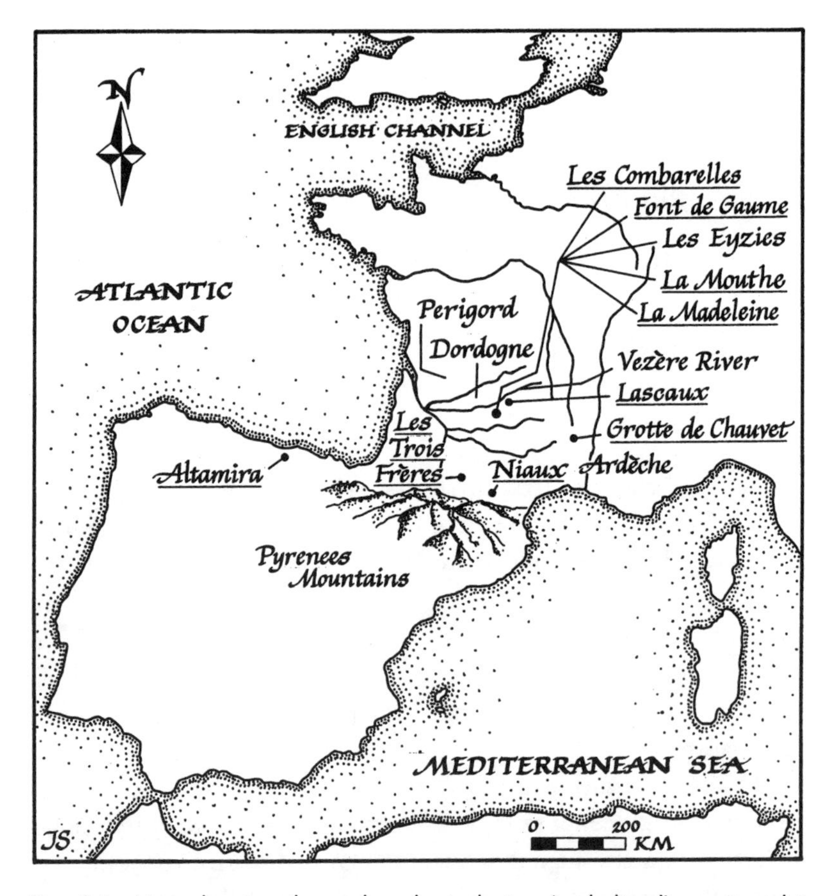

Fig. 2.1. Map showing the archaeological sites (underlined) mentioned in chapter 2.

the cave walls like a slide show. Perhaps, as scientific analyses of paintings in Niaux Cave in southwestern France suggest, the quester then made a hasty sketch, either while in a trance, or after returning to normal consciousness. The resulting painting or engraving was a recreated vision, a living, breathing work of art.

Other cave art, such as the large, impressive images in the famous bison chamber at Altamira in northern Spain, may have performed a special function. People hallucinate what they expect, or have been led to expect. Thus, the art may have been placed close to the entrance of this chamber in order to prepare vision questers for the hallucinations they would experience deep underground. And in the depths of the newly discovered Grotte de Chauvet, it is possible that the bison-skin-clad shaman waiting on the wall near a great frieze of dangerous animals is watching the entrance where the audience will approach.

It is difficult to decode the full meaning of Cro-Magnon art, separated as it is from us by more than 3,000 centuries.* Anatomically modern and intellectually more developed than their famous predecessors the Neanderthals, the Cro-Magnons were some of the first artists in the ancient world. Thanks to the devoted labors of nineteenth-century anthropologists and inspired archaeological research a century later, we have been able to lift a corner of the veil of passing millennia and enter a Cro-Magnon world where powerful images and dark caves allowed shamans to pass in trance from the living realm into a potent spiritual world. Perhaps such journeys were at the very core of human existence some 30,000 years ago.

LES COMBARELLES AND LASCAUX

The Vezère River flows through a land of deep limestone gorges, lush water meadows, and dense woodland. In late Ice Age times, 200 millennia ago, Europe was in a deep freeze, but the narrow gorges with their high cliffs and huge rock shelters were an oasis of pine and birch forests for animals and humans alike. The Cro-Magnon people who lived here dwelled in a circumscribed environment of river valleys, sheltered and hemmed in by high cliffs with a segmented view of the sky overhead. Theirs was a world of horizontal layers set in a vertical plain: the sky above, the valley floors and uplands, and, below the

*Many groups with different material cultures inhabited late Ice Age Europe, so I use the generic term Cro-Magnons here rather than confuse the reader with a proliferation of technical, cultural terms. The Cro-Magnons (named after a rock shelter of that name found during the construction of the Les Eyzies railroad station in 1868) were round-headed, anatomically modern people, who flourished in western Europe during the late Ice Age, between about 32,000 and 11,000 years ago. Their genes still contribute to modern Europeans: recent DNA studies have confirmed a genetic link between a living Somerset man from southwestern England and a 9,000-year-old skull from a cave in the same region.

ground, an unknown, and potent subterranean world accessible only by dark caves. For thousands of years they painted and engraved on the uneven walls of these sacred places.

I first visited Les Combarelles in 1955, when only a handful of visitors passed through the cave every summer's day. Famous for its 20,000-year-old Cro-Magnon wall engravings, Les Combarelles lies in the corner of a limestone cleft, which opens into the woods above the Vezère River, near the small village of Les Eyzies in southwestern France. Les Combarelles is a deep cave, with small, awkward passages that twist and wind for hundreds of yards underground. Only a few people can explore it at a time.

The guide and six of us walked along the narrow passage deep into the earth, our only light delineating the outline of the guide's shoulders. At times we squeezed through narrow passes, or bent almost double as the roof hung low above us. A hissing acetylene lamp cast yellowish shadows on the walls and threw the dark folds in the rock into high relief. Our guide's soft voice warned us of low ceilings. An oppressive stillness pressed on my ears. I thought of the dripping June world outside, where cascades of mist swirled around the high limestone cliffs and trees in the gloom. Two hundred yards from the entrance, we discerned walls covered with intricate engravings. Mammoths, wild oxen, horses, and reindeer danced in the soft acetylene light.

Held close to the wall, the flickering light shone on the incised lines of a wild aurochs. In the yellow luminescence the figure became so real that for a moment I imagined myself in the midst of a bellowing Ice Age herd in an open valley, eyeing a huge, pawing male about to charge. Its head crouched in defiant profile, the lines of the engraving shimmered. As the aurochs charged, the light brightened and moved to another figure. Then suddenly, the guide turned out the lantern. A few exclamations were heard and then, as we stood in the dark chamber, except for occasional soft drips from the ceiling, complete silence. The darkness was so intense I could feel the weight of it: the silence of death, of unexplained forces casting their compelling spell in this subterranean place. Bright flecks of light passed in front of my eyes. Ocher, yellow, and black images pressed in on me. Pulsating dots became lines—changing shape, coming alive.

Moments later, the soft hiss of acetylene impinged on my consciousness. Transitory images faded into the background, and I was back in the twentieth century.

I was lucky to see Les Combarelles by carbide light. Such lanterns, like flickering candles or pitch torches, heighten the illusion of powerful animals lurking in the gloom. Today, electric lights illuminate the once-dark passageways. Les Combarelles has become a static display, an Ice Age art gallery frozen in time, not a place of living performance and gyrating lights as it was when the ancient artists wended their cautious way through the defile, armed only with small animal fat lamps and firebrands.

A few days later, I visited Lascaux Cave near Montignac, the greatest of all the painted caves near the Vezère River. An iron door protected the paintings. Inside, electricity bathed the chamber in a soft, uniform light, highlighting the white, yellows, and grays of the cave walls and reminding me of the flickering acetylene light at Les Combarelles. Four small boys who were hunting rabbits found Lascaux in 1940, when their dog became trapped in an underground cavern. They fetched a rope and, to their astonishment, found themselves in a chamber adorned with huge, prancing bulls, bison, horses, and stags. *Bos primigenius*, the primordial aurochs, galloped along the walls, great horns curved high (figure 2.2). This nimble, fierce beast was formidable prey for ancient hunters.

When I first saw the two black and red bison in Lascaux facing each other, and the row of stags, their heads slightly raised, swimming across a stream, I felt thunderstruck. The powerful, haunting beauty of the yellow-brown wild horses prancing in abandon and the intricate engravings of exuberant animals covering the walls looked vibrant and fresh—even after 15,000 years; it was an awesome sight.

Then and now, people call Lascaux the "Sistine Chapel of Prehistoric Art," the ultimate masterpiece of ancient artistry. Perhaps it was the crowds and the soft electric lights, but it struck me how, for thousands of years, late Ice Age men and women had visited Lascaux's painted galleries, as we were doing. But they did not come to admire the paintings. They came for more profound, and still not thoroughly understood, reasons. Were the painted caves used for initiation rites and communal ceremonies? Or did shamans venture

Fig. 2.2. One of the Lascaux aurochs (*Bos primigenius*). Photograph from Artworks.

alone deep into their remote passages in quest of revelation and enlightenment?

Seven years later, when I returned to Lascaux early in the tourist season, I was there almost alone and had more time to explore. The main chamber with its expanses of more-or-less smooth walls had made an ideal surface for painting large animals like the aurochs. During this visit I noticed greater detail: deftly executed manes, the use of soft hues, tilted heads and well-defined eyes, and the expert use of bulges in the rock to accentuate relief. The more I looked at the animals honed over by generations of repetitious painting, the more the paintings assumed a power over me, the experience becoming much more than merely a visual one.

Today, the painted caves have become prehistoric art galleries, sanitized and illuminated in the name of mass tourism. Now a visitor to Lascaux has to be content with an exact replica built nearby. I had been lucky to have experienced the cave paintings moving in darkness, picking out the images with candle or flashlight. The scale of the cave, its dimensions, and the darkness focus your attention on the images' menacing power—the animals come alive, set amidst stage props of dark, light, and shadow.

The very act of painting and engraving reveals the close identification of these people with their natural world. The sight of the magnificent bulls, which seem to shimmer and move, makes one wonder what Lascaux had been like when aurochs were still alive, when shamans chanted and people crowded close together in half-darkness surrounded by depictions of their prey. What compelled Cro-Magnon artists to penetrate deep underground for their canvases? For generations, archaeologists have grappled with this question and have utilized various research methods in their attempts to answer it.

COPYING CRO-MAGNON ART

Scientific research into Cro-Magnon rock art is an immense challenge to archaeologists' ingenuity and scientific abilities. The challenge begins with the need to make accurate, scientifically neutral copies for comparative purposes.

A century ago, researchers had only rudimentary photographic technology, so they relied on their artistic skills. The legendary French archaeologist and Catholic priest, Abbé Henri Breuil, began his copying in the early years of this century, and became the doyen of Cro-Magnon rock art studies for more than 50 years. Slim and agile, he often worked on his back, wedged into narrow defiles for hours on end, tracing faint images onto florist's or rice paper (the most translucent drawing surfaces then available) taped to the rock or held by an assistant, who was roundly cursed if he fidgeted. At Altamira, Breuil lay on sacks filled with straw and ferns with only candles to illuminate the bison paintings above him (figure 2.3). Since these cave artists had used a pasty pigment, he was not able to place paper directly on the figures. Instead, he made rough sketches, then measured the figures. At first he tried watercolors as his medium, but the paint would not dry in the damp atmosphere. He then switched to crayons, but had no black, so he lit a fire outside the cave and equipped himself with burnt sticks and corks. All the time, inquisitive

Fig. 2.3. An Altamira bison. Photograph from Scala/Art Resource, NY.

Spaniards bombarded him with questions. Later, Breuil worked his tracings and sketches into full watercolors, using black-and-white photographs to check the details.

As an artist, Breuil thought in terms of friezes. The Cro-Magnon artists approached their work differently and for other-than-artistic reasons, using the same rock "panel" again and again. Many painted surfaces became palimpsests for mazes of animal figures and lines, indecipherable signs, swirls, and meandering dots. Only occasionally did human figures appear; one of them, a dancing man wearing a reindeer headdress, is found at Trois Frères Cave in the Ariège (figure 2.4). What, then, did the art mean? This question has intrigued scientists for a century.

Breuil's paintings are superb renderings, even if modern research shows some of his reconstructions to be somewhat imaginative. Like all artists, he had his own distinctive style, which tended to make figures from different sites look more similar than they actually are. His paintings have been reproduced again and again, to the point that modern researchers visit these sites with preconceived notions of what cave paintings look like.

Fig. 2.4. Abbè Henri Breuil's rendering of a shaman wearing a reindeer headdress, from Trois Frères. Photograph from Musée de l'Homme, Paris.

In contrast to Abbé Breuil's inadequate lighting and materials, today's rock art copyist has an arsenal of vastly superior materials at hand—supple plastics and acetates, and many forms of pens and markers. Direct tracing is now scientifically unacceptable, so artists sometimes set up a sheet in front of the wall and trace from a short distance away, all the while checking their renderings with photographs and measurements. (Microphotographs taken many years later show how Breuil's direct tracings slightly damaged some of the paintings. He even flicked dust off paintings with his handkerchief!)

The advent of color photography and the 35mm camera after World War II allowed systematic recording of entire friezes in a fraction of the time needed by an artist. Individual figures or friezes can be enlarged to full size, then traced before being checked against the original. Photography and tracing complement one another, for they allow accurate recording without impacting the images, and also enable one to disentangle figure superimpositions by tracing each separate image from the photograph, often under laboratory conditions. Color photography also allows for the statistical recording of paintings throughout entire cave systems within economic time frames. This approach was the basis for André Leroi-Gourhan's quantitative analyses of Cro-Magnon art (described later in the chapter).

Usually photographs are taken perpendicular to the wall, but photomontages of paintings or engravings in narrow defiles or other awkward spots are often used. Today's researchers use both color and black-and-white film, diverse light sources such as lamps and electronic flash, and filters to enhance contrast. Infrared film or light makes red ochers transparent, to the point that you can see other pigments under red figures and even identify different paint mixes—which is extremely beneficial when reconstructing the sequence of painting. Ultraviolet (UV) radiation causes calcite and living organisms on cave walls (but not manganese and other ochers) to fluoresce, which allows the researcher to assess damage to paintings caused by wall growths. Thus, a UV light source, whether from a lamp or film, can highlight painting detail buried under calcite layers. Ultraviolet lamps were invaluable during a recent research project at Niaux (also described later in this chapter).

Today, the process of recording goes hand-in-hand with accelerator mass spectrometer radiocarbon dating (see section on p. 39), direct observation, and analyses of paint composition. The photographer employs carefully controlled light—and, at times, infrared film—to photograph friezes and figures from different angles, and to identify faded portions of the paintings. The scientist also uses a binocular microscope for close-up views of the painting, especially when issues of homogeneous composition and paint analysis are involved.

Other new processes include casting technologies that use elastomer silicones and polyesters—quick and easy-to-apply materials that produce exact, durable casts of even the finest engraved lines on cave walls. The casts are readily carried away for study in the laboratory, and can easily be displayed to the public. The researchers can coat the cast surface with a mixture of ink and water that is then wiped away to expose engraved lines, a technique that is, of course, unthinkable with regard to original paintings.

Today's researchers are also using computers to scan photographs and then enhance and fine-tune the pictures for accuracy. Digital storage of photographs is also coming into use, and photogrammetry, the same technique used for generating contour maps from aerial photography, offers the prospect of contoured, three-dimensional images of clay figures and other delicate art works. Since many of these approaches are still fairly new, their full potential is yet to be seen.

PAINTINGS AS ARTIFACTS: "SYMPATHETIC HUNTING MAGIC"

Miles Burkitt was the first archaeologist I ever met, as a freshman at Cambridge University in the mid-1950s. Master raconteur and teacher extraordinaire, Burkitt had enthused generations of undergraduates about archaeology. In his youth before World War I, he had explored the Vezère Caves alongside archaeologists like Breuil, when the paintings were fresh and had not suffered from a surfeit of visitors.

A letter of introduction from Miles Burkitt enabled me to meet Abbé Breuil in the museum at Les Eyzies, high above the Vezère River. He was short, gruff, now stout, and rarely without a cigarette in his mouth. I had been warned that Breuil did not suffer fools gladly and was well set in his archaeological ways. In the early years of this century, he had been one of those who had developed a cultural sequence for late Ice Age cultures in southwestern France and northern Spain based on numerous rock shelter excavations and the different artifact forms found in successive cultural levels in

Cro-Magnon sites. Like his contemporaries, he tended to think in linear terms, of late Ice Age cultures that followed one another in a ladderlike, almost geological sequence. He himself had subdivided 30,000 years of French Stone Age culture into different cultural stages using antler spear points and barbed harpoons as "type fossils."

We started in the museum study collections, sifting through drawers of artifacts from the nearby La Madeleine rock shelter, which was occupied by people living on the Vezère's banks more than 12,000 years ago. His pudgy hands flitted from antler harpoon to antler harpoon, as he categorized them item by item, slipping them into a six-part subdivision of the "Magdalenian culture," which he had devised as long ago as 1906. "Magdalenian IV," he pronounced, fingering a short, forked-based antler spear point. A moment later, he pointed to a barbed antler harpoon: "Magdalenian VI." I peered into a drawer of double-barbed harpoons with fine lines engraved on the barbs. I wanted to savor each delicate artifact, but Breuil moved on, secure in his classifications. It was as if, in his mind, the Magdalenians had been factory workers, stamping out hundreds of identical double-barbed harpoons, each conforming to a "Magdalenian culture" specification.

Breuil's briefing was an archaeological tour de force. I wondered privately whether he thought of the paintings in similar impersonal and linear terms. I soon found out. We visited the Font-de-Gaume Cave near Les Eyzies, famous for its frieze of mammoths with long, curving tusks. To show me the masterpieces, Breuil used a flashlight to amplify the electric lights, shining it at an angle to decipher the superpositions for me. I saw two engraved and painted reindeer head-to-head, the doe kneeling. This time, there were no tricks with lights. I was reviewing a scientific, linear analysis of changing styles, pointed out with a walking stick. On this occasion, Abbè Breuil the archaeologist overruled Abbè Breuil the artist. When I asked him about the meaning of the art, he stated that they were "Sympathetic hunting magic"—paintings created to ensure the success of the hunt and the continued abundance of game and animal life.

Breuil and his contemporaries viewed late Ice Age art in the same way that they viewed late Ice Age cultures—as a single continuum. The caves and rock shelters of southwest France and northern Spain contained layer after stratified layer of late Ice Age occupation, each with its own characteristic artifacts (table 2.1). Archaeologists of the day believed there had been a uniform, orderly development of Cro-Magnon culture, in which society followed society with all the regularity of geological strata (see information in box, p. 37). Breuil believed he could trace the evolution of Cro-Magnon art in the same way, by deciphering the superpositions of paintings and engravings in major caverns like Altamira, Font-de-Gaume, Lascaux, and Niaux. He argued that the earliest and crudest paintings and engravings would be at the base of the friezes and stated that Cro-Magnon art began with the Aurignacian people (see table 2.1). They executed engravings and outline paintings in which the necks, bellies, and other anatomical features of the animals were grossly exaggerated. Their heads and feet were often turned toward the viewer. After many thousands of years, according to Breuil, these artists became expert at depicting their prey, especially on small objects. They used polychrome hues and shading, and even modeled in clay to produce bas-reliefs and small sculptures.

Clearly, Breuil thought of the paintings as enigmatic artifacts, to be analyzed and treated in much the same way as the stone and antler tools found in the dense occupation layers in Les Eyzies's rock shelters. Faced with the seemingly impossible task of explaining the meaning of Cro-Magnon art, Breuil instead turned to familiar archaeological research tools of artifact classification, such as the frequency of animals in different caves—the most common animal being the preferred prey. In so doing, he missed a fundamental point: Tools and art are pieces of a single continuum, in that they comprise a part of the entire constellation of ancient cultures in which artifacts, painting, ritual, and the daily business of obtaining food were all integral components of living.

A generation later, we have moved a long way from the paintings-as-artifacts perspective. Careful observation, high-tech photography, and computer inventories of paintings at dozens of sites provide comprehensive data for studying the placement of the art within the caves. They also reveal many inconsistencies in the Breuil "hunting magic" theory. For instance, in the case of the famous Niaux Cave in southern France's Ariège, bison are the most common

TABLE 2.1: CRO-MAGNON CULTURAL SEQUENCE IN WESTERN EUROPE

Date in Years	Culture Name	Named After	Characteristic Artifacts
ca., 32,000	Aurignacian	Aurignac rock shelter, Ariège, France	Notched and sharpened blades Split-based antler points
ca., 32,000	Perigordian	Perigord, France	Fine blunt-backed stone points Antler/bone points
ca., 23,000	Solutrean	Solutré rock shelter, Perigord, France	Leaf-like stone points
ca., 19,000 to 10,000	Magdalenian	La Madeleine rock shelter, Dordogne, France	At least six stages or variants. Rich antler/bone technology. Many specialized tools

The original classification of Cro-Magnon cultural traditions developed by Breuil and others was entirely linear. During the 1930s, the linear sequence gave way to a more elaborate scheme of two parallel cultural traditions, the Aurignacian and Perigordian, which came together in the late Ice Age to form the famous Solutrean and Magdalenian cultures. This branching sequence is the basis for the present cultural framework shown in table 2.1, which has been greatly elaborated due to more sophisticated excavation methods, and statistical analyses of artifacts from individual occupation layers. Modern researchers have categorized these cultural traditions into numerous subdivisions—at least eight for the Magdalenian alone—each based on artifact forms and statistical groupings of stone and antler tools. Whether these have any cultural significance is a matter of continued debate.

animals on the walls (figure 2.5). The people who visited Niaux lived at the La Vache Cave across the valley, where ibex bones are more commonly found than those of bison, thus making it apparent that these people hunted primarily ibex, not bison.

ROCKSHELTER EXCAVATION METHODS

No one can blame the archaeologists of a century ago for their simplistic, linear interpretations of the Cro-Magnons. Their crude excavation methods prevented them from understanding the full complexities of rock shelter occupations. Working with small, adzelike handpicks, Breuil and his contemporaries dug into deep rock shelters along the Vezère River in arbitrary levels of 10 to 20 centimeters, searching for finished antler, bone, and stone artifacts. They took little account of animal bones and subtle evidence of sporadic human occupation. Using "type fossils," they developed entirely linear cultural sequences, as if one society had followed another in an orderly manner.

Eventually, both excavation methods and artifact analyses became more sophisticated and cultural sequences more elaborate. Rock shelter excavation is now among the most sophisticated of all archaeological investigation. Modern-day Cro-Magnon excavations move slowly, but attempt to place more than 20,000 years of late Ice Age foraging within a complex environmental setting. Today's fieldworkers spend months refining the linear systems: areas that were once thought to be continuously occupied are now accurately chronicled as favored locations that were repeatedly visited—sometimes long term.

The process is meticulous and time consuming. The excavator lays out a horizontal meter-square grid over the occupation deposits, then digs each natural layer with brushes, dental picks, and trowels, following them from square to square and observing them both horizontally and vertically. Every square centimeter of cave earth passes through fine screens, as the diggers search for tiny artifacts, beads, rodent bones, plant remains, and environmental evidence. In this way, they isolate individual hearths, and even transitory visits to the site that may have lasted but a few days. The excavation is a multidisciplinary exercise, involving botanists, geologists, pollen experts, soil scientists, and zoologists, among others. Sample columns of cave deposit are subjected to flotation to recover minute plant remains (see chapter 4). Pollen and soil samples yield valuable information on climatic change in a region where the environment changed swiftly within individual centuries, requiring rapid cultural response.

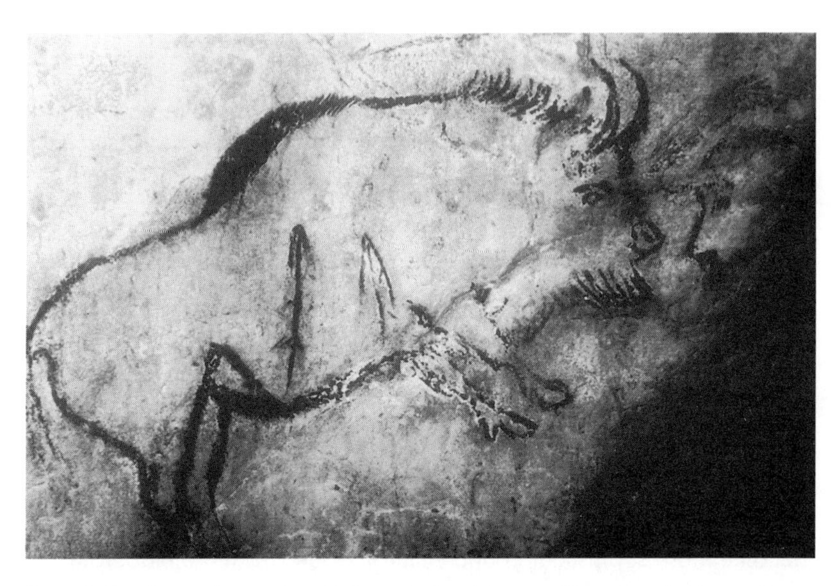

Fig. 2.5. Bison from the Salon Noir, Niaux. Photograph from Photo Researchers.

Another expert, French archaeologist André Leroi-Gourhan, working in the 1960s and 1970s, refused to speculate about the motives of the art. He used computers to analyze individual paintings, their relationships to one another, and their positions on cave walls at dozens of painted sites. Leroi-Gourhan subdivided each cave into three sections: the entrance area, the main friezes, and the deepest zones. For each of the friezes, he assigned sexual value to the animals and signs, either male or female, believing there was a fundamental duality in late Ice Age thinking that was reflected in the paintings. Thus, he concluded in his book, *Treasures of Prehistoric Art* (1964), that, in the celebrated Salon Noir at Niaux, finely painted female bison are associated symbolically with male horses.

Leroi-Gourhan was deeply influenced by the anthropological humanism of the 1930s, which proclaimed that art was a "universal essence." He believed that Cro-Magnon art was one of the great artistic movements of the human past, and inseparable from a coherent religious system; therefore, one could not study the art without taking religion into account. Leroi-Gourhan's theories are controversial. The complex reality of the paintings and engravings at

Lascaux, Niaux, and elsewhere defies easy analysis, and there are many exceptions to his gender duality. For example, all one can conclude about the paintings at Niaux is that bison are the major animal present. Their relationship to other species such as horses and ibex is unknown. The main reason that Leroi-Gourhan was unable to analyze the paintings thoroughly, however, was because he worked before accelerator mass spectrometry (AMS) revolutionized radiocarbon dating (see p. 41). He had no means of dating individual paintings accurately, or of establishing the length of time it took to build up the associations in individual friezes.

In recent years, scientists have turned from generalized descriptions like Leroi-Gourhan's to sophisticated analyses of the artists' paints and to new radiocarbon dating techniques that allow the precise dissection of individual friezes, such as those at Lascaux or Niaux. These approaches, combined with research thousands of kilometers away in South African rock shelters (see chapter 3), promise new interpretations of late Ice Age art.

ACCELERATOR MASS SPECTROMETRY RADIOCARBON DATING

For half a century, Breuil and his successors assumed that Cro-Magnon art evolved steadily over many thousands of years during the late Ice Age. A few experts thought this theory too simplistic. They compared the development of the art to a slowly growing tree with many branches, believing that the art was a result of many spurts in growth and individual episodes of genius, among, perhaps, the artists in a single group. However, no one could date the paintings, except on those very rare occasions when a slab of painted rock fell into a contemporary occupation deposit in the cave (Breuil attempted to do so, by using a crude technique of direct comparison of art styles on walls with those on antler and bone artifacts found in archaeological sites).

Then, in 1949, physicists Willard Libby and J. D. Arnold of the University of Chicago developed the radiocarbon dating method, which dated ancient organic materials like charcoal and bone. Radiocarbon dating is the most common way of dating archaeological sites younger than 40,000 years old. The method was developed from nuclear research during World War II. Willard Libby theorized that because living vegetation (and animal tissue) builds up its own organic matter by photosynthesis and by using atmospheric carbon dioxide, the proportion of carbon in it is equal to that in the atmosphere. Libby knew that cosmic radiation produced neutrons, which entered the earth's atmosphere and reacted with nitrogen. The product of this reaction is carbon 14 (radiocarbon), a carbon isotope with eight rather than the usual four neutrons in the nucleus. With these additional neutrons, the nucleus is unstable and is subject to radioactive decay. Carbon 14 (radiocarbon) is believed to behave exactly like ordinary carbon from a chemical standpoint, and enters into the carbon dioxide of the vegetation in combination with carbon. When a plant dies, the intake of carbon (and carbon 14) ceases and the process of decay begins. Thus, argued Libby, the amount of radiocarbon in ancient organic materials such as burned bone and charcoal was a direct function of the length of time the sampled organism had been dead.

Libby calculated that it took 5,568 years (the "half-life" of carbon 14) for half of the carbon 14 in any sample to decay, allowing him to develop a time clock for the past 40,000 years. He tested radiocarbon dating with samples from ancient Egyptian mummy cases, whose ages were already known, cleaning the samples and burning them to create a pure carbon dioxide gas. Then he used a lead-shielded Geiger counter to record the radioactive emissions from the gas, free of outside contamination. A simple formula allowed him to turn his counts into ages with appropriate statistical limits of error.

With this new technique, archaeologists could measure rates of cultural change and date important developments like the origins of farming in widely separated parts of the world for the first time. However, old-style radiocarbon dating required large samples, for only a small number of carbon 14 atoms break down over the many hours of the sample. Back in the 1960s, I remember collecting in plastic bags handfuls of charcoal from ancient hearths, the rule being, the larger the sample the better. In those days, dating a late Ice Age painting *directly* meant destroying the entire image to obtain an adequate organic sample. No one would countenance such drastic action. So cave art experts took *indirect* radiocarbon dates from charcoal samples in layers lying on the cave floor below the art. By the same token, scientists could not date small objects like seeds, maize cobs, or tiny wood fragments embedded in the sockets of prehistoric bronze spear heads: the samples were too small. Such objects are especially hard to date, and very difficult to pin down chronologically, for tiny finds like seeds can easily move upward or downward into older or younger occupation layers, either by human means such as trampling, or through natural phenomena such as burrowing animals. The development of a new radiocarbon method based on AMS in 1983 revolutionized radiocarbon chronologies, for it allowed the dating of individual seeds or of tiny charcoal flecks removed from an ancient rock painting without damaging the figure itself.

Using an accelerator mass spectrometer, researchers estimate the remaining carbon 14 in a sample by directly counting carbon 14 atoms rather than counting decay events (beta counts). In doing so, they can date tiny samples. The development of small, high-energy mass spectrometers solved a major problem—that of background noise from ions or molecules of similar mass to carbon 14, masking the presence of carbon 14. The new instruments filter out background, as a proportion of the sample's atoms are propelled through an accelerator toward a detector. Ionized carbon atoms from the sample are pulled in beam form toward the accelerator. A magnet bends the beam, so the lighter atoms turn more sharply than heavier ones, and move toward the inside of the diverging beam. A filter blocks the passage of all charged particles except those of atomic mass 14. The accelerator pushes the stripped beam through a second beam-bending magnet that filters out any last non-carbon 14 particles. A magnetic lens focuses the beam as a carbon 14 detector counts the number of remaining ions, thus calculating the age of the sample.

Today, archaeologists use radiocarbon dates calibrated with tree rings, to give actual dates in years (see information in box, chapter 8, p. 175). Originally, Willard Libby assumed that the concentration of radiocarbon in the atmosphere remained constant as time passed, so that ancient samples would have contained the same amount of radiocarbon as today. In fact, changes in the strength of the earth's magnetic field and alterations in solar activity have varied the concentration of radiocarbon both in the atmosphere and in living things. Fortunately, tree-ring dates obtained from ancient trees are extremely accurate (see chapter 8), so radiocarbon experts have developed calibration tables that convert radiocarbon dates into actual dates in calendar years. So far, tables exist for the period A.D. 1950 to about 6,800 B.C., which convert most sites described in this book. This calibrated time scale will expand considerably in coming years.

Since AMS works with small samples, an archaeologist, instead of extrapolating from generic dates derived from occupation layers, can now date individual figures within a cave painting frieze and work with blocks of AMS dates from dozens of samples. In a decade or so, AMS radiocarbon dating will give us finely detailed chronologies for religious sanctuaries and sacred art. It will also enable us to study the relationships between individual material objects and the places in which they are found, a useful tool for establishing the general nature of ancient beliefs. AMS radiocarbon dating of individual antler artifacts and cave paintings may, one day, allow direct and simultaneous comparisons between art on walls and engraved and carved objects in different occupation levels and neighboring sites. Perhaps then we may even be able to identify the work of individual artists.

NIAUX CAVE AND CHAUVET, THE CAVE OF BEARS

AMS dating is so new to cave art research that French and Spanish archaeologists have obtained dates from figures in only nine caves. In general, these dates confirm earlier estimates. Three of the famous Altamira bison, located in northern Spain, date to about 14,000 years ago. The most interesting results, however, come from the Niaux Cave in the foothills of the Pyrenees Mountains in southern France, and from the Grotte de Chauvet in the Ardèche region of the southeast, discovered only recently in December 1994.

Niaux is a deep cave, famous for its Salon Noir, a high-ceilinged cul-de-sac 700 meters from the entrance and measuring 20 meters across, adorned with magnificent paintings of horses, bison, and other animals, executed with fine, black shading. Niaux's deep passages also contain isolated friezes and figures, far from the great salon. For years, everyone believed Niaux dated to about 14,000 years ago, about the time of Altamira. But AMS dates in the Salon Noir tell a different story. One bison figure dates to about 13,850 years ago, another to around 12,890 years before the present; thus, the Niaux paintings span at least a millennium.

Those at the Grotte de Chauvet extend over an even longer period of time. The Grotte de Chauvet lies deep in the gorges of the Ardèche in southeast France, far east of the painted caves of Les Eyzies and the Pyrenees. Some cave art was known from the region, but nothing prepared scientists for the magnificent figures in the depths of Chauvet. On December 18, 1994, three local speleologists with an interest in archaeology, Eliette Brunel Deschamps, Jean-Marie Chauvet, and Christian Hillaire, crawled into a small opening in the Cirque de Estre gorge. The entrance was a mere 80 centimeters high and 30 centimeters wide, but led to a narrow vestibule with a sloping floor.

The three explorers felt a draft flowing from a blocked duct. They pulled out the boulders that blocked it, and saw a vast chamber 3 meters below them. After returning with a rope ladder, they descended into a network of chambers adorned with superb calcite columns. Calcified cave bear bones and teeth lay on the floor, and they noticed shallow depressions where the beasts had hibernated. Suddenly, Deschamps cried out in surprise. Her lamp shone on two lines of red ocher, then on a small mammoth figure.

The group then penetrated into the main chamber and came upon further paintings: hand imprints—both positives and negatives—and figures of mammoths and cave lions, one with a circle of dots emerging from its muzzle. As they gazed at the paintings, the three explorers were "seized by a strange feeling. Everything was so beautiful, so fresh, almost too much so. Time was abolished, as if the tens of thousands of years that separated us from the producers of these paintings no longer existed" (Chauvet, Deschamps, and Hillaire 1996, 42). In their book, *Dawn of Art: The Chauvet Cave,* they noted how, like the excavators of the Egyptian pharaoh Tutankhamun three quarters of a century before, they felt like intruders: "The artists' souls and spirits surrounded us. We thought we could feel their presence" (ibid.).

The Grotte de Chauvet had lain undisturbed since the late Ice Age. Hearths on the floor looked as if they had been extinguished the day before. Flaming torches had been rubbed against the wall to remove the charcoal so they would flare anew. One such rubbing lay on top of a painting. On a second visit later the same evening, the explorers found an extraordinary frieze of black horses, wild oxen with twisted horns, and two rhinoceroses facing one another. There were lions, stags, engravings of an owl, and animals never before seen in cave paintings, covering an area of more than 10 meters. The horses had half-open muzzles, their eyes depicted in detail. One horse outline had been scraped to make it stand out better. The artists made use of the contours and crevices of the rock to produce illusions of relief and perspective. A little farther on in the chamber lay a slab that had fallen from the ceiling. A bear skull had been set on top of it, and the remains of a small fire lay behind it. More than thirty calcite-covered and intentionally placed bear skulls surrounded the slab.

Realizing they had discovered a cave that rivaled Altamira, Lascaux, and Niaux, the three speleologists returned on Christmas Eve. Acutely conscious that the cave was undisturbed, they unrolled plastic sheets over the imprints of their footprints from their previous visit to ensure that later visitors stayed on the same route. In an end chamber, they came across another 10-meter frieze of black figures dominated by lionesses or lions without manes, rhinoceroses, bison, and mammoths. Far to the right they discerned a human figure with a bison head. They wrote that it "seemed to us a sorcerer supervising this immense frieze" (ibid., 58). To their eternal credit, Deschamps, Chauvet, and Hillaire took precautions to prevent trampling and unauthorized visitation. Within a few days, the cave was sealed with an iron door.

When archaeologist and top rock art expert Jean Clottes arrived at the cave, he had his doubts as to whether the paintings were genuine. But his skepticism soon vanished when he saw that several paintings lay under a layer of calcite, which had formed over many millennia. Piles of charcoal from burnt Scots pine branches lay intact. The undisturbed floors with their numerous bear bones and undisturbed hearths left no doubt in his mind that Chauvet was a priceless archive of Cro-Magnon painting preserved in its original context, artifacts and bones still in the exact places where they had been dropped.

Clottes identified at least a dozen flint artifacts and realized that analysis of the edge wear might yield clues as to their use. He estimated that at least three hundred animal figures and numerous signs would be found once the cave was fully studied and explored. But he noted that rhinoceroses were most common, followed by lions, mammoths, wild horses, and bison. The artists were masters of perspective, overlapping the heads of animals to give the effects of both movement and numbers (figure 2.6). They even scraped some of the walls before painting them, to make the figures stand out better; and spread the paint with their hands over the rock, to obtain values that showed dimension or color tonality.

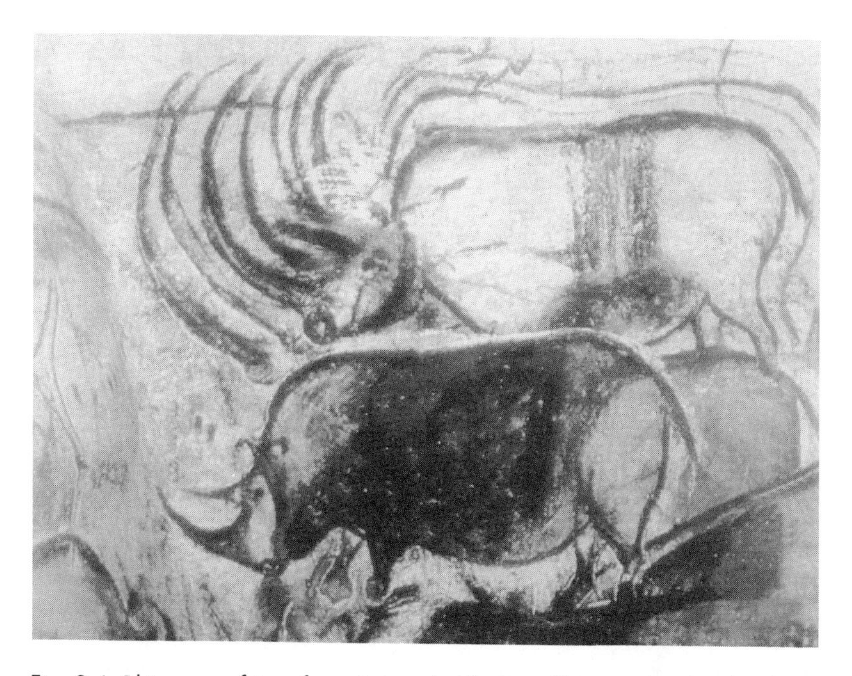

Fig. 2.6. Rhinoceros frieze from Grotte de Chauvet. The artist has used multiple lines of horns to give the impression of a herd. Photograph from Photo Researchers.

After careful stylistic comparisons, Clottes dated the Chauvet paintings to between 17,000 and 21,000 years ago. However, AMS dates from two rhinoceroses and a large bison produced dates from a 1,300-year period around 31,000 years ago—making them the earliest dated art in the world. Two more samples from torch smears on the walls were found, by AMS, to date to around 26,500 years ago, while two charcoal samples on the floor gave readings from about 24,000 years. Clottes claims humans visited Chauvet on several occasions over at least 6,000 to 7,000 years. Whether they painted over that long period is still unknown, but AMS dates will ultimately produce some answers.

The AMS data gathered from Chauvet show us that Breuil and his contemporaries erred in thinking of a gradual evolution of Cro-Magnon art. In reality, as proven by AMS, the art flourished and reached high peaks of artistic achievement in many places at different times, starting from the earliest moment when modern humans, our remote ancestors, settled in western Europe.

Interestingly, many of the animals on the cave walls represent dangerous members of the late Ice Age bestiary—bears, lions, mammoths, rhinoceroses, bison, and even, occasionally, the nimble and ferocious aurochs. But Grotte de Chauvet was a bear cave, a place where these powerful animals hibernated. Did bears use the cave before humans? What was the relationship between bears and people? Clottes pondered the paintings and wondered if human visitors to the cave, with its claw marks, hollows, prints, and scattered bones, came to the chambers to acquire the potency of the great beast whose smell probably lingered in the darkness. Another approach not yet attempted at Chauvet—that of highly detailed paint analysis—promises some indirect clues as to the ways in which the paintings were executed, and, perhaps, for what purpose.

ARTISTS' RECIPES

In 1902, a French scientist named Henri Moissan scraped flakes of Stone Age paint from the walls of Font-de-Gaume and La Mouthe caves near Les Eyzies. He identified a number of constituents, including red ocher and charcoal, but lacked the analytical tools to carry out more precise analyses.

New analytical techniques that allow scientists to analyze cave paintings extensively are now available. The process begins in the caves themselves. Archaeologists first photograph the paintings with great care and examine them meticulously. They study ways in which the paint was applied to the walls, and record any superpositions of one painting over another. They also check the figures for homogeneity of composition, a vital step, for it establishes how many paints were used. A binocular microscope provides a close-up view of the painting in situ, and a much better chance of observing details, such as hair shading, than conventional observation. Next, technicians photograph the figures with the aid of three optic fibers that guide light onto the painting. All of these observations and photographs provide the basis for research that determines how many paint samples will come from individual figures.

Back in the laboratory, the analysis begins with the chemical constituents of the paint. The Cro-Magnon artists used pigments such as charcoal or red ocher, combined with an organic binder or an extender to ensure cohesion and fluidity. A scanning electron microscope attached to an X-ray detector allows visual identification of the morphology of the constituents and a basic analysis of the main chemical substances in the paint. Then, X-ray diffraction analysis provides information on the crystalline nature of the minerals. The exact chemical composition comes from proton beam analysis, using a linear accelerator. The paint analyses provide information about the chemical composition, and, more important, about the quantities and proportions of the paint constituents, and the amount they were ground or otherwise processed. Spectrographic analyses reveal trace elements, which sometimes can be used to link original paint materials to sources some distance from the cave.

Identifying organic binders requires careful selection of paint samples, for only those where no pollution has affected the paint can be tested. Even touching a painting once can contaminate the sample with invisible oily deposits. Gas chromatography identifies lipids like triglycerides, the main constituent of oil or grease, and sterols, the molecular structure that allows scientists to determine whether samples are of animal or vegetable origin. Mass spectrometry provides confirmation of the identifications. The new paint analysis techniques are expensive and time consuming, so research moves slowly.

Niaux Cave in the foothills of the Pyrenees mountains lies amidst a cluster of Magdalenian occupation sites and painted caves. The great cave enters the pyramidlike, 1,189-meter-high Cap de le Lesse massif at the foot of a steep cliff. Niaux's cave system extends nearly 2 kilometers underground, a labyrinthine system of chambers and side galleries. The setting is spectacular, set into a conspicuous peak, that might itself have had spiritual significance.

Niaux is famous for its black animal figures, mainly centered in the Salon Noir (figure 2.5). Visited for more than a millennium, Niaux—with its many images, some far from the open air—is ideal for paint analysis research. Jean Clottes has taken seventy-five paint samples from the cave in recent years. In the Salon Noir, both observation and laboratory analyses showed that very often the artist made a preliminary outline sketch in charcoal, a "cartoon" as it were, before executing the actual painting. Ostensibly the painter intended to create a specific, well-thought-out composition. The Salon Noir is relatively close to the entrance, where the carefully prepared figures are painted on the walls, in a chamber often visited by the Magdalenians. The paintings in deeper chambers show no signs of careful preparation. Rather, the artists painted the animals directly on the walls without preliminary drawings, evidence that the Magdalenians made only short expeditions into the remotest dark chambers.

The paint recipes used at Niaux were based on hematite for the reds and manganese oxide with or without finely ground charcoal for the blacks. The artists also used an extender: either potassium feldspar or a combination of potassium feldspar and a large quantity of biotite. All these minerals are readily found near the cave and easily extractable. Extenders saved on paint and produced a more homogeneous mixture that adhered better to the cave walls. The binders used in the Ariège region contained water and, in two caves, animal and plant fats, which allowed the artists to use real "oil paints." The Niaux artists used at least two "recipes," replacing the earlier potassium feldspar extender with the biotite combination, which was more easily obtained and had better adhesion qualities.

The discovery of at least two paint recipes in one cave destroys Breuil's earlier theory, which he had based on stylistic comparisons, that assumed that all the Niaux paintings had been executed as a contemporaneous whole and painted over a very short period of time. The new paint analysis approaches allow individual figures to be considered. One of the Niaux panels at the extreme left of the Salon Noir was painted with only the potassium feldspar/biotite recipe. Almost certainly, this panel was conceived of as an integrated whole. But the artists ran out of paint in the middle of the panel. Trace elements in the paint samples reveal that two separate, slightly different batches were used. Other Niaux panels consist of paintings that used both recipes and were painted over a considerable period of time. Clottes believes the accessible locations were chosen because of convenience. By using paint analysis, he has been able to assign paintings in remoter chambers to specific moments in the Salon Noir sequence.

The Niaux analyses reveal that an apparently homogeneous sample of wall art has at least two different stages, so an apparent unity may conceal a complex reality. How long, then, did stylistic themes and conventions endure? We do know that the two paint recipes were used in succession, giving us a rough chronology. But could the recipes have had different social or ritual meanings? Perhaps the same recipe was used for horses and bison, while different constituents were employed for signs and animals, or by men or women exclusively, or even at different seasons of the year.

The history of art tells us that new paint recipes that are technically superior or more economic will sometimes rapidly supplant those in current use. Acrylics have replaced oils in some instances. Oil pastels are preferred over chalk, and ink over watercolor, according to personal taste. Today, some people even "paint" with graphics programs on computers. Thus, a recipe can acquire a chronological significance. Thanks to AMS dating, we can study individual ancient sites, and groups of sites as part of a much bigger cultural landscape.

Cro-Magnon artists sometimes painted images by drawing outlines in charcoal. On other occasions they used the paint like a wash or even blew it on to the rock surface. As we have seen, paint manufacture involved the use of local ingredients, but the painting of large images, like those in the Hall of the Bulls at Lascaux, required large quantities of pigment that must have taken many people to prepare.

I have not seen the Chauvet paintings: few outsiders have yet had the chance. Nor should they, for our primary concern must be to protect them from harm. But color pictures of the friezes give an impression of movement, of a subterranean, animal-filled realm where supernatural potency lurks behind the rock faces, in a remote cosmos of layers. The caves were the places where humans and the spiritual world met and illuminated the darkness.

The famous polychrome bison at Altamira, which stand and crouch on the rocky ceiling, are modeled around natural protrusions in the rock (figure 2.3). Maybe the artists believed that spirit-animals could empower them through a transference of power that occurred during the act of painting animal figures and from touching them. This activity was probably confined to a few individuals who as shamans and artists acquired spiritual powers during solitary vision quests in the depths of dark, sacred caverns. In the same manner, through identifications at Chauvet, we may be glimpsing one of these ancient people, dressed in a bison skin, facing outward toward the entrance to the chamber.

A century after the first authentication of Cro-Magnon art, we have begun to take halting scientific steps toward understanding an artistic tradition deeply embedded in an ancient hunting culture. We may never fully discern the motives behind the paintings of Chauvet, Lascaux, Niaux, and other caves. But recent archaeological and ethnographic researches in southern Africa (described in chapter 3) have given us some fresh and provocative insights into the ancient spiritual world where artistic traditions play a significant part in ritualistic performance.

chapter three

SAN ARTISTS IN SOUTHERN AFRICA

As a young archaeologist, time and chance allowed me to work in southern Africa, where I had many opportunities to explore the ancient art tradition of the San people. The San (sometimes, pejoratively, called Bushmen) are the indigenous hunter-gatherers of southern Africa. Today, small numbers of them survive in the Kalahari Desert of Botswana, where they have been in contact with farmers and herders for many centuries. Their ancestors have lived in southern Africa for at least 10,000 years and have been studied extensively through numerous archaeological excavations of caves, rock shelters, and various types of open-air sites from Zambia to the Cape of Good Hope. San artists, who sometimes painted in well-lit caves, but mostly in open, sun-filled rock shelters, were still at work a little over a century ago. Thus, their art is separated from the Chauvet paintings by more than 25,000 years. The education I gained from studying San art was priceless in helping me to understand the motives behind the much older Cro-Magnon paintings.

My first sighting of African rock art was in late afternoon. The Matopo Hills in western Zimbabwe (figure 3.1) shimmered in the soft, yellow winter light. Great weathered granite domes and vast boulders nestled in dense woodlands. I climbed high on a rounded peak where a deep overhang cast a cool shadow on a smooth-walled concavity in a rock, and when I looked up, I could see painted red giraffes with their elongated bodies outlined in white. Groups of San hunters, drawn with sticklike precision, prepared for the hunt below

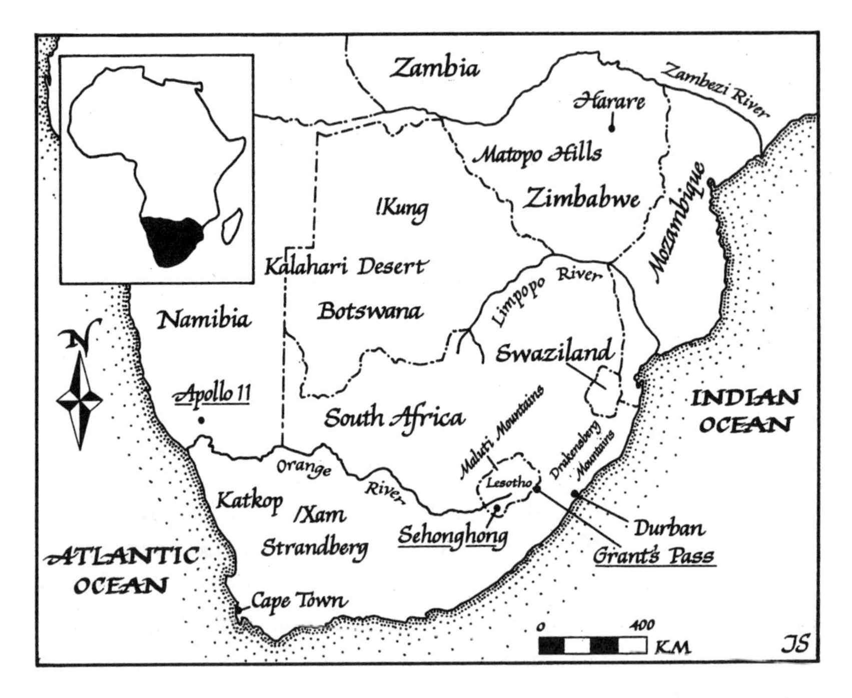

Fig. 3.1. Map showing the archaeological sites (underlined) mentioned in chapter 3.

them. The intricate frieze ran for several meters; a wild maze of reds, oranges, browns, whites, and yellows that included animals, humans, and a multitude of dots, oval lines, and symbols. As the sun went down, the paintings stood out clearer on the light-colored wall. Defying immediate interpretation and understanding, I nevertheless stood enthralled, caught up in the colors, lines, and images.

A few days later, I went exploring by myself in a remote corner of the Matopos, visiting several more painted rock shelters. Everywhere I went, the paintings were out in the open, not hidden in dark caves like those of the Cro-Magnons. Late that afternoon, I was exploring a small cleft between two enormous boulders when I came across a single portrait of a red hunter. He had been depicted as walking calmly by himself, carrying a digging stick and bow and arrows on his shoulder. Quite alone, with no animals nearby, the hunter was making his way across the weathered rock face toward a vertical crack. His manner, to me, appeared purposeful, as if he were intent on his destination. Strangely, there were no other paintings anywhere nearby. Apparently, someone long ago had painted this solitary figure walking toward a crack with a specific reason in mind. Later, I learned I was the first person to have seen him in many a year, certainly the first archaeologist to admire him—and admire him I did.

My Cambridge University lecturer, Miles Burkitt, had visited South Africa in 1927. He too had admired the rock paintings. However, he considered them to be decorations for the walls of oft-visited caves—"prehistoric wallpaper," as he called it. My solitary hunter convinced me that he was wrong. Why, after all, would anyone paint a picture all by itself, by a crack in a rock, just for the sake of it? Unbeknownst to me, two linguists and a magistrate had stumbled across some of the mythology behind San art more than a century ago.

EARLY RECORDS OF SAN FOLKLORE

When I taught for a couple of months at the University of Cape Town in 1960, I spent hours in the Jagger Library at the university reading about the San. The literature was enormous, scattered in dozens of specialized, and often long-discontinued, scientific journals. I read early anthropologist George Stow's *Native Races of South Africa,* published in 1905; and discovered an 1874 article of his in the prestigious *Journal of the Royal Anthropological Institute*, in which he describes a meeting with a group of southern San who did not themselves paint, but had heard of distant neighbors who did. Stow remarked that this "was spoken of in the past tense as if the practice had been discontinued at the present time" (1874, 245).

I delved briefly into folklore, myth, and linguistics, but soon tired of the complexities of San dialects. For a few hours I dipped into Wilhelm Bleek and Lucy Lloyd's *Specimens of Bushmen Folklore*, a long-forgotten work that appeared in 1911. I read of /Kaggen,* the

*Khoisan languages such as San dialects make extensive use of palatal, tongue, and lip clicks, which are set down with exclamation marks and other standard scholarly conventions. For example, !Kung is a word that begins with a palatal tongue click.

insect known as a mantis and trickster, who assumed the form of various animals; and of the children sent to throw the sleeping sun into the sky. Soon, however, the exotic myths palled, and I returned to the more familiar world of Stone Age archaeology. I had no idea that the Jagger Library housed thousands of pages of research notes on San folklore that would prove to be of great value to archaeology.

The existence of these archives is a minor scientific miracle for several reasons. Thousands of San perished as a result of white settlers' encroachment on their homelands during the eighteenth and nineteenth centuries. Many were hunted down by farmers when the hunters raided settlers' cattle herds. Others retreated into the vast mountain and desert landscapes, far from settled lands. Additionally, only a handful of travelers and scholars mastered the intricacies of San dialects. German-born linguist Wilhelm Heinrich Immanuel Bleek (1827–1875) was one such scholar.

Bleek began his linguistic research with the Hebrew language, then turned to the "so-far unexplored African languages." He came to Natal in 1854 and spent 2 years recording Zulu languages and traditions, living among the people for weeks on end. Two years later, Bleek became chief interpreter to the British High Commissioner in Cape Town, a post that allowed him ample time to complete his famous monograph, *A Comparative Grammar of South African Languages* (1862), in which he classified both Bantu and Khoikhoi (Hottentot) languages. Bleek found himself increasingly fascinated by San dialects, but was unable to visit any speakers of San in the field, for the nearest groups lived far in the interior. By chance he discovered in 1870 that there were twenty-eight San convicts working on the breakwaters of the Cape Town harbor. He interviewed the men in the breakwater jail. The atmosphere was uncongenial, so he persuaded the authorities to release the better informants into his care. They lived at his house, working closely with Bleek and his sister-in-law Lucy Lloyd on a compilation of vocabularies and grammars. Almost immediately, Bleek and Lloyd realized that they were collecting a valuable body of mythology and folklore. Their vast repository of material has proven to be a priceless source of information for students of rock art trying to interpret scenes depicting common San myths.

Most of Bleek and Lloyd's San informants came from the Strandberg and Katkop mountain areas of the east-central portion of the Cape, where they still lived a partially nomadic life. The informants included young men and older individuals, among them //Kábbo, who lived with Wilhelm Bleek and his family from 1871 to 1873. He was about 60 years old, and was apparently a medicine man who was also a rainmaker. //Kábbo looked after the garden; upon finding the ground too hard in October of 1871, he went into a trancelike state, during which he said that he spoke to the rain, asking it to fall. He also claimed to have gone on a "journey" to his homeland and spoken with his wife and son. //Kábbo was at first hesitant to speak with Bleek and Lloyd, but soon became an excellent informant, waiting patiently while Lucy Lloyd laboriously transcribed his words. He contributed over 3,100 pages of material to Lloyd's notes. His descriptions of rainmaking rituals have all the immediacy of firsthand accounts.

Between 1866 and 1874, Wilhelm Bleek filled 27 notebooks with San linguistics; while Lucy Lloyd collected no less than 10,300 pages of folklore between 1870 and 1884. She set down the material in /Xam, for which Bleek developed a phonetic script. Then she translated it and read the story back to the informants for verification. Unfortunately, the mythology and vocabularies were too exotic for popular taste and most of her collection remained unpublished, buried in her notebooks, which, after her death in 1914, resided in the University of Cape Town Library.

Both Bleek and Lloyd were interested in San paintings, but they had few copies of them for their informants. Lloyd occasionally showed them a picture, but was more interested in having the informants identify the subject matter than in probing the meaning of the paintings. The informants always accepted the art as the work of their own people. Some of them were also familiar with the practice of painting, and had some of the same basic conceptual patterns as the artists of the Drakensberg Mountains of Natal in eastern South Africa.

In 1873, J. M. Orpen, chief magistrate of St. John's Territory, hired a San guide named Qing to accompany him on an official expedition to explore the Maluti Mountains of Lesotho, a short distance from the Drakensberg Mountains. Qing had never encountered a white man before "except in fighting," but had a passion for horses and hunting, so the two men got on well. Orpen set down Qing's stories while they were sitting around campfires. He always used interpreters, which must have weakened the impact of the stories. He also complained that the stories were "fragmentary," and, therefore, he strung Qing's stories together consecutively. Although the magistrate's research methods were primitive by modern anthropological standards, there are striking similarities between the Lloyd and Orpen material. For example, both stress the importance of the mythical trickster /Kaggen. Above all, both the Lloyd and Orpen mythologies placed major importance on the eland. Both said that /Kaggen created the eland and cherished it, but hunters killed it without his permission, thereby angering him.

Orpen began his work with Qing by asking questions about the rock paintings that the San pointed out to him as they journeyed toward the Maluti Mountains. Orpen made copies of paintings from four caves, which he sent to the editor of the *Cape Monthly Magazine*. The editor in turn sent them on to Bleek, but without Orpen's accompanying account, which recorded Qing's interpretations. Bleek shared the pictures with an informant, Diä!kwain, who gave a ready explanation. When the two interpretations were compared, Bleek discovered remarkable similarities between both accounts—by San living many hundreds of kilometers apart. Both agreed that one scene from Sehonghong Cave depicted a rainmaking ceremony. Men have attached a rope to the nose of a rainmaking animal and are leading it over parched ground to break the drought. Since this was a hazardous task, the medicine men had first charmed the animal with *buchu*, which are sweet smelling herbs (another scene from a different site in figure 3.2). Although Qing spoke in more complex metaphors, perhaps based on firsthand experience of trancelike visions, Bleek was convinced that his explanations of the paintings showed how San paintings "illustrated" their mythology. Generations of later researchers, however, dismissed these ethnographic accounts as being of dubious value, and instead focused their efforts on systematic recording (see information in box, p. 58).

Fig 3.2. J. M. Orpen's copy of a painting from Mangolong Cave in the Maluti Mountains, South Africa, of men leading a rainmaking animal. Photograph from San Heritage Center, University of the Witwatersrand.

SAN PAINTINGS AS THOUGHT PATTERNS

Generations of scholars have tried to date southern African rock art, but have experienced problems similar to those encountered by the archaeologists who studied the Cro-Magnon paintings. These difficulties were compounded by a relative lack of portable art objects that could be excavated from dated layers and compared to wall paintings. Painted slabs have come from the Apollo 11 rock shelter in Namibia, which are said to be about 26,000 years old, contemporary with the Grotte de Chauvet in France. However, the majority of sites date to more recently than 10,000 years before present; in fact, most surviving paintings were probably painted within the past 500 years. Some San paintings date to historic times and depict European activities: red-coated British soldiers, full-rigged ships, and cattle raids. AMS dating is only now being applied to South African rock art, but dates from individual paintings will soon delineate the chronology of San artistic traditions.

South African archaeologist David Lewis-Williams is a rock art expert at the University of the Witwatersrand in Johannesburg. He

SOUTH AFRICAN ROCK ART COPYING

As in Europe, scholars in southern Africa have experimented with various rock art copying methods. In the 1890s, Abbé Breuil was the first archaeologist to make color reproductions of South African rock art, using butcher's paper. Another early scholar, Walter Battiss, painted in watercolors. The beginnings of a revolution in San rock art studies came with the development of affordable color photography in the 1950s. A South African rock art expert, Alex Willcox, photographed thousands of paintings, especially in the Drakensberg Mountains, where some of the finest cave paintings in Africa are to be found. Willcox was somewhat of a romantic. Captivated by the beauty and variety of the paintings, he waxed lyrical about the leisurely, prosperous life of the ancient San. He wrote of expert artists who took great joy in their depictions of animals and people. This, he said, was "art for art's sake." In reality, the paintings were an invaluable source of information about ancient San life and hunting practices. Patricia Vinnecombe also worked in the Drakensberg region and compiled a remarkable statistical record with drawings and color photography. A 1970s scholar, Harold Pager, photographed the paintings in black and white, measured the drawings, and then returned to the site to color in the photographs. Another photographer, Neil Lee, used color film, shooting the art from an overall perspective, then moving closer and closer to take detailed close-up photographs. This approach allowed him to study the painter's technique, the draughtsmanship, the types of brushes used, and the different paint types.

violently disagreed with the "art for art's sake" theory and was struck by the uniformity of the paintings over a large area of South Africa. They were, he believed, the product of shared beliefs and behavior. But what were these beliefs and behaviors? He combed through Bleek and Lloyd's long-forgotten notes at the University of Cape Town Library for answers and became convinced that San groups in the Drakensberg Mountains, the Cape, and elsewhere shared many common spiritual concepts. Both the art and the oral traditions preserved in Bleek and Lloyd's collection over a century ago were different expressions of a single belief system.

In 1974, Lewis-Williams and anthropologist Megan Biesele gathered together a small group of !Kung San informants from the Kalahari Desert at a camp in Botswana "to talk" and formed a group of six !Kung men and women who took the scholars' work seriously. Lewis-Williams showed copies of some rock paintings to the informants, who were able to identify the animals in the paintings; however, they were most interested in the more complicated scenes. The !Kung recognized different species of eland, their sex and position, and groupings of males and females characteristic of different times of the year. Lewis-Williams left Botswana convinced that the !Kung and other San groups share a basically common cognitive culture; and that present-day accounts of eland hunting rituals, puberty rites, and trance performance can be used, cautiously, to amplify much earlier informants' remarks.

The eland, which are predominant figures in some areas where San art appears, are the largest and fattest of all African antelope—lumbering animals that can be easily run down by an agile hunter on foot. One eland could feed a San band for many days, which meant that they assumed great importance in environments where food supplies could be irregular. Wilhelm Bleek's informants had dictated several myths in which they associated eland with honey, a substance with a strong, sweet smell similar to that which rises from a dead eland when it is skinned.

Lewis-Williams has examined hundreds of eland paintings in the Drakensberg and elsewhere (figure 3.3). At Grant's Pass Cave, he found a painting of an eland with deeply sunken eyes, staggering in its death throes. Dancers are cavorting around the animal, one decorated with cloven antelope hooves crossed like those of the dying eland. White dots depict sweat drops falling from a dancer "dying" in trance. Lewis-Williams believes that these dancers are acquiring the potency released by the death of the eland, a process shown by the antelope heads, feet, and hair on the dancers. He found the same scene in dozens of other eland paintings, where it appears that the whole being of the medicine man and the people become merged with the most potent of all animals. From his extensive studies,

Fig. 3.3. Eland scene from Fetcani Glen, Barkly West, Cape Province, South Africa, showing intentional superpositioning of eland and figures. Photograph from San Heritage Center, University of the Witwatersrand.

Lewis-Williams realized that painting after painting linked society with the supernatural; and that the medium "responsible" for this link was the medicine man, who entered the spiritual world during an altered state of consciousness—a trance. San shamans induced trance, not by ingesting psychotropic drugs, but by intense concentration, prolonged rhythmic dancing, and hyperventilation.

Bleek's informants spoke of at least four categories of medicine men, three of whom were associated with game, rain, and curing illness. To achieve their ends, the medicine men went into trance, during which they manipulated a supernatural potency, which was possessed by animals like eland and hartebeest, by rainfall, and by girls at puberty. Medicine men also used the same powers in curing rituals and to travel through supernatural realms outside their own bodies. Today, San in the Kalahari Desert will dance next to the carcass of a freshly killed eland. A medicine man who has special control over eland potency will enter a trance during this dance, and then cure everyone of ills by removing "arrows of sickness" that may be directed against them. I have seen medicine men among the Kalahari

!Kung activate their *n/um,* a supernatural potency used for curing. As they activated their potency, the medicine men trembled, then sweated, then bled from the nose, as the potency took hold of them—an eerie sight to behold. Many white hunters have seen dying eland, trembling with wide-open mouths, sweating profusely, with melted fat gushing like blood from their nostrils. Perhaps the San likened this phenomenon to "death" in a medicine man's trance.

By combining careful observations such as these with his anthropological research data and nineteenth-century ethnography, Lewis-Williams believes he can "read" some of the rock painting friezes not as art, but as meaningful scenes. He claims that the art in one rock shelter in central South Africa depicts San shamans in the midst of a trance dance. Their attenuated bodies convey the sense of altered consciousness (figure 3.4). In some paintings, a line of dots along the spine of the central figure portrays the "boiling sensation," when potency—supernatural power—rises up the spine and "explodes" in the head. Perhaps the power comes from animals such as

Fig. 3.4. Stages of trance performance. At left, a medicine man collapses to the ground, bleeding from the nose. Another figure, also bleeding, bends toward him. Fetcani Glen, Barkly East, Cape Province, South Africa. At right, a man bleeding from the nose and wearing a skin cloak, sits out the dance to control the trance. Halstone, Barkly East, Cape Province, South Africa. Drawing from San Heritage Center, University of the Witwatersrand.

the eland, which in the original painting is situated to the right of the frieze.

Lewis-Williams's research has led him to believe that the San paintings are visual representations of the people's back-and-forth thought patterns—thoughts of the mind in both the unconscious and conscious states. He cites the example of a girl's puberty ritual, where the young girl is said to have "shot an eland." In the Eland Bull Dance, miming and sounds make the eland appear real before the eyes of the participants. As the medicine man dances, he hallucinates and "sees" the eland standing in the darkness beyond the glow of the fire. As the dance continues, the dancers become one with the eland spirit, and the transfiguration is complete—they have become the eland. Afterward, the shaman-artists remember their trance experiences, and paint what they hallucinated on the walls of rock shelters. These representations complete the transference process. The creative act of "seeing" has served the function of ritual, in which the object and the subject become interchangeable in the subtle web of San thought and belief. These visions of the unconscious are then painted and thus transferred to the world of the conscious.

ALTERED STATES OF CONSCIOUSNESS

Altered states of consciousness can be brought about in many ways: through the use of hallucinogens, from sensory deprivation or intense concentration, or even hunger, pain, or migraine. Although the San used hyperventilation, rhythmic music, and intense concentration, their hallucinogenic experiences were very similar to those induced by LSD (lysergic acid diethylamide), peyote, or other hallucinogens.

The nervous system controls trance states. Neuropsychological research using volunteers revealed that there are three general stages in the sequence of mental imagery during altered states of consciousness. In the first stage, the subject experiences entoptic phenomena (mentally generated images). These luminous visions take the form of incandescent, shimmering geometric forms, curves, and spirals. The patterns move, rotate, and at times enlarge themselves; and appear independent of any light source. The imagery flows rapidly; so rapidly that the subject cannot keep up, although both training and experience make it possible for the subject to observe and describe the images more accurately. Since all humans share similar nervous systems, we know that San shamans would have experienced entoptic images similar to those of modern subjects. Some of the geometric images that research subjects recorded are identical to those on panels of rock panel engravings (though rarely in painted rock shelters), so it can be concluded that the San shamans who used an altered state of consciousness to see into the spiritual world and for other purposes turned the basic entoptic forms that they had experienced in the first stage of trance into shapes and objects that had an expressed value (figure 3.5).

During the second stage of altered consciousness, the subject tries to make sense of the entoptic images—to turn them into something recognizable. Just as in normal consciousness, the brain of a person in an altered state receives a constant stream of sense impressions supplied by the nervous system. The subject's brain matches the images against previously stored experiences in his or her brain, in an attempt to identify them. Thus, an enigmatic round object seen in altered consciousness may become a cup of water if the subject is thirsty, a female breast if he or she is in a state of heightened sexual excitement, or some other object, depending on the subject's "state of being."

A common entoptic motif found on shelter walls is a series of nested, "U" shapes. Some San shaman interpreted these as curved honeycombs, even painting dozens of bees, complete with wings, around them. Lewis-Williams believes that the hive-curve association may have resulted from a humming sound often heard by people in an altered state of consciousness. The Kalahari !Kung believe that bees possess great potency, and, therefore, dance when bees swarm, as a way of harnessing their power.

In the third and final stage of altered consciousness, the laboratory subjects experience a vortex that seems to surround them, the walls of which are marked by lattices of squares. These walls are like television screens, and carry spontaneous images of people, animals, and other iconic images from memory that are associated with powerful emotional experiences. In this stage of altered consciousness, these images appear startlingly real in a strange world of nonreality.

Fig. 3.5. A San curing dance involving trance, from the Nyae Nyae region of Namibia, southern Africa. Photograph from AnthroPhoto.

Some of the San paintings take us into this realm. I saw a cave in South Africa's Free State that shows a rainmaking bull, associated with torrential thunderstorms. A zigzag enters or leaves its eye, and white dots adorn parts of the body. Rain shamans surround the beast, most with zigzags leaving their bodies.

The zigzag is a common entoptic that is associated with hallucinations. Zigzags often accompany human figures in San rock art. Some paintings with zigzags appear to represent shamans in stage-three trance, with appendages on their arms that appear to depict feathers. This may also portray a trance state, for, according to San folklore, the mantis, itself a shaman, sometimes obtains feathers when it enters trance and flies away.

Dozens of San paintings provide evidence of religious experience. These are not representations of a shaman's solitary vision quest in which power is sought, but rather the product of hallucinations experienced at deeper stages of trance, long after power is acquired. San rock art may seem serene and tranquil, but it was conceived in the turmoil of powerful hallucinogenic experiences as a trancer explored the inner recesses of the spiritual world.

The dances and trancing left lasting impressions on those who observed them. When anthropologist George Stow showed some rock painting copies to an elder San couple, the woman began to sing and dance. Her husband begged her to stop because the old songs saddened him. But she persisted. Eventually, the old man joined her. Stow watched as the aged couple became lost in the dance, exchanging looks of deep happiness, the present forgotten.

IMAGERY AND INGREDIENTS

San rock art combined the real and supernatural worlds. The artists often painted observable reality such as dancers and dancing rattles, ostrich hunts, or campfire scenes. However, they also acquired imagery in trance, a spiritual reality that was "seen" only by shamans, creating an art with an ever-shifting reality, the exact details of which are inaccessible. During trance dances, shamans might draw everyone's attention to things they can see, such as spirit-eland standing in the semidarkness beyond the fire. The participants look in the same direction and share the same vision. When everyone has returned to a normal state of consciousness, the visions are then described. The shaman is then able to manipulate these visions for the audience, using such things as a dance, flickering lights, or paintings, to direct his narrative.

The San attached great importance to the ingredients in their paintings. In the early 1930s, anthropologist Marion Walshaw How conversed with a 74-year-old Sotho (Bantu) man named Mapote, who had learned to paint with neighboring San in his youth. He told her red painting pigment was called *qhang qhang.* It glistened and sparkled, and came from nearby basaltic mountains. Mapote said that the pigment had powerful supernatural associations and was heated red hot, then ground between two stones into a fine powder. This work was done by women, carried out at full moon.

When asked to paint himself, Mapote asked for another vital ingredient—"the blood of a freshly killed eland." Only fresh blood would coagulate and soak into the rock, implying that much painting took place after a successful hunt, a time when many communal trance dances took place. Another informant, known as "M," was an old woman of partial San descent, whose father had been a shaman-artist. When "M" was interviewed in the 1980s, she recalled how a young girl would accompany the hunters and point an arrow at their prey that had been smeared with a "medicine" prepared by the shamans. The hunters would then kill the animal, which was "led" back to a place near the rock shelter where the paintings were to be made. The artists would then mix fresh eland blood with ocher, making the resulting painting a storehouse of potency.

Two types of blood were used in painting: that of the dying eland, and that of the shamans, who, as mentioned previously, often bled from the nose when they entered trance, just as dying eland sometimes bleed nasally as well. The preparation of paints, therefore, offered a way of expressing consistent symbolic messages, by simply changing ingredients to reflect different meanings known to the shamans. Thus, the San rock paintings were far more than pictures, and the processing of pigments, which can be determined by chemical analysis, was far more than a material technique.

Whatever the motives for the San art, it is certain that a high level of artistic ability was needed, and that only a relatively small number of shamans were painters. In some areas, those individuals exhibiting an expertise at rock painting may have been given a higher status in society. Since it would have been impossible for artists to have painted while in a trance—due to their trembling and being sometimes in an almost unconscious state—Lewis-Williams believes that they painted while in a state of normal consciousness, recalling their vivid images of the spirit world and then recreating, through their paintings, those visions and the animals that were their sources of potency. He writes: "Like Wordsworth's observation on poetry, San rock art should be seen as powerful emotion recollected in tranquillity" (1981, 76). Today, Kalahari San shamans recount their powerful spiritual experiences as everyone listens intently. The ancient paintings parallel the modern verbal reports, giving a vivid impression of the variety of things these modern-day shamans see in the spiritual world.

COMMON ARTISTIC GROUND

Westerners think of paintings as finite works of art to be admired and cherished over generations, centuries, and even thousands of years. The academic literature on ancient paintings is laden with references to "galleries of prehistoric art," as if the original artists painted the walls, then held open houses for visitors to admire their work. The southern African wall art lies in full view, in shadow and sun, changing character with shifting sunlight and clouds as if it was, indeed, on display. But the paintings were far more than art to the San; they were objects of significance, in and of themselves—images with potent ingredients of ocher and eland blood. In many San paintings, a figure or an animal enters or leaves a crack, climbs an uneven rock surface, or emerges from the shelter wall. These figures may actually be artistic representations of a specific San belief: that an underground journey takes the shaman to the spirit world. Perhaps the rock shelter was seen as an entrance into the spirit world, the wall itself being a kind of curtain between the living and supernatural realms. Thus, the shaman could coax an animal and other inhabitants of the spirit world from behind the rock, and perhaps paint their images on the rocky veil itself. Whether this was dangerous work, we do not know, nor can we discern if the shaman-artist prepared himself for this complex role. But the possibility of a rock shelter, a rock wall, alluding to another world is a compelling one to contemplate.

Lewis-Williams believes that the San paintings continued to have supernatural power associated with them long after their execution. His informant "M" took him to a cave, arranged his hands on a depiction of an eland, and explained that the potency of the eland would flow into a person in this manner, giving him or her special powers. The San both looked at and touched their paintings. We know this because some friezes display patches of paint that have been worn entirely smooth, most likely through hand contact. In some rock shelters, the walls bear imprints of human hands: the San covered their hands with paint and then placed them on the wall. Clearly these were not mere paintings of hands, but involved ritual touching of the rock.

"M" showed Lewis-Williams how San people danced in the painted rock shelter to which she took him—how they raised their hands during the dance and turned to the paintings when they wished to intensify their potency. As they looked at the paintings, the potency flowed from the images and entered into them. Thus, the painted images helped to form and constrain the stream of mental images that flowed through the dancers' minds as they hallucinated, and therefore, along with the dancing, clapping, and singing, the painted figures controlled the spiritual experiences of the shamans and other members of the group.

Why, then, the jumble of paintings at many sites? In time, some rock shelters may have acquired exceptional potency. Here, the artists returned again and again, adding new images on top of old ones, building up palimpsests of paintings.

San rock art was painted by shaman-artists living in a semitropical savanna homeland quite unlike the bitterly cold late Ice Age world of the Cro-Magnons. But can one use the San experience to throw more light on the engravings and paintings in the deep chambers of Altamira, Lascaux, or the Grotte de Chauvet? Here one treads on shaky scientific ground, for any comparisons have to bridge between fourteen and thirty millennia. However, Lewis-Williams points to some general similarities that emerge from the scientific investigation of Cro-Magnon caves. His research, with its emphasis on shamans and trance, suggests that much Cro-Magnon art was a form of shamanistic expression. He points to the placing of images in the depths of dark caves, the presence of human-animal figures, and the close relationship between the figures and the surfaces upon which they were painted as evidence for the presence of shamans. The combination of representational and geometric images also carried an undertone of altered consciousness. Furthermore, the Niaux Cave paint analyses tell us that the Cro-Magnons attached great importance to the ingredients in their paintings and that they painted over existing figures with meticulous care, just as the San did.

Thus, like the San, the Cro-Magnon artists may have acquired their images in trance. The ceremonies may have taken place in the open air, inside rock shelters, or in large chambers not far from the depths of deep caves. They may have attracted large numbers of people, who witnessed shamans obtaining their visions. The shamans and many of those present may have experienced a range of altered states of consciousness. Perhaps those in the deepest trances used psychotropic drugs; others may have been caught up in the ecstasy of the dancing and music. But everyone shared some of the insights of the shamans. Archaeologists can assume, based on scientific analysis, that altered states of consciousness and painting or engraving played a major role in the lives of both late Ice Age Cro-Magnon people and San hunter-gatherers of the past few millennia.

chapter four

FERTILITY AND DEATH

For tens of thousands of years, Stone Age hunter-gatherers lived in tiny family bands, moving across large hunting territories in established patterns according to the season. At times of plenty, the nomadic bands gathered. During these periods, marriages were arranged, initiation ceremonies were performed, and exotic goods like marine shells and commodities such as fine toolmaking stone were exchanged, before everyone dispersed for the lean winter months.

In the course of a lifetime, the average Stone Age person encountered few people by today's standards, spending most months with only his or her own family. The life of a hunter-gatherer was very flexible, and always in a state of flux. If two individuals quarreled, they could settle their dispute by simply moving away to another band. Thus, human societies of the Stone Age split and fissured regularly, surviving by virtue of this flexibility, even though there was usually constant tension within the small groups.

Farming altered human life beyond recognition. Once humans began to farm, they changed from nomads into permanent settlers, occupying small villages. Their lives unfolded according to new rhythms and realities. No longer did the movement of game herds or the seasons of wild plants determine lifestyles. Instead the continuous cycles of planting and harvest, life and death, governed human life.

Crowded in small villages, closely tied to their lands, early western Asian farmers dwelled in a constricted world, bounded by the family, household, and fellow kin. The villagers focused their energies on their families and on the task of raising crops and tending herds; beyond the village fields lay the unknown. Almost immediately, the dynamics of human society changed beyond recognition, as the relationship between people and their land, and between the living and their ancestors, assumed a central role in human life. This relationship persists to this day in many village farming societies.

I have visited small Egyptian villages by the Nile that are reminiscent of these ancient farming villages. The mud-brick houses crowd together around small courtyards, set in a maze of lanes and paths. Cleared fields surround the villages with a sea of brown and green. The sun beats down inexorably, and the constant north wind blows fine dust in the air, stirred up by people's feet. The crowded houses appeared to crumble before my eyes, trickling into the soil, repaired again and again with bricks fashioned from the same earth. Come sunset, the warren comes to life. Herdsmen and their cattle return from the fields and crowd the narrow alleys. The village pulses with renewed life, the rhythm of evening. As the sun sets, black-clad women prepare the evening meal, as itinerant vendors fill the quiet dusk with their throaty cries. There is a timelessness about the fading light, as day passes into the night.

FLOTATION: DOCUMENTING THE CHANGEOVER

In the 1950s, most experts believed that humans changed over from hunting and gathering to farming sometime before 4000 B.C., this date being determined from less than a handful of carbonized seeds and a few radiocarbon dates from the Fayyum Depression west of the Nile River. No one had found an early farming village elsewhere in western Asia. At this time, University of Chicago archaeologist Robert Braidwood searched for early farming villages in the foothills of the Zagros Mountains. He pioneered the multidisciplinary approach to such sites, taking botanists, a geologist, and animal bone experts with him into the field. However, for all his expertise at excavation and teamwork, Braidwood did not have the technology to acquire large samples of ancient plant remains. He relied entirely on finds of carbonized seeds in hearths and storage pits—at best rare discoveries. Inevitably, he could obtain only a very incomplete picture of the earliest farming economies.

Thanks to new recovery technologies, we now know that the economic changeover took place over a few centuries around 8000 B.C., and may have resulted from a combination of cooler, drier climatic conditions, and the need to feed growing hunter-gatherer populations by supplementing wild products with limited food production.

When Yale University archaeologist Andrew Moore excavated a large occupation mound at Abu Hureyra in Syria's Euphrates River Valley in the late 1970s, he discovered a hunter-gatherer settlement lying under a very early farming village. Using sophisticated flotation techniques, he acquired enormous samples of plant remains from the dry, ashy deposits.

Like so many important scientific methods, the technique of flotation was developed from a simple idea. In the 1960s, archaeologist Stuart Streuver, confronted with the problem of recovery of tiny seeds from ancient villages at Koster, Illinois, tried to screen the deposits through water. He discovered that seeds and other fine plant remains floated on the surface, while heavier sediment sank to the bottom. Streuver and other American excavators then began to use jury-rigged flotation machines, made up of fine screens and oil drums, to recover thousands of seeds, which allowed botanists to determine that Indian communities of 5,000 years ago exploited fall nut harvests and gathered large quantities of native grasses. More important, the new samples provided evidence that demonstrated deliberate cultivation of local plants, such as goosefoot, by 1000 B.C., long before maize came to eastern North America. Over the past quarter century, flotation technology has evolved rapidly. Mechanized flotation machines process samples much faster, allowing for the collection of enormous seed assemblages, which can then be analyzed statistically.

In the Abu Hureyra excavation, researchers took large column samples of the occupation levels. They poured each sample into a large flotation tank set at a higher level. An inlet pipe pumped air into the body of the tank at a constant rate, while a small amount of detergent was added to the water to help separate the seeds from the soil. The fine elements floated to the surface and were carried away into 2 gossamer-fine screens that caught the finest residues. Meanwhile, the heavier elements sank to the bottom of the tank and were flushed out onto a fine mesh screen, thus enabling the operators to recover small beads, tiny stone tools, and other minute objects from the fine sludge. As a result of flotation, Andrew Moore and his colleagues recovered 712 seed samples from Abu Hureyra, some of them containing more than 500 seeds from over 150 different species. Botanist Gordon Hillman had so many seeds to work with, that he was able to study the ancient landscape almost as easily as if he had been there in person. Thus, a chronological history of the economic transition was pieced together.

Abu Hureyra overlooked the Euphrates floodplain and extensive grassland steppe. In 9000 B.C., a time of more abundant rainfall than today, open forests rich in nut-bearing trees lay within easy walking distance. Today they are at least 120 kilometers west of the site. Hillman's flotation samples showed how the first inhabitants exploited hackberry, pistachio, plum, and medlar trees on a large scale; as well as wild wheat and rye, which grew close to the site. Within a few centuries, people stopped gathering nuts from the forest fringe, possibly because it was no longer within easy reach. Much drier conditions descended over the Euphrates Valley. In response, the Abu Hureyra people increased their exploitation of wild cereal grasses, which would have flourished on the floodplain as the forest retreated in the face of much drier conditions. Soon even these resources diminished, as Abu Hureyra lay in the grip of a prolonged drought. The people abandoned their settlement by 8000 B.C. Three centuries later, a new village arose on the same mound—a closely knit community of rectangular, mud-brick houses. The flotation samples from this occupational layer yielded large quantities of domesticated wheat and barley, and far fewer wild plant forms.

Flotation documents a dramatic change in human subsistence caused in large part by drought, rising local populations, and the need to feed the growing masses. One logical solution was to cultivate wild grasses, to extend the range of existing cereal stands. Within a few centuries, foragers had become village farmers, with much closer ties to the land, and a radically changed social environment. Interestingly, biological anthropologist Theya Molleson has shown how women's skeletons from these new farming villages show clear signs of malformation resulting from long hours spent on their knees grinding grain.

Excavations like Abu Hureyra offer hope that one day we will be able to document the dramatic changeover from foraging to agriculture in such detail that we will also be able to chronicle the social, and even ritual, changes that resulted from sedentary settlement and the resultant closer bonds to farming land.

ANCESTOR CULTS

Within a few centuries, farmers transformed the environment of western Asia through the clearing of fields, the planting and harvesting of crops, and the grazing of herds. Deforested hillsides and floodplains turned into a patchwork of small fields that surrounded permanent villages close to rivers large and small. The nomadic hunter-gatherer of centuries earlier became the sedentary farmer, living side by side with other households, usually in small, flat-roofed mud-brick houses separated by narrow alleyways.

Although farmers inherited the botanical and zoological knowledge of their hunter-gatherer predecessors, they focused their efforts on a few acres, passed down from one generation to the next. The stream of time flowed onward, season after season, in an endless rhythm of winter, spring, summer, and fall, where the land itself lived and died, just as human life flowed and ebbed in a constant passage of one generation to the next. Homer himself wrote of this renewal in the *Iliad* (Fagles 1990, 200):

> Like generations of leaves, the lives of mortal men,
> now the living timber bursts with the new buds
> and spring comes round again. And so with men:
> as one generation comes to life, another dies away . . .

Within the span of a few short years, the living became the dead—the revered ancestors, who had once farmed the same lands in earlier years. The cycle of time unfolded from one generation to the next, with each generation assuming that their descendants would inherit the same earth, the same world, following the pattern that they and their ancestors had enjoyed. Such a cyclical view of life, of time itself, fostered a close relationship between the living and their ancestors.

The farmers' sacred places reflected this intimacy with the dead, now the guardians of the land. Like modern-day subsistence farmers, they clothed their surroundings with symbolic meaning. Thus, their sacred places moved into the household, and ancetor worship—which was a family ritual for these people—was usually conducted within the privacy of one's home.

Ancestors, land, and farming go hand-in-hand. From the very earliest days of agriculture, we find ancient ancestors in small cache pits, buried under Jericho house floors, also at Abu Hureyra. Apparently, ancestor worship was a family ritual, conducted within the privacy of one's home.

ANCESTORS AT JERICHO

The 9,000-year-old plastered skull with the aquiline nose stared at me through slightly hooded eyes, with a gaze that penetrated the inner recesses of my consciousness. This ancient came from Jericho in the Jordan Valley (figure 4.1). Resting on a museum table, timeless and serene, an inhabitant of a very different world from that of the Cro-Magnons, he had emerged from an archaeological excavation over forty years ago.

Jericho is one of the great archaeological sites of western Asia, home of famous Bronze and Iron Age cities, and one of the earliest farming communities in the world. In the 1950s, British archaeologist Kathleen Kenyon approached Jericho's deep city mound with a disciplined, stratigraphic approach that started by cleaning up still-open and eroded trenches dug by earlier excavators. She then sank a deep vertical cutting down to the base of the great mound, and began excavating each occupation level separately and tracing the transitions in architectural and pottery styles. Kenyon collected animal bones and plant remains, as well as charcoal samples for radiocarbon analysis. Her combination of disciplined excavation, precise stratigraphic

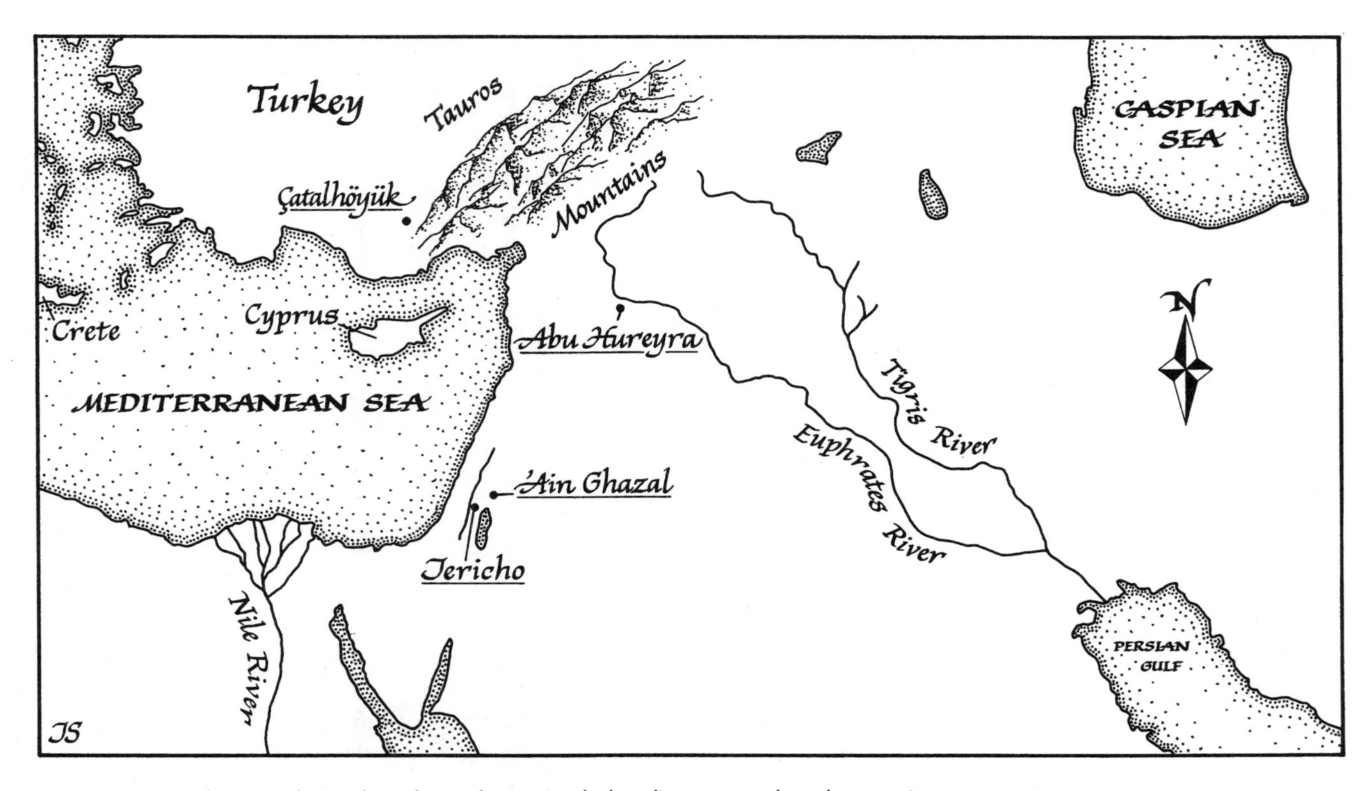

Fig. 4.1. Map showing the archaeological sites (underlined) mentioned in chapter 4.

observation, and radiocarbon dating produced startling finds: farming settlements going back to at least 8000 B.C., the earliest town walls in the world, and evidence of ancestor worship.

The plastered skull first appeared as a smooth cranium surface projecting from a trench wall. Kathleen Kenyon looked at the 9,000-year-old skull cap for days, but refused to move it from the cutting face deep in the Jericho mound. Archaeologists dig cuttings with vertical walls, so that they can record the complex occupation layers of past centuries—the archives of long-forgotten settlements. The skull was left in place for the remaining weeks of the 1953 season. Eventually, Kenyon drew the stratification in her notebook and reluctantly gave permission for its removal. The site supervisor set to work with trowel and dental picks, expecting to uncover just a human skull, but he soon encountered a layer of fine clay, which appeared to surround the cranium. Nothing could have prepared the archaeologists for what emerged from the wall: a human skull covered with clay plaster, that was molded to form the features of its owner, the eyes inset with shells (figure 4.2). Two more plastered skulls could be discerned at the back of the small hole.

As so often happens, one of Jericho's most spectacular finds came to light at the very end of the field season, when nearly everything was packed up. Kenyon and her colleagues stayed on in considerable discomfort for another week, and excavated what turned out to be a cache of plastered skulls. For 5 days, they worked their way into the vertical wall, exposing 7 plastered skulls crammed in a tumbled heap, surrounded by stones and hard earth. The skulls lay on top of one another in a small pit, their bones so friable that they had to be hardened in place. When all were removed, they provided a unique portrait gallery of Jerichans from more than 9,000 years ago.

Kenyon suspected at once that she was dealing with an ancestor cult. The skulls lay in a special pit under a house floor and had been treated with great respect after being detached from their parent bodies. Each head was packed with solid clay. The modeler had enveloped the lower part of the skull, from the temples downward, with clay plaster, leaving the crown of the skull bare. Only one skull had retained its jaw: the others had a mandible modeled over the upper teeth, giving them a somewhat chubby appearance. Flattened

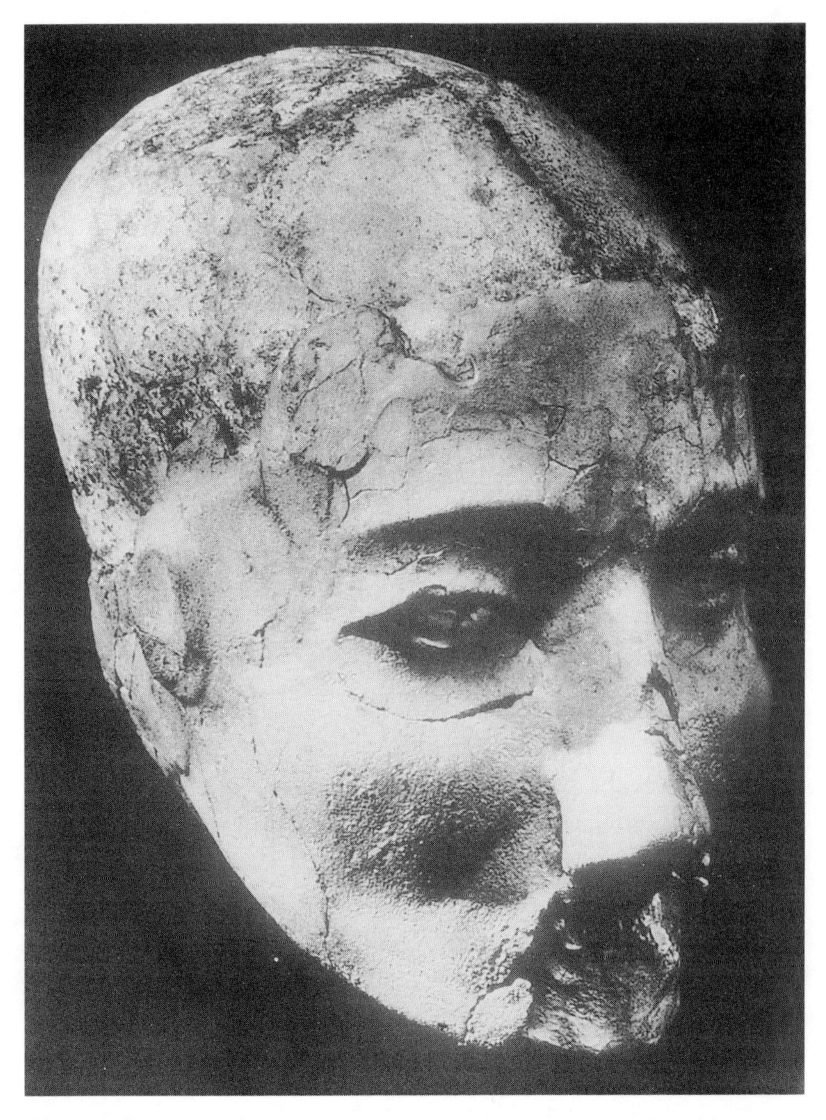

Fig. 4.2. Plastered skull from Jericho. Photograph from The British School of Archaeology at Jerusalem.

clay covered the base of each cranium. One skull was colored a reddish-flesh hue. Another bore bands of brown paint, as if it once wore a headdress. Their cowrie-shell eyes, which lay in fine clay that filled each eye socket, glared unblinkingly at the beholder, giving an impression of inner wisdom. Each head was a naturalistic, individual sculpture of an ancestor, with nose, mouth, and ears molded with remarkable delicacy.

Three years later, in 1956, Kenyon unearthed two more plastered skulls from underneath a slightly earlier residence at the same general location. This earlier dwelling also yielded about thirty skeletons packed into a small area, many of them with missing skulls. Most of the skulls had been removed from their bodies after they had sufficiently decayed, so that separation of the limbs from the trunk, and the skull from the neck and jaw could be achieved successfully. In later seasons, Kenyon uncovered layer after layer of houses. Almost all of them covered burials of skeletons, many of them with the skulls removed.

The Jericho plastered heads come from one of the earliest farming settlements in the world. The ancient biblical city, famed for Joshua's siege chronicled in the second book of Kings, was located at a strategic caravan route that linked the Jordan Valley with areas far inland and with the Mediterranean coast. The city itself lay close to a perennial spring and the Jordan River. This fortified town flourished only a few centuries after the people of the Jordan Valley turned from hunting and plant gathering to cereal agriculture and animal domestication.

Kathleen Kenyon caused an archaeological sensation in the late 1950s, when she dug down to the very bottom of the great Jericho city mound and uncovered a tiny shrine by the spring, radiocarbon dated to nearly 8000 B.C. Immediately above, she found a small town, protected by a massive stone wall of boulders carried in from half a mile away. On the outer side lay a rock-cut ditch 9 meters wide and 3 meters deep. A solid stone watchtower 9 meters across lay on the inner side of the wall, with a stone-built staircase of 20 steps in its center, leading to a horizontal passage. The passage was tightly-packed with skeletons, which had been deposited there long after the tower itself fell into disuse. No one knows whether they were war casualties, or, more likely, placed there after the watchtower stopped being used.

Close to where she had found the plastered skulls, Kathleen Kenyon unearthed a small room, which had been formed by dividing off a section of a modest residence. The builders had placed a small niche in the end wall, and had set a rough stone in the base to serve as a pedestal. A volcanic rock, imported from 13 kilometers away, lay close by. It had once stood in the niche, chipped to a pointed oval. Kenyon surmised that this was a cult object, perhaps personifying a supernatural force. Herein lies a basic archaeological principle: Context is everything. Had the volcanic rock been found in isolation, it would have been recorded as merely a fragment of shaped, imported rock. But it came from the collapsed rubble of a building containing what appeared to be a family shrine. Therefore, Kenyon interpreted it as a religious object. In yet another structure, Kenyon discovered 3 figurines—a bearded man, a woman, and a child. These 3 figurines were perhaps symbolic of family life, of the continuity of generations.

Jericho is not alone in its ancestor cults. At another early farming settlement called 'Ain Ghazal, also in Jordan, archaeologist Gary Rollefsen recovered a series of human statuettes made from plaster, which may once have stood under a special shelter. The sexless figures with modeled features apparently represented some form of ancestor cult (see information in box, p. 81).

For these early farming communities, ancestors were the very essence of life and death: guardians, protectors of the living, denizens of the spiritual world. The ancestors acted as intermediaries between the living and the dead, for individuals and society as a whole. Ancestor worship was centered in the household and comes down to us in the form of venerated clay figurines, or carefully rendered sculptures of earlier generations modeled in clay plaster on their skulls. The artifacts of the ancestor cult at Jericho and elsewhere offer powerful clues regarding rituals that were as old as farming itself.

THE MOTHER GODDESS

The cycle of birth, death, and rebirth lies at the heart of ancient thinking regarding creation. In many farming societies, the earth is female, the source of life and rich harvest. Worshiped from the earliest days of farming, and perhaps much earlier, the myth of a universal Earth Mother flourished long before archaeological evidence had given such a goddess validity. References to an age-old universal Mother Goddess can be found in the classical literature of Homer.

THE 'AIN GHAZAL FIGURINES

'Ain Ghazal ("the Spring of the Gazelles") on the outskirts of Amman, Jordan, is, like Jericho, one of the earliest farming villages in the world (figure 4.1). Nine thousand years ago, only a few centuries after farming began, the inhabitants lived with their tethered goats in a crowded settlement of oval houses with carefully plastered floors. In 1985, a bulldozer operator carving out a suburban road uncovered some crumbling plaster statues 18 meters below the surface. The humanlike figures lay facedown in a pit below a white-plastered hut floor. Archaeologist Gary Rollefsen and conservator Lynn Grant lifted the figurines and the surrounding soil in a single earthen block for laboratory examination.

The Smithsonian Institution's Conservation Analytical Laboratory in Suitland, Maryland, undertook the delicate task of disentangling the incomplete figures from their matrix. The extraction of the figures unfolded like an archaeological excavation. Technicians used X-ray diffraction and electron microscopes to identify the baked clay fragments in the heart of the block. The conservators divided the boxed earthen block into a small grid and excavated the figure fragments, recording the precise position of each piece before lifting it. Then they hardened the brittle clay with special chemicals as the body parts emerged from the surrounding soil. The "dig" completed, the conservators reconstructed the figures from a three-dimensional jigsaw puzzle of arms, limbs, and body fragments, using informed guesswork and precise measurements to fill in the gaps between the original plaster. They employed acrylic putty, shaped on an epoxy form, and rods to prepare the figures for display. This epoxy can be dissolved in the future for further analysis or attempts at different reconstructions.

The compelling human images that were pieced together by the technicians' expert handiwork had stared at me with serene, almost ghostly confidence from a display case in the Smithsonian's Sackler Gallery. Androgynous and near life-size, the 9,000-year-old plaster figures gazed wide-eyed across the centuries, as if possessed with boundless wisdom (figure 4.3). It had felt as if their eyes were following me around the room—their impact upon me lingers still.

To create each of these figures, the 'Ain Ghazal artists started with armatures of twine-bound reeds, which survived as impressions on the interior surfaces of the plaster. They then molded fine plaster over the reeds to form the body and legs, and even sculpted fatty folds at the base of the thighs. Both arms rest by the sides. The long neck perches atop the body, supporting a head with a calm, expressive face. The mouth smiles slightly; cheekbones stand out in delicately modeled, pinkish clay; and the carved eyes, inlaid with bitumen, stare unnervingly. The figures are remote, almost otherworldly, emissaries from a vanished society. Several of the figures have two heads attached to a rough torso, perhaps symbolizing husband and wife, possibly twins, or some form of long-forgotten dual deity.

Something seemed familiar about the figures. They reminded me of shop-front mannequins waiting to be dressed. Their stylized heads, rough torsos, and realistic legs could have once worn garments, perhaps skirts or ceremonial gowns. This would also explain the odd shapes of the heads, which are ideal for wearing headdresses, or even flowing scarves. Pure speculation, but I can think of several historical analogies. The ancient Egyptians and Sumerians often clothed or painted statues of deities; and the cult statue of Athena in the Parthenon at Athens wore a magnificent saffron-covered robe that took nine months to weave. Even today, effigies in Catholic churches wear robes that change with the religious season. Perhaps the 'Ain Ghazal figurines were adorned with different regalia for separate festivals, when the ancestors assumed different roles. We can only guess.

When archaeologists discovered clay female figurines in early farming villages throughout western Asia, the Aegean Islands, and southeast Europe, they recalled the myth of the primordial Mother Goddess, and applied it to their interpretations of these finds. Since farming and fertility went together, it seemed logical to add an Earth Mother, a Mother Goddess, to the equation.

Even Kathleen Kenyon was tempted by the idea. She unearthed a unique shrine at Jericho, a substantial rectangular chamber, measuring 6-by-4 meters, with a small, fire-scorched basin in the center. Finely burnished plaster covered the floor, walls, and basin. Rounded annexes with domelike walls lay at both ends of the building. Kenyon

Fig. 4.3. 'Ain Ghazal figurines. Photograph from The Arthur Sackler Gallery, Washington D.C., and reproduced with the permission of the Jordan Department of Antiquities.

believed that this was a religious structure, for inside were found 2 small female figurines, each with a long, flowing robe gathered at the waist. They are both in sitting positions, with their arms akimbo, their hands placed under their breasts. Kenyon wrote: "They are strongly suggestive of a Mother Goddess or fertility cult, very natural for a community closely dependent on the productivity of the soil for their existence" (1957, 59).

In recent years, the "Mother Goddess" theory has gained new followers thanks to the work of the late Marija Gimbutas of the University of California, Los Angeles, who considered the great Mother Goddess the "cosmic giver and taker of life, ever able to renew Herself within the eternal cycle of birth, death, and rebirth" (Gimbutas 1991, 86). Gimbutas was a redoubtable scholar with a distinguished career in central European archaeology. An expert on early central European farmers, Gimbutas placed women and their fecundity at the heart of early farming life. She developed her theory by studying hundreds of clay figurines from ancient agricultural villages in western Asia and Europe, focusing on the style and decoration of individual figures. Gimbutas joined her figurines in a "threaded necklace" of archaeological finds that traced the Mother Goddess through dozens of societies from Jordan to Czechoslovakia, over a span of more than 20,000 years.

In later years, Gimbutas became preoccupied with her theories of an "Old Europe," where matriarchy prevailed, even as a new generation of computerized, pattern-based artifact studies undermined her comparisons. Gimbutas traced the beginnings of her primordial cult back to an explosion of artistic creativity in the late Ice Age. Some Cro-Magnon artists of 20,000 years ago painted "several kinds of abstract and hieroglyphic symbols" along with animal subjects, in which Gimbutas discerned "an iconography of the Goddess." These symbols persisted, so she claimed, long after the Ice Age, culminating in a great flowering of the Mother Goddess cult and a period of great prosperity within a peace-loving, egalitarian, matrilineal society, which began in the eastern Mediterranean and quickly spread to eastern Europe. This "Old Europe" lasted until about 3500 B.C., when a warlike, patriarchal culture from the Black Sea displaced the placid peoples of earlier times. "Paradise Lost," as a cynical reviewer in the *New York Times* put it.

When I read Gimbutas's books, it was as if she herself took me by the hand, and, like a prophetess with an archaeological key to the Sacred Mysteries, led me back into a matriarchal past. We passed from site to site, and image to image with effortless ease, following Mother Goddess symbols through the millennia from the late Ice Age to modern Czechoslovakia. I waded through page after page of descriptions and interpretations of figurines, art motifs, shrines, and dwellings that were remarkable for their uncritical subjectivity. A typical example: "More than half the figurines of Old Europe appear to be nude above the hips, hence we presume they represent goddesses or priestesses as they enact rituals."

The entire Mother Goddess thesis rested on interpretative subjectivity rather than critical analysis. Gimbutas's stirring saga of matriarchy falls into New Age literature and attracts feminist students of patriarchal institutions in Christianity and Judaism. Radical ecofeminists examining women's roles in ecological and environmental issues often embrace the Mother Goddess cult as an example of ancient religion, with its emphasis on the close relationships among fertility, motherhood, and earth.

Ancient figurines of all kinds have two statuses, one as "art," the other as "artifact," with precise contexts in time and space. Gimbutas focused on the figurines as art, perhaps because many such objects came from poorly documented archaeological contexts, thus leaving her only the objects to work with. She tended to ignore chronology, except in general terms, and rarely studied the precise locations of figurines in dwellings, middens, or archaeological layers. A very different picture emerges, however, when the context of early western Asian female figurines is examined.

INTERPRETATION VIA SCIENTIFIC CONTEXTS

The study of ancient religion is, fundamentally, the study of ancient iconography, for only durable symbols of intangible beliefs survive through the ages. This means that one must look at religious iconography in three scientific contexts: First, any objects that represent this iconography must be dated as accurately as possible, just as were the Jericho portrait heads. Second, the excavator must establish the precise spatial context of each find. The law of association states that all objects found within the same occupational level are broadly contemporary: some are absolutely contemporary (e.g., grave furniture deposited with a burial), and others are more generally contemporary. At Jericho, for example, the male, female, and child figurines that Kenyon unearthed were in significant juxtaposition, as if there was an association between their positioning and the idea behind the grouping. Spatial context also has a broader significance. For example, portrait heads like those found at Jericho occur at other sites in the Jordan Valley, but they do not appear in Egypt or Turkey. This pattern of distribution hints at a local cult. But if you find similar artifact associations at other sites, can you claim that the local phenomenon represents a universal custom?

Our third scientific context is that of social association. What was the social organization of the society in which the inconography of, for example, female figurines occurred? Was it a simple, egalitarian culture with little evidence of social ranking; or an elaborate, state-organized society with a hierarchy of social ranks, as well as a pyramidlike ranking of deities; or was it something in between, where there was a form of centralized authority that organized the building of communal tombs or small temples? In egalitarian societies, shrines tended to be simpler, and most often confined to individual households. This is because these types of shrines did not require the resources that were necessary to build elaborate sacred places. It was probably no coincidence, therefore, that most ritual activity at Jericho took place in private houses.

If we examine the early female figurines of western Asia and Turkey utilizing these three scientific contexts, we arrive at a very different interpretation from Marija Gimbutas's of the universal Mother Goddess cult. An obscure, but telling study of early Greek farmers' figurines illustrates this.

MOTHER GODDESSES IN GREECE

Over the past 50 years, archaeologists have recovered the legs of 18 female figurines from 5 early farming settlements in Greece's northern Peloponnese. The villages lie in rugged terrain, and are between a half to several days' journey apart by foot and boat. Nevertheless, there are enough common artifacts among them to show that they formed part of an active trading network for many centuries. The 18 leg fragments are well made and carefully modeled. They all measure between 4 and 10 centimeters long and are usually broken at the waist. Almost all of them have protruding buttocks, a painted or incised pubic triangle, scars and incisions indicating a hand on the upper thigh, and painted linear decoration. Not surprisingly, their finders labeled them Mother Goddesses (figure 4.4).

Archaeologist Lauren Talalay further investigated these clay limbs and discovered that they were not pieces of complete Mother Goddess figurines, for no female torsos have come from any of the settlements. Thus, the legs were never intended as complete figurines. Talalay found that each of the legs was once attached to another; indeed one specimen from Franchthi Cave in the eastern Pelopon-

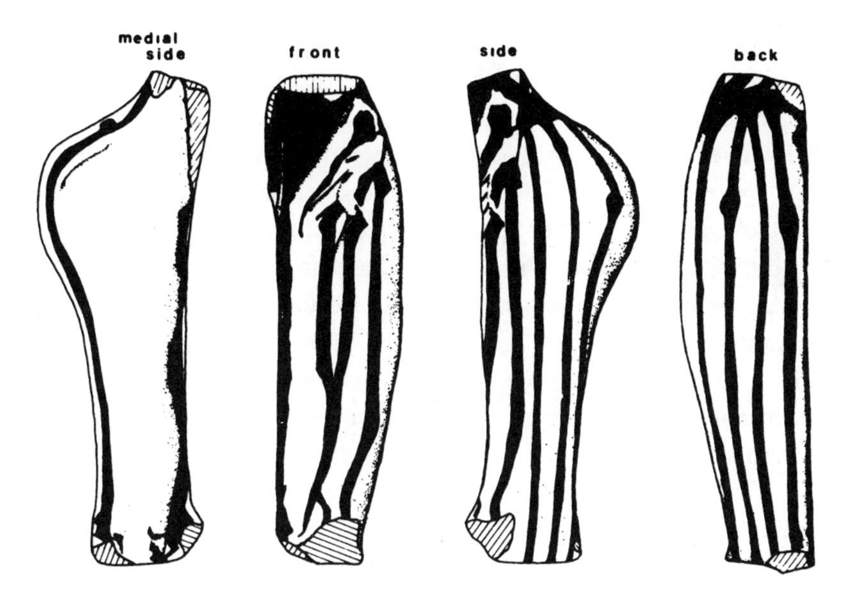

Fig. 4.4. Leg fragment from Franchthi Cave, southern Greece. One-half full size. Courtesy Lauren Talalay.

nese still retained part of a second leg. She admired the flat and well-burnished inner surfaces of the legs, a finish that made little sense, for the matching leg would have been placed alongside this inner surface. Had the makers intended for the legs to remain attached, she reasoned, then they would probably not have treated them in this manner, since burnishing and smoothing the surface would have weakened the joint between them. Additionally, each side had a distinctive painted design. She argued, therefore, that the legs were designed so that the two halves could be easily separated.

What, then, was the context in time and space of the legs, she asked? Unfortunately, all of them came from general habitation debris or disturbed layers. None had any unmistakable association with a shrine or other kind of sacred place. Therefore, to answer the question regarding context, Talalay turned to another source: classical literature. She remembered that the intentional splitting of objects sometimes served as a means of identification, or agreement, between two parties. The Greek word σψμβολα ("symbol") means a tally. Another symbolic meaning of this word, however, is an object broken into two halves by two parties as evidence of the identity of the presenter to the receiver. She soon found that the intentional splitting of objects has a long history in many different circumstances and cultures. The Greek writer Herodotus refers to split tokens being used to seal agreements between a man in the community of Miletus and Glaukos, an inhabitant of Sparta. Chinese, Japanese, and Romans also used halved tokens or split objects as bases for identification, in economic transactions, or to convey the bona fides of messengers, much as some people use the two halves of a dollar bill for this purpose today. Therefore, tokens have several functions. They are used in marriage contracts, as emblems of membership in kin groups, or as a means of identifying trading partners, to mention only a few.

Talalay wonders if the classical Greek σψμβολα had a deep antiquity in humble tokens like the eighteen split legs of much earlier times. In an attempt to narrow down the social context of the legs, Talalay compared the painted designs on the legs with contemporary pottery decoration, and found close similarities. She was struck by the homogeneity of the painted vessels found in the five sites. The pots from each settlement bore painted designs that were too complex to be learned by anyone other than through extended face-to-face lessons. Talalay therefore concluded that the five farming communities were exogamous—that people sought wives in other villages, since local populations were too small to be biologically viable through the sole practice of internal marriage. The women of these communities were potters, and carried their carefully learned skills with them when they married, sometimes into settlements several days apart. At the same time, close ties of reciprocity between widely separated villages would have been essential to ensure adequate food supplies in years when rains failed and famine loomed in some areas but not in others. Thus, the five communities, and others, were, of necessity, in frequent contact, bartering grain and exotic materials that were used for toolmaking, such as the obsidian (fine volcanic glass used for stone tool manufacture), marble, and nonlocal flint that were found in each site. Perhaps, says Talalay, convenient small tokens such as clay legs may have served as identifying tokens in intervillage trade. Far from being Mother Goddesses, the small limbs were an integral part of daily life, to the point that they were simply discarded with the trash when no longer needed.

The moral of Lauren Talalay's research comes across loud and clear. The credibility of the Mother Goddess thesis depends on archaeological context, not mere stylistic resemblances. Unfortunately, almost no early farming sites have yielded more than a handful of clay figurine fragments, many of them sexless, and most of them broken pieces found in occupation layers or rubbish dumps. Only one site—a large early farming community in Turkey—brings us face to face with intact sacred places, and provides evidence of both the worship of goddesses and the worship of ancestors.

THE SHRINES OF ÇATALHÖYÜK

Çatalhöyük on Turkey's Konya Plain—a major center for obsidian trade—was a town before 6000 B.C. (figure 4.5). The town's sun-dried brick houses with flat roofs rose in terraces above one another, the bare walls forming an outer wall for the settlement. People entered their houses from the roof, climbing down a wooden ladder into a well-plastered main room with benches, a hearth, and a wall

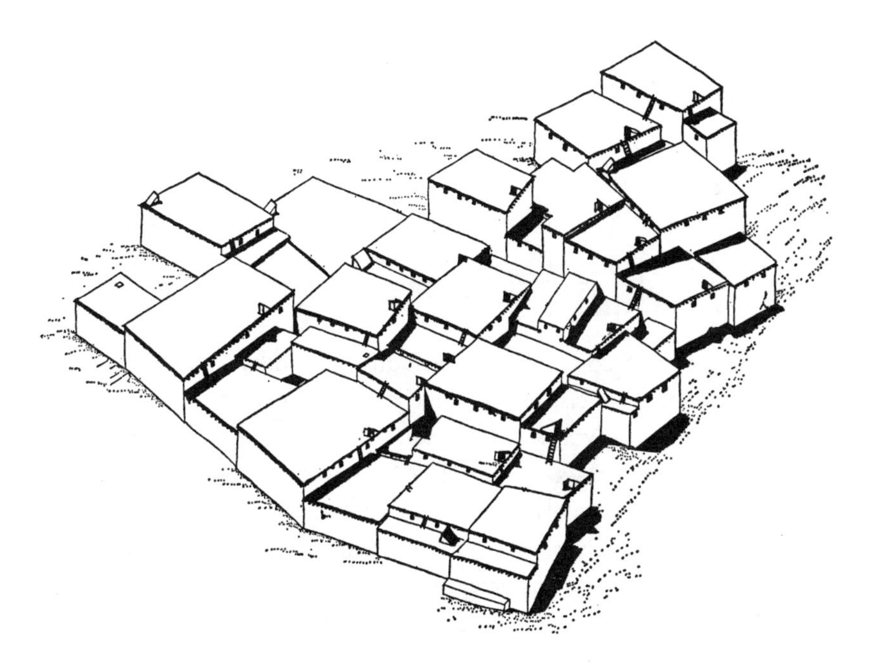

Fig. 4.5. Schematic reconstruction of terraces of houses and shrines from level VI at Çatalhöyük. Courtesy Grace Huxtable.

oven. Most houses had their own storerooms, often fitted with clay grain bins. Here, as at Jericho, life unfolded at the household level; each family fed itself, with the women grinding grain on querns set into the floors of their homes.

Çatalhöyük was a well-planned community. Its builders were aware of the need for orderliness within such a cramped environment, so all of the houses had the same general floor plan. Even doorways and bricks were measured to standard sizes using adult hands and feet as units of measurement. The architects were conservative, always constructing rectangular houses, sized according to the household, the outer walls forming a defensive perimeter for the town. A potential attacker would have to climb on the roofs and would be at a tactical disadvantage from defenders in the houses underfoot. The greatest danger was not enemies, however, but fire. Fast-moving blazes regularly destroyed much of the town. Eight thousand years later, archaeologists give thanks for these disasters, which buried priceless information regarding the town's shrines under layers of ash, charcoal, and mud-brick.

Çatalhöyük was no ordinary settlement. At 13 hectares it was a much larger village than early Jericho. Its citizens prospered off cereal agriculture, stock raising, and, above all, long-distance trading in lustrous, black obsidian obtained from the peaked cone of Hasan Dag and other volcanoes about 130 kilometers to the east (see information in following box). For ten centuries, between 6000 and 5000 B.C., Çatalhöyük flourished, accumulating sufficient food surpluses to

---

OBSIDIAN SOURCING

Obsidian, a fine volcanic glass, was highly prized by ancient stoneworkers for artifacts, mirrors, and other ornaments. This resplendent material was widely traded in many parts of the world, including the eastern Mediterranean, Mesoamerica, and the southwestern Pacific. Fortunately for science, each obsidian source has its own distinctive "signature" in the form of trace elements in the stone that can be identified by spectrographic analysis, thereby allowing archaeologists to "source" obsidian fragments hundreds of miles from their original location. Such characterization studies began in the

1960s, when British archaeologist Colin Renfrew and others pinpointed obsidian sources in central and eastern Turkey. The concentrations of the main elements of the volcanic glass—silicon, oxygen, calcium, and so on—are broadly similar everywhere. However, the trace elements, measured in a few parts per million, vary from source to source and can be measured using optical emission spectrometry or inductively coupled plasma atomic emission spectrometry.

The outer electrons of the atoms of every chemical element emit light of a certain wavelength and color when heated to a high temperature. The resulting light is composed of several wavelengths, which can then be separated into a spectrum, by passing them through a prism or equivalent device. The presence (or absence) of various elements can be established by searching for the spectral lines of their characteristic wavelengths. The sample is atomized and excited in a stream of argon plasma to produce high temperatures that reduce the potential of interference between different elements. Trace element analyses in western Asian obsidian have also relied on neutron activation analysis, in which the nuclei of the atoms of a sample's elements are bombarded with thermal neutrons. This process produces radioactive isotopes of the elements, the intensity of which can be measured accurately.

Such sourcing methods have determined that most western Asian obsidian came from Turkish sources. Central Anatolian obsidian, like that from Hasan Dag, passed south and westward down the eastern Mediterranean coast to the Jordan Valley. The eastern Anatolian sources provided obsidian to trade routes through the Zagros Mountains and into Iran. A pattern of exponential falloff emerged from an analysis of the distribution of obsidian at dozens of excavated sites. Plotting the quantity of obsidian against distance on a logarithmic scale showed a straight-line falloff from a central supply zone within 320 kilometers of the source where people collected raw material for themselves, to a rapid drop in quantities as distance from source increased.

---

maintain a large settlement that was still essentially a village, where both secular and religious life were based in the household.

British archaeologists have a long tradition of excavations at Çatalhöyük. James Mellaart worked there in the 1960s, and Ian Hodder of Cambridge University began a long-term field project in the 1990s. Together, they have pieced together a complicated mosaic of sacred places—household or kin group shrines—within the confines of the settlement.

Mellaart's excavations exposed about an acre of Çatalhöyük, a total of 139 rooms, approximately 40 or more of which were what he called "shrines,"—elaborately decorated spaces adorned with exotic figures and figurines, which tended to blend into residential spaces with their plastered sleeping benches, hearths, and ovens. Mellaart developed a set of rigorous scientific criteria for shrines, which included detailed wall paintings with obvious ritual significance; plaster reliefs of deities, animals or animal heads; cattle horns set into benches; the presence of cult statues; and human skulls set on platforms (figure 4.6). By applying these criteria, and stressing archaeological context, he tried to decipher the significance of religious beliefs and rituals at Çatalhöyük.

Mellaart began with the orientation of wall decorations. Çatalhöyük's shrines bore wall paintings or plaster reliefs, executed according to well-defined rules. Death scenes were always depicted on the north and east walls, with the dead buried beneath them. Birth scenes appeared on the west walls, whereas pictures of bulls appeared only on north walls, facing the distant Taurus Mountains. Animal heads associated with red-painted wall niches adorned the east walls, but goddess figures, and bulls' and rams' heads embellished any wall. The south wall, usually the kitchen, was rarely decorated.

The wall paintings themselves were not mere decoration. They were painted, then "erased" with a layer of white paint after use. Later, another frieze would adorn the same wall, only to be painted over soon afterward. Clearly, these wall paintings had a profound, transitory significance, perhaps as an element in powerful ritual performances that unfolded in the shrine.

The artists drew plain and geometric patterns, flowers, plants, and other symbols, as well as human hands framing geometric and

Fig. 4.6. Reconstruction of the eastern and southern walls of shrine VI at Çatalhöyük. Bulls' and rams' heads are modeled in relief, whole horn cores (bucrania) adorn a bench. The entrance to the shrine was from the roof. Courtesy Grace Huxtable.

naturalistic designs. The naturalistic paintings included goddesses, human figures, bulls, birds, leopards, deer, and vultures. There were elaborate scenes of deer hunts, and of bulls cavorting with human figures. The artists also drew landscapes, including pictures of volcanoes erupting, and town and mortuary structures. Like Cro-Magnon and San art, some of the Çatalhöyük wall paintings are difficult to interpret, especially those that appear to represent fields of flowers or netlike patterns.

Three shrines have decorated walls where vultures are attacking human bodies, as if they are cleaning the newly exposed corpses of the dead before the family members recover the defleshed bones and bury them. In one case, the legs of the vultures are human, leading one to wonder if the scene depicts a ritual performed in vulture garb. Human burials lie below this painting. Several incomplete skeletons found in the excavations show that the Çatalhöyük people disposed of their dead in exactly this way. They exposed the deceased in some form of mortuary outside the town, where vultures cleaned off the corpses, leaving only the bones and dry ligaments behind. Then the relatives collected the bones, taking care to maintain the articulation of the body, and buried the dead wrapped in cloth or skins below the platforms of houses or shrines.

In another intriguing scene, hunters with their weapons dance, accompanied by some headless human figures wearing leopard skins. Most of the dancers advance to the left, where a large bull stands, but others cavort with stags. Mellaart believes that the headless figures in the hunting scene and others associated with vultures may represent ancestors (figure 4.7).

A unique landscape scene depicts the rectangular, huddled dwellings of Çatalhöyük in the foreground, while a twin-peaked volcano erupts in the distance. Fire spouts from the summit, and lava streams flow from the base. Dots cover the mountain, perhaps the dust that rained down from the eruption. The mountain must be Hasan Dag, the only twin-peaked mountain in Turkey, the source of the magical obsidian that brought the town its prosperity. Obsidian's volcanic origin may have linked it to the otherworld beneath the earth, the place of deities and ancestors. This potent material was a gift of earth, of the land that brought forth crops and life itself.

Looking at these scenes, with their combinations of humans, animals, and dots or other symbols, I was irresistibly reminded of the vivid rock friezes painted by San shamans in South Africa—recollections of trance experiences that joined the living and spiritual worlds in a powerful synthesis of potency. The paintings found at Çatalhöyük, however, are different, and thus we must assume that the meanings behind them are also different. Perhaps these people, in an altered state of consciousness, cavorted with headless ancestors and animals. Thus, their art—which depicts vultures eating human corpses—might actually be symbolic representations of the journey of the dead as they turn into ancestors.

ÇATALHÖYÜK'S DEITIES

The Çatalhöyük excavations also provided evidence for important deities. Bulls' heads and goddesses adorn Çatalhöyük's shrines. One large shrine from level VII, relatively late in the occupation, contained

Fig. 4.7. An enormous vulture attacks two headless figures, one crouched like a corpse, on its left side. Part of a frieze in Shrine VII, Çatalhöyük, Turkey. Courtesy Grace Huxtable.

a massive bull's head with meter-long horns, molded nostrils, and mouth set in a red-painted muzzle. The head sat across from the shrine's west wall, where a painting depicted vultures attacking headless corpses. Another elaborate shrine in somewhat earlier level

VI yielded rows of wild bulls' horns placed on benches and platforms. Mellaart believes this particular shrine was dedicated to a male deity.

A level VII shrine contained a painted relief of a pregnant goddess on its east wall. She wore a dress brightly painted with red, black, and orange patterns. The dress extended like a veil between her upturned arms and legs. Another shrine held the seated figure of a goddess, who also wore a painted dress. She sat between two posts, as if she were appearing to worshippers in a doorway. In yet another sanctuary, the goddess figure was modeled in bold relief with a small bull's head above her left hand. Another goddess figure is in profile, her long hair floating behind her in the wind. The artist outstretched and foreshortened her arms and legs to give the impression of rapid motion. She seems to be running, dancing, or just whirling in uncontrolled movement.

The people of Çatalhöyük also collected stalagmites from deep caves within the Taurus Mountains and brought them back to lie in the shrines alongside cult statues. Many of them resemble clusters of breasts, udders, or even human figures, perhaps collected because their shapes recalled important elements in local ritual. Statuettes of goddesses carved from stalactites reinforce the notion of a powerful subterranean otherworld.

Unlike other archaeological sites, the small cult statuettes found at Çatalhöyük come entirely from shrines. They are very different from the small, stylized clay figurines of animals and schematized humans stuck between house bricks or thrown out in rubbish pits. Many of them stood in groups, carved out of the same material, and often bearing stylistic similarities, perhaps because they were carved by the same sculptor. They depict different aspects of male and female deities, including ritual marriage, pregnancy, birth, and command over wild animals, as if they each tell a story in the life of a god or goddess.

The Çatalhöyük shrines reveal a strong duality between human and animal. James Mellaart believes that the people thought of their deities in human form, endowed with supernatural attributes from a familiar animal world. For example, bulls or large rams were impressive symbols of male fertility. The formidable leopard symbolized the power of both animal and human life, as did the fierce wild boar

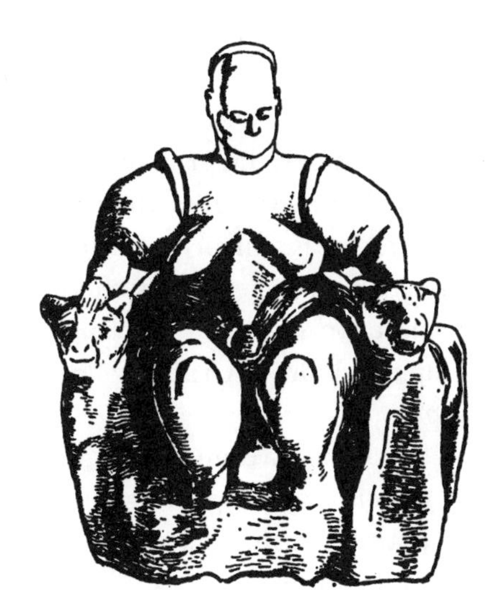

Fig. 4.8. Clay statuette of a Çatalhöyük goddess giving birth, supported by leopards. Courtesy Grace Huxtable.

and the flocks of griffon vultures on the walls of the shrines. In many paintings, leopards support the goddess as she gives birth (figure 4.8). Sometimes she wears leopard skins, while leopard cubs rest on her shoulders. In another tableau, two goddesses of different ages stand behind leopards patting their backs. In yet another, a boy-god rides another leopard, as if a divine family were master and mistress of wild animals.

The archaeological associations between wild animals and the Çatalhöyük Goddess may also reflect her ancient role as patroness of the hunt, for her statue alone appears in one hunting sanctuary, while maimed and wounded animal figurines occur in pits near other shrines. Her associations with agriculture are even stronger: floral and vegetable patterns appear in her shrines and are painted on female cult figurines. In one shrine, Mellaart found goddess statues lying in what were once piles of grain and petaled flowers. A goddess giving birth lay in the grain bin associated with another shrine.

Life and death lay at the very core of the Çatalhöyük Goddess cult. She is often depicted as pregnant or giving birth, and in Çatalhöyük's many layers, the same shrine associations and motifs repeat themselves throughout the centuries. Such reccurring themes include those of a goddess with upturned arms and feet, often in a position indicative of childbirth, or giving birth to bulls or rams. On other occasions, the grim-visaged goddess accompanied by a bird of prey, perhaps a vulture, symbolizes approaching death. Sculptures of mothers' breasts incorporate fox or weasel skulls, or the lower jaws of boars with huge tusks, symbolic of the scavengers who preyed on the dead. In one shrine, a pair of red-painted clay breasts contain complete skulls of Griffon vultures, their beaks protruding from the nipples.

Although we are able to reconstruct only the symbolic world of the Çatalhöyük people in general terms, we do know that there also was a strong duality between male and female. This duality may have been a form of divine family personified by two deities: perhaps a great mother/daughter goddess, and a lesser father/son paramour. Somewhat similar symbolism occurs in later Minoan religion (see chapter 10). In farming societies, women played a major role—through planting and harvest—in the creation of food, which has symbolic associations to fertility and abundance, and life and death. Therefore, it is not surprising that Çatalhöyük's sacred places, with their impeccable archaeological contexts, hint at a religion that reenacted the mysteries of birth, life, and death, where the female breast, pregnancy, and horned animal heads celebrated the duality of male and female, the conservation of life itself.

Was the Çatalhöyük Goddess the Earth Mother, the Mother Goddess? Without question, she was a creation deity, a symbol of the endless cycles of the new farming life, of the passage of life and the seasons. At Çatalhöyük the archaeological context is impeccable (see information in box, p. 99).

There can be no doubt that the earliest of village religions commemorated both the people's ancestors—the guardians of the soil and of human existence itself—and the continuity of life, as measured by the seasonal rhythms of planting and harvest. The earth became the mother, the womb of existence, and the place where the ancestors lived when the winter of life on earth arrived.

Thus, the mysteries of life and death, embodied in an Earth Mother and revered ancestors, were an integral part of the life of these western Asian farmers—a life spent bound to the soil. In these new farming communities, with their focus on agriculture and animal husbandry, ancestor cults and the worship of female deities symbolized these cultures' concern with fertility, child bearing, and the planting and harvesting of grain—with the continuation of human life.

A UNIVERSAL MOTHER GODDESS?

Can we use the excavation at Çatalhöyük to proclaim the existence of a universal Mother Goddess 8,000 years ago? Archaeologists have found hundreds of human figurines in western Asia over the past century: figures from Mesopotamia with no signs of divinity; anthropomorphic male and female depictions from 6,000-year-old Egyptian villages; the figures from Çatalhöyük; and a range of male, female, and sexless figurines from Turkey, the Aegean Islands, and Greece. In 1968, British archaeologist Peter Ucko studied every known early female figurine from the eastern Mediterranean region and found that few of them had precise cultural contexts. Many were thrown out with the domestic garbage; some were children's toys, others trade objects, mourning figures in tombs, ancestral spirit figures, or tokens to encourage pregnancy.

Despite Ucko's rigorous, critical analysis that showed that these figurines had several meanings and uses, archaeologists and others have persisted in writing about universal Mother Goddess cults. However, evidence points to the contrary: that there was never a *supreme* Earth Mother in the world of the first farmers—indeed the ancient Egyptians considered the earth male.

---

chapter five

POWER AND THE ANCESTORS

What made, and still makes, reverence for those who lived before such a powerful part of human existence? Fortunately for archaeology, there are still many living societies in the non-Western world who enjoy a close relationship with their ancestors and consider them to be a powerful force in their daily lives. At a general level, such societies provide rich analogies for interpreting the archaeological record of much older ancestor cults. They also highlight the complex relationship between the living and spiritual worlds that are mediated by shamans or spirit mediums, men and women with special ritual powers.

Ancestors and their mediators provide timeless messages passed down from one generation to the next. Moral values, good crops and rains, strong children: the continuity of human life resonates through the millennia. Before we explore more complex sites and sacred landscapes than Çatalhöyük or Jericho, it is helpful to look at how one still-flourishing subsistence farming society, the Dande of central Africa's Middle Zambezi Valley, governs its relationship with those who have gone before (figure 5.1). The intricacies of Dande ancestor rituals provide a salutory lesson into the difficulties of inferring religious beliefs and rituals from archaeological data that is worth remembering.

Once a low-lying floodplain hemmed in by rugged escarpments, the Gwembe Valley is now a vast lake impounded by the Kariba Dam, built in the late 1950s, which straddles the Zambezi.

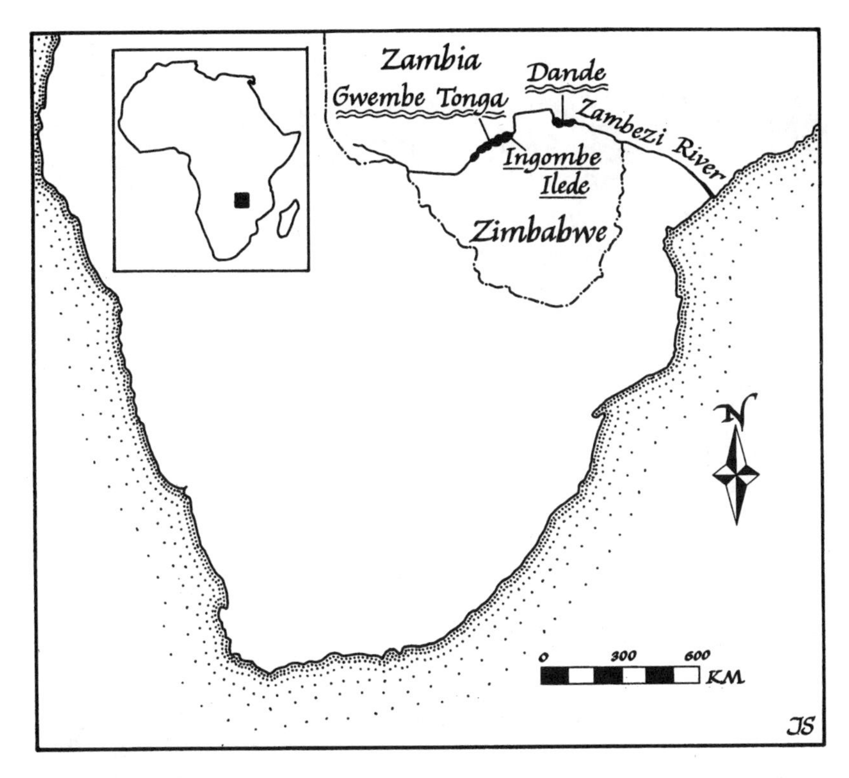

Fig. 5.1. Map of Gwembe Tonga and Dande country, Central Africa. Archaeological sites are underlined; peoples, double underlined.

Until rising lake waters caused forced resettlement and the abrupt end of a centuries-old way of life, the Gwembe Tonga and their neighbors farmed in this low-lying, intensely hot environment.

Waves of invaders have entered the valley over the past millennium: Shona peoples from the southern highlands, Islamic and Portuguese traders, and European colonists. None has changed the age-old rhythm of farming life in this brutal environment. The cycle begins in December, when rains soak the parched valley floor. Lush grass sprouts, and tributary rivers and streams flow with hillside runoff. The people plant their crops and weed their fields, praying for good rains. With April comes the harvest and the end of the rainy season (figure 5.2). Lush vegetation withers and dies, and streams dry up on every side. The hot sun and cloudless skies suck moisture from the thin soils, and the earth, in essence, dies.

Fig. 5.2. Gwembe Tonga women storing grain in the late 1950s. Courtesy Nigel Watt, Zambia Ministry of Information.

Come September, the men clear the gardens of trees and shrubs, burn off the vegetation, and allow the ash to fertilize the fields. Women hoe out the furrows, then sow the seeds and weed. They scan the sky for telltale signs of impending storms. Everyone watches bird movements, the behavior of insects, and the vagaries of winds sweeping across the hot valley. Temperatures rise and tempers flare in the heat. Then the rains come and life begins anew.

ARCHAEOLOGY IN THE GWEMBE VALLEY

The Tonga and other Zambezi River peoples have deep roots in the remote past. The modern people's oral traditions go back only a few generations, but archaeology reveals tens of thousands of years of prehistoric occupation in the low-lying valley.

Although the construction of the Kariba Dam and the flooding of the Gwembe Valley on the frontier between the nations of Zambia and Zimbabwe in the late 1950s was a spectacular achievement of modern industrial enterprise, it was a massive human and archaeological tragedy. Lake Kariba inundated many square kilometers of Zambezi River floodplain, displacing thousands of Gwembe Tonga villagers from their homeland of many centuries. The archaeological destruction was immense, but drew little attention at the time. A small team of archaeologists, headed by J. Desmond Clark, fanned out over the low-lying, remote Gwembe Valley, searching for Stone Age settlements and later agricultural villages before rising floodwaters destroyed the archaeological record forever. Although the fieldworkers achieved miracles with minimal resources, hundreds of valuable sites vanished under Lake Kariba.

Clark and his colleagues discovered evidence for human occupation along the Middle Zambezi going back at least half a million years. After 10,000 years ago, ancestors of modern-day San people lived in the valley, before being replaced by farming peoples some 1,800 years before present. The Gwembe Tonga themselves settled in the valley at least 1,000 years ago, perhaps even earlier. The Middle Zambezi Valley became a strategic trading route after A.D. 1000, as traders from the Indian Ocean coast penetrated far upriver in search of gold dust, elephant ivory, and other raw materials. The river was a natural highway to the interior, with many paths leading to the mineral-rich plateaus on either side of the Zambezi.

The most spectacular archaeological discovery came after the flooding of Lake Kariba. In 1960, the construction of a water tank on a low hill named Ingombe Ilede, near Lusitu and downstream of the

Kariba Dam, unearthed nine richly decorated skeletons of men. Archaeologist James Chaplin was able to rescue the graves before they were destroyed. His excavations, however, were confined to the summit of the hill, where the richest burials lay. He found traces of a substantial mud-and-thatch house that may have covered the graves, but its foundations had been demolished by a bulldozer's blade before he arrived on the scene.

The nine men had died in the fifteenth century A.D., just before historical records document the arrival of Portuguese ships on the distant Indian Ocean coast. The skeletons lay in cotton robes, with finely wound copper wire bangles on their arms and legs. One individual wore a necklace of 18-carat gold beads and multicolored, imported Indian glass beads, as well as nine *Conus* shells from the Indian Ocean coast—all objects of great rarity and prestige, indicative of the status of a chief. Long sheaves of copper bangle wire, tools for manufacturing it, ceremonial hoes, and iron gongs—which were an important symbol of chiefly office in Africa, and were sounded when a chief entered a village—lay in the graves.

The artifacts found with the skeletons suggest that the nine men were engaged in the ivory and gold trade, and had powerful authority over the local people. Both leaders and traders, these men probably lived in some isolation from the rest of the settlement. They used fine clay vessels and thin-walled bowls quite unlike any of the other coarser pots made and used elsewhere in the settlement. These men were also expert artisans, who could convert bundles of square-sectioned trade wire imported from the nearby plateaus into fine wire bangles that adorned their own limbs. Historical records tell us that the same kinds of bangles were used as trade objects between neighboring African villages for many centuries. The copper salts in the bangles preserved large pieces of cotton fragments. Most of the cloth was relatively coarse grained and woven locally, judging from the many clay spindle whorls on site. The textile fragments with a fine weave were perhaps from imported cotton fabric that came from as far away as India.

I carried out later excavations both to date the burials and to search for more graves. The dig at Ingombe Ilede still required urgent fieldwork in the face of pipeline and water tank construction. My trenches examined areas away from the central precinct and were designed to establish the chronology of the site and to study the way of life of the inhabitants as a whole. We showed that Ingombe Ilede was most likely occupied, on several occasions, between the seventh century A.D. and A.D. 1450. The spectacular burials on the summit belonged to the latest occupation. We also unearthed forty-two burials of humbler folk from the same period, which lay on the edge of the village. The dead lay crouched in shallow graves, deposited one atop the other in a haphazard way. Two-thirds of the skeletons were from infants who were less than a year old at death. In contrast to the central burials, almost all of these forty-two people were buried without any body decoration or accompanying grave furniture. Only a handful of skeletons wore small strings of imported glass beads or freshwater shell necklaces. A vast economic chasm separated leaders and farmers at Ingombe Ilede.

My excavations showed that trading activities and wealth were confined to a few people. We were never able to establish the identity of the richly decorated men, but their cultural links appeared to lie to the south, in modern-day Zimbabwe. Though sourcing methods were still in their infancy, we were able to show that the copper in the central graves came from nearby plateau areas (see information in box, chapter 9, p. 195). These Ingombe Ilede burials provided dramatic proof of the wide range of the Zambezi trade.

Before I was able to carry out my rescue excavations, I had been required to obtain permission from the local community. I attended a village meeting, bringing gifts for the chief and the local spirit medium. The small thatched house was dusty and hot in the late afternoon sun, and we were crowded together on hard wooden benches. An elderly man stood up and started clapping. In unison, all stood, removed their hats and shoes, and joined in the ritual clapping. Two hours later, the same ritual ended the meeting. Later, I learned that this ritual was performed in order to invoke the blessing of the ancestors, so that the decisions made by the gathering had the sanction of the "original owners of the land." Gradually, I gained an understanding of the spiritual association between the Gwembe Tonga people and their patchwork of gardens along the north (Zambian) bank of the Zambezi River.

While archaeologists studied endangered sites, a group of anthropologists headed by Elizabeth Colson and Thayer Scudder documented the intimate spiritual ties between the Gwembe Tonga, the ancestors, and their land. Colson and Scudder studied these subsistence farmers a few years before the resettlement. They described how the Tonga dwelled in tiny hamlets. These villages were sometimes no more than a few mud-and-pole huts—a community composed of one man and his wives and children. Larger communities were comprised of a group of households, each with its own small garden. The villages clustered near tributary river banks, or by the Zambezi itself. Endemic tsetse fly infestations made cattle herding impossible; therefore, everything depended on the land, the rains, and ancestral blessings. Each Zambezi-side territory and each garden lay under the control of a particular social lineage, and every lineage had its own ancestors. The living, the dead, and the land formed a triumvirate. Colson and Scudder have shown in later studies conducted after the resettlement that this powerful triumvirate still endures in remoter areas.

"THE OWNERS OF THE LAND"

The Dande people, who live across the river from the Gwembe Tonga, provided me with further insights into the nature of lineage ancestors as the *varidizi vepasi*, "the owners of the land." Named after the ancestors, lineages own land—specific territories, usually with natural boundaries, that are headed by a chief. Ancestral history of the ruling lineage provides the spiritual experience of "owning" the land, and chiefs' graves link the revered dead to the acreage. Because the ancestors lived on the land, every tree, hill, and wet-season pond has ancestral associations.

Ancestors are the disembodied spirits of people who lived in the past. These spirits—or shades—are men and women who provided for, and protected, their families during their lifetimes. After they died, they continued to care for their descendants from the otherworld. Cultural anthropologist David Lan writes: "The richness of their personalities and the depth of their experience do not come to an abrupt end and dissipate for all time" (1985, 31). He likens death to a weir across a river, which blocks the flow of life. The stream backs up on itself during the processes of dying and burial. When the weir gates open, however, the flow of life continues. Thus, the dead, with all their strengths and weaknesses, become *midzimu*, ancestors.

Ancestors have no material form. They are *mweya* (air), and hence, ubiquitous. They can see and hear, cure every ailment, and know the future. Thus, death enhances human life beyond recognition. As generous providers, these ancestor spirits protect; they are never evil. In order to acquire the goodwill of the ancestor, the descendants perform the requisite ceremonies. When a person dies, their spirit leaves the body and hides in the remote bush. Descendants dance on the grave at the funeral. Within a year of the death, the ritual dance leads the ancestor back into the household, and thus the ancerstor becomes a protector of the living. From time to time, the living brew beer and distribute it in the ancestor's name, lest he or she be forgotten.

Among the Gwembe Tonga, descendants of the same sex and lineage inherit newly deceased shades, the inheritor being chosen during the funeral. Thereafter, the shade can wander freely and visit the homes of descendants and fellow lineage members. When the inheritor dies, the shade passes to another member of the same lineage until it is finally forgotten. Thereafter, the ancestor merges into the amorphous body of unknown shades of the ancient dead, which maintain a tenuous connection with the lineage. The shades of the most recent dead—only two or three generations removed from the older living members of society—are the ones most actively involved in the affairs of the living. The recently deceased also provide a link between the present generation and much older ancestors.

The dead loom large in Dande life. People want to live on their ancestral lands, and be buried there, even if they have spent years elsewhere. Human existence thrives in the fullest sense when living alongside the home of the dead, near the places where the ancestors once lived.

Some Dande royal lineages extend back as many as fifteen generations, linking members to the original "owners of the land." A chief is "father" to his people; his ancestors are the "grandfathers."

While alive, Dande chiefs look after the welfare of their people—they provide grain in time of scarcity and enforce the law. When they die, each becomes a *mhondoro* (lion), a royal ancestor and a source of the fertility of the land. The royal ancestors provide rain and protect growing crops. If their descendants obey the laws and perform the requisite ceremonies, they will live in prosperity. Each *mhondoro* rules over a spiritual realm, the specific territory he once controlled. These territories vary greatly in size. The senior *mhondoro* exercises supreme control over all of them.

Each territory is an autonomous spiritual unit where heads of households participate in annual rituals at special shrines to bring rain. They must also make offerings of the first grain from the harvest to the senior royal ancestor. Newcomers who settle in a spiritual territory must make an offering to the *mhondoro* and obtain his permission to cultivate the land.

When an ancestor feels the need to communicate directly with its descendants, it chooses a man or woman and uses his or her mouth to speak. These individuals thus become spirit mediums—someone "grabbed by an ancestor." The ancestors "grab" people at random, regardless of social status. Dande people do not become spirit mediums voluntarily. Possession by an ancestor is a hardship and a trial, during which these people lose control of their bodies. Mediums become mere receptacles of the shades: they have no special qualities and are powerless to influence the will of an ancestor. However, they utter the words of the dead, thus making them as close as one can become to being divine.

Almost all Dande mediums begin their careers at a moment of personal crisis. The subject is usually young, for the old are almost ancestors themselves. When an individual falls ill, and Western and local medicines are unsuccessful, the patient either gets worse or does not recover. It is at this point that suspicion falls on a shade. A traditional healer may then recommend an attempt at possession, so that the ancestor can reveal its identity, its message, and its demand.

Possession ceremonies are elaborate performances, usually held in the weeks after the harvest when beer is plentiful. The patient's father and the healer prepare and supervise the ritual, which begins soon after sunset with drumming. The drums summon the ancestor.

The patient sits on a mat, his or her head bowed and covered with a white cloth. Soon female kin and neighbors dance in a semicircle facing the drummers. They move back and forth, approaching and retreating from the patient, singing the same one- and two-line refrains again and again. Meanwhile, mediums who have experienced possession will many times join the dancing as the crowd presses ever closer. They brandish their ritual axes and spears, fall into trance, and dance more aggressively. If their spirit starts to speak, they break away from the crowd and crouch in the darkness as their ancestor shouts its message. The patient remains seated in front of the drummers, until possessed. At this point he or she joins the dancers, encouraged by the other mediums.

At dawn, close members of his or her lineage and the mediums lead the candidate into a sealed hut. They question him or her, and try to identify the spirit from the answers. Many times, the spirit is reluctant to speak, sometimes refusing to deliver a message at all. If it does speak, the words come falteringly at first, then with increasing fluency and force. Sometimes, long harangues catalog the ancestor's demands, accompanied by groaning, weeping, and agonized cries. The questioners gauge success or failure from the answers. Sometimes the spirit identifies itself as the medium's own clan ancestor, a *mudzimu*. Less frequently, the shade comes from a different clan, raising the possibility that it is a *mhondoro*, a royal ancestor. Then the aspirant becomes a *mhondoro* medium in his or her own right. He or she goes to live in the spiritual province of the appropriate royal ancestor, and has to persuade an established medium to validate him or her.

Five to ten years may pass. The *mhondoro* or *mudzimu* spirit may disappear for months on end, never return, or return with increasing frequency. Gradually, a successful medium acquires a reputation for unusual powers, an ability to forecast the future, and skill at punishing people who do not believe him or her. When he or she is confident of his or her abilities, the aspirant is ready for the final test, a public display of ancestral knowledge.

The Dande believe that wisdom acquired by the living, by "those who eat," comes only from experience gained during their own lives, and from learned fragments of historical lore. But having a body and biological needs prevents living people from acquiring the most important forms of knowledge. True erudition, which gives power over life and death, and over the forces of nature, belongs to the ancestors. The dead have different, superior knowledge, even when there is an implicit continuum between the knowledge that is possessed by the living and the knowledge of their forebears. This belief in continuity of knowledge, however, has critical relevance to the selection of mediums, because it means that the elders have enough knowledge of the past to judge whether the aspirant's claims are valid. To gain acceptance, the candidate must utter a commonly accepted and believed version of the past. The *mhondoro* or *mudzimu* spirit must now show its trust of the medium by sharing the genealogy of the royal ancestors.

The medium goes into a trance, and recites the ancestral genealogy and details of the ancestors' history: where they lived, whom they married, the battles they fought, and other topics. An accurate recitation by the passive medium convinces the other mediums and the crowd that the ancestor has indeed spoken through the aspirant. Then, the candidate faces a final, and crucial, test. He or she is confronted with a pile of carved staffs and is asked to select the staff that was used by the previous medium, or that was carried by the royal ancestor when he was alive. If the aspirant does not select the correct staff, he or she departs, returns to the community, and lives an ordinary life. Cheating is impossible: if an aspirant fakes possession, it is believed that the spirit will kill him or her.

Mediums are people marked by inexplicable illness, who have accepted the diagnosis of a traditional healer, and achieve possession. They don ritual dress and abstain from sex for many years. Spirit mediums wear two lengths of black and white cloth. One surrounds the waist and falls to the ankles. The other, tied to the right, drapes around the shoulders. They also wear sandals and carry a carved, ceremonial ax at all times.

Once established, mediums practice in their spiritual territories for the rest of their lives. They tread a fine line, knowing that ancestors have the welfare of the people at heart above all else. Everything depends on their maintaining their reputation of authenticity.

Accusations of greed, selfishness, or using one's office for illegitimate ends flow easily from disgruntled clients. However, in actuality, few mediums ever acquire great wealth. They use gifts given to them to cover the expenses of the shrine, or hospitably share them with their many visitors.

A CIRCLE OF SHARED POWER

In 1609, Portugese explorer João dos Santos recorded an instance in which a Shona chief presided over major religious rituals, and communicated directly with the ancestors. The chieftain was a political and religious leader, and, just as many other chiefs of central and southern Africa, had mystic powers over the land. Undoubtedly many rulers throughout the world based their political power on their supernatural abilities as intermediaries and rainmakers. This role reached great sophistication among the Mayan lords of Central America, and the divine kings of Sumerian and ancient Egyptian civilizations, who were seen as deities, with a special relationship to the spiritual world.

The links between chiefly and ancestral power are not always so direct. In Dande society, there are separations of power. The living chief supervises law and politics, but is not a medium at all. The ancestral chiefs of the past direct moral behavior, and protect the crops, guard over wild animals and plants, send rain, and cure the sick. They also select and install living chiefs by expressing their opinions through a *mhondoro* medium, a living person who is a passive agent, without opinions of his own. As David Lan states: "The influence of the past bears down especially heavily on the present when a chief dies" (1985, 57). With many claims and counterclaims to adjudicate, the *mhondoro* medium treads a delicate path, since he or she must ensure that royal ancestors speak directly to their descendants, while maintaining complete disinterest. Every time a medium allows his or her body to be filled with the royal ancestor, he or she travels back to the beginning of the world and crosses the boundary between life and death. A successful medium presents the illusion that he or she is not merely the medium of a *mhondoro* —a chief of the past—but the chief himself returned to earth. For this reason, Dande mediums have a great deal of covert influence over political activity. A living chief, however, has no authority to make changes in the ritual domain. Successful Dande mediums acquire great skill in balancing consensus and innovation.

Each Dande medium has a *mutapi*, who manages the shrine. He acts as an intermediary between the *mhondoro* medium and those who wish to consult him or her. The *mutapi* receives visitors, and summons the *mhondoro* to possess the medium. Then he interprets or explains the statements uttered by the *mhondoro*. The *mutapi* is the active counterpart of the passive medium and, thus, a highly respected individual in the local community. Each *mutapi* is a lineage head. Invariably, he possesses an intimate knowledge of the spiritual territory in which he resides—the history that is handed down over many generations. A medium, usually an outsider, has no credibility without the patronage of a *mutapi*, who is willing to organize rituals and validate his or her authority.

The *mutapi* is more than a medium's assistant; he also serves as the minion of the *mhondoro* spirit, and is, thus, an intermediary between the spiritual realm and the chieftaincy. He has an eye on the present and on the past, acquiring his authority from his prestigious position within the chiefly hierarchy and from his knowledge of the ways of the *mhondoro* ancestor.

Thus, the Dande operate within a complex living and spiritual world, where the representative of the *mhondoro* on earth is the chief. He derives his authority because of his descent from the royal ancestor, who chooses him from all available candidates as the true heir to the chieftaincy. One day, he will become a *mhondoro* himself. The chiefs and the *mhondoro* mediums are part of a single system, where the royal ancestors and the living engage in constant dialog through the medium, with the *mutapi* acting as a royal messenger. In the final analysis, however, it is the *mhondoro* spirits who supply rain to the fields. As long as they do, their living descendants have legitimate claim to the land, confirming their vital role in chiefly leadership. Through their ability to control rain, the *mhondoro* show again and again that their propitiating function is their most auspicious gift.

THE PAST AND THE PRESENT

Lan calls the medium's performance "the great spectacle of the past . . . Rights to ownership in the present derive from rights to ownership in the past and with each royal ancestor returned to life and living in his spirit province, the past is available for inspection right now, right here, right in front of your eyes" (ibid., 68). Dande mediums present a continual historical spectacle. The dead ancestors of the present chiefs return to earth. These heroes of the past appear to give advice and to pass on wisdom learned after death through the mediums. With their traditional chiefly clothing and daily rituals, mediums assume many attributes of chieftainship, as if they are ancestral chiefs returned to earth.

The Dande conceive of life in biological and spiritual dimensions. Biological life is red, blood-drenched, and associated with women as marriage partners. Women reproduce the human heart with all its wetness, softness, and blood. Men control the social and intellectual life-in-death of the royal ancestors. Males reproduce human life through the agency of the mediums. The set of rigid prohibitions that surround mediums separate these two dimensions and reinforce the historical spectacle that unfolds on every side. No *mhondoro* medium may see blood, or have anything to do with burial. He or she cannot witness the fundamental biological aspects of life such as birth and death—reproduction and decay. Additionally, no one wearing any shade of red may attend a possession ritual, for if a *mhondoro* spirit sees red, the medium will die.

The world of the medium is dry and light, a symbol of increasing age when men and women come closer to the ancestors and become increasingly desiccated. The lineage's age and authority are associated with dryness, which is why the corpses of chiefs—future royal ancestors—lie on platforms until only the bones remain, to be buried in the earth with their forebears. For obvious reasons, mediums receive the same burial as a chief.

The phases of the moon symbolize the cycles of life and death that take a person from the insubstantial and temporary world of the living to an eternal and unchanging universe. A *mhondoro* spirit cannot enter its medium and speak to its descendants on dark, moonless nights. Thus, the moon's cycles regulate possession, as well as agricultural activities. People rest on the day after the first moonless night; no one hunts or works in the fields. Two days later, another rest day commemorates the return of the moon when the ancestors reappear and the heavens gleam with renewed light. Without the moon, the *mhondoro* spirit cannot reappear.

The "great spectacle of the past" incorporates both Dande myths and oral histories, drawing together individual lineages and traditions into historical narratives that change from generation to generation. Local possession ceremonies legitimize the oral traditions and rituals of households, villages, and larger communities. As time passes, these many local ancestor rituals combine to link together families, communities, entire lineages, and secular and spiritual territories. Those who head the senior "royal" lineages may be conquerors, or simply perceived as "owners of the land." However, no political leader, whether conqueror, lineage leader, spirit medium, or chief, can claim greater power than that of the ancestors, for they become the source of the fertility on which the lands depend.

Archaeologists attempt to recapture the subtle nuances of the intangible. Listening to the passionate chants and drumming that summon the ancestors helps us to look beyond earthworks, paintings, sepulchres, and scatterings of artifacts and food remains. As we shall see in the next two chapters, sacred landscapes such as Avebury and Stonehenge provide material evidence of dynamic relationships between the living and the dead. From communal graves, sacred enclosures, and the close juxtaposition of burial mounds and stone circles, the intangible survives—but still only hints at much earlier grand spectacles of the past. These unfolding spectacles, when staged in carefully controlled settings, could conquer and act as powerful instruments of social control.

chapter six

AVEBURY: LANDSCAPES OF THE ANCESTORS

I made my way to Avebury, driving along quiet English country roads, twisting and turning down narrow lanes, over low chalk ridges brown with freshly harvested fields. Blackbirds scavenged in the corn stubble, ever-watchful as they ate. Black-and-white cows grazed placidly in a lush field. Long hedgerows rolled toward the far horizon where distant ridges looked dark against the pale September sky. Soft, white clouds were chased by the keen southwest wind. Long grass rustled and cascaded in a nearby field in an endless tracery of sudden gusts.

The contours of Wessex, southern England's chalkland, showed few signs of human habitation. The domesticated landscape has been cultivated for more than 6,000 years, yet, after all this time, the ancients will not allow their presence to be forgotten. Their burial places watch over the land, over Avebury (figure 6.1).

As I traveled, I came upon the silent burial mounds, stark against the blue heavens arching overhead. Their conical shapes looked green as the road climbed closer. I stopped the car to gaze at the Wessex countryside. At one time, the barrows had gleamed brilliant white with fresh chalk, signposts for sacred places where ancestors slept. Even before that, a dense woodland had covered much of this rolling chalkland. The primordial forests, however, have long vanished in the face of the oxen and the plow. Small gardens and fields of the Stone Age lie under the landscape. Mere shreds and patches of the ancient farming countryside survive; and within these patches can be found

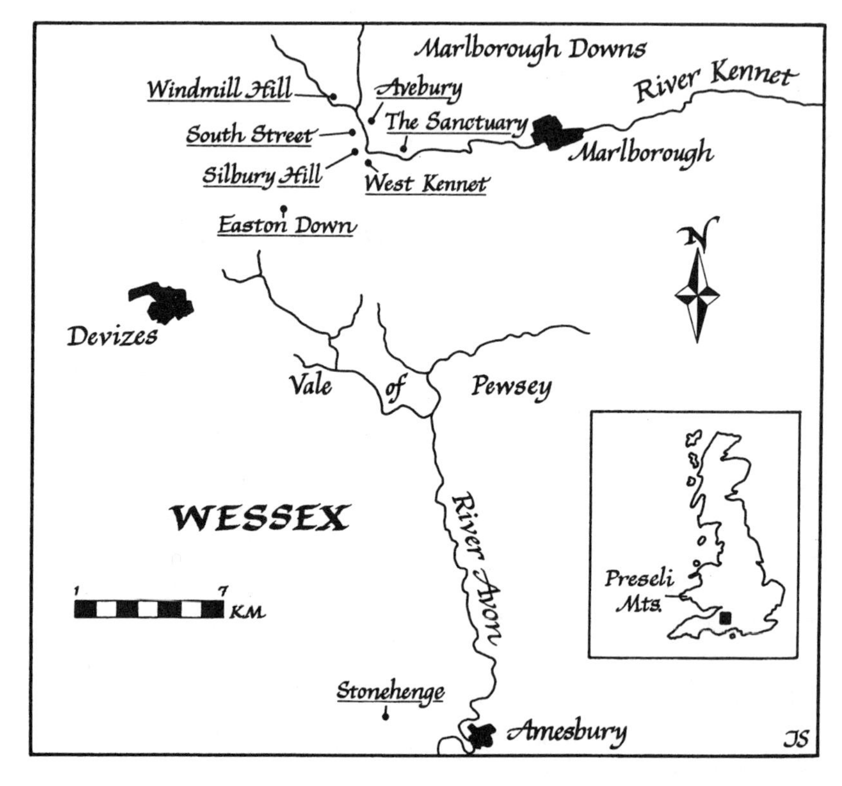

Fig. 6.1. Map showing the archaeological sites (underlined) mentioned in chapters 6 and 7.

mollusca and plant pollens that preserve the undisturbed turf line, earthworks, and burial mounds.

The natural woodland, pasture, and cleared fields of Stone Age times perished under the insatiable plow. However, barrows, long-vanished earthworks, ditches, and camps show up in aerial photographs. The silent bulk of an enigmatic mound named Silbury Hill and the stone circles of Avebury represent some of the great archaeological sites of western Europe. They beckon to us with tantalizing clues of human deeds and actions that stretch back beyond memory.

THE DISCOVERY OF AVEBURY

In the year 1649, antiquarian and country squire John Aubrey rode into the middle of Avebury's stone circles while out fox hunting. The self-promoting Aubrey tells us he was "wonderfully surprized at the sight of these vast stones of which I had never heard before" (Malone 1989, 21). Perhaps the "surprize" was no first-time discovery, for Aubrey had an eye for publicity and royal preferment. He returned to sketch and explore. King Charles II read his descriptions, announced his intention to visit the site, and instructed Aubrey to dig for human bones. Aubrey never excavated at Avebury, but he did complete the first plane table survey of the earthworks and stone circles, recording the position of many stones that have since vanished. King Charles II himself visited the site with Aubrey in 1663 and together they climbed Silbury Hill.

John Aubrey, a typical intellectual of his day, was interested in everything from archaeology to natural history and classical mythology. He traveled widely, sketching many ancient monuments for a never-to-be published masterwork, *Monumenta Britannica*. He wrote: "Surely my stars compelled me to be an antiquary, I have the strongest luck at it, that things drop into my mouth" (ibid., 21).

Aubrey realized Avebury's stone circles dated to pre-Roman times. He wrote: "It does as much exceed in greatness the so renowned Stonehenge, as a cathedral doeth a parish church" (ibid., 22). He considered the builders of the stone circles "Ancient Britains," "savages" who lived in a shady dismal wood." He compared them to native Americans: "They were two or three degrees I suppose less savage than the Americans" (ibid., 21).

Another antiquarian, William Stukeley, read Aubrey's account and visited Avebury in 1719. An eccentric and romantic man, Stukeley was preoccupied with ancient Celtic priests—the celebrated Druids that Roman general Julius Caesar observed in 55 B.C. Also an accomplished landscape archaeologist, Stukeley rode kilometers into the country tracing earthworks, ditches, and stone alignments. He was the first observer to place Avebury in the heart of a much larger sacred hinterland of burial mounds, enclosures, and stone alignments. All serious research at Avebury has been built on his original fieldwork, completed before mechanized agriculture transformed the Wiltshire countryside.

STUDYING CLIMATE AND LANDSCAPE CHANGES

William Stukeley studied Avebury's environs from the back of a horse. Twentieth-century archaeologists use aerial photography to obtain a much higher, bird's-eye view of the same countryside (figure 6.2). After seventy years of sporadic aerial reconnaissance, thousands of photographs taken at every season of the year document familiar and hitherto invisible features of the Avebury Stone Age landscape (see information in following box). The aerial photographs reveal

---

AERIAL PHOTOGRAPHY

The overhead view has fascinated scientists since the days of the French balloonist Joseph Montgolfier, who floated over Paris in 1783. More than a century passed before archaeologists serving as artillery spotters on the Western Front and in Mesopotamia during World War I photographed Roman forts and other sites from the air. They soon expanded their research after hostilities ended. In the 1920s, the famous landscape archaeologist O. G. S. Crawford and a wealthy marmalade manufacturer's heir, Alexander Keiller, photographed Avebury and its surroundings from the air. Decades of thorough aerial photography in southern England now provide a chronicle of hundreds of sites, photographed at all times of the year, under a wide variety of weather conditions, and vegetational covers. Still the discoveries continue to be made, especially when conditions are unusual. For example, a new stone circle appeared in photographs taken inside Avebury during a recent severe summer drought.

Aerial photographs provide an overhead view of known earthworks, but, more important, reveal sites invisible from the ground due to natural erosion or intensive plowing. "Shadow sites"—locations that have reduced topography resulting from plowing and other modern activity that has flattened the original banks, ditches, or other features—show up clearly from the air. The rising or setting sun can help set off deep shadows, emphasizing the relief of almost-vanished banks or ditches at such places. "Crop marks" and "soil marks" reveal earthworks as differential growth resulting from the amount of moisture the plants can derive from the soil and subsoil. Lusher vegetation grows if the soil depth has been increased through the excavation and filling in of ditches or pits, or through the erection of banks. Conversely, paved surfaces that are below the surface, such as roads, inhibit plant growth.

Most aerial photography uses black-and-white or color film, the former being especially useful with its wide variety of filters. Infrared film has layers sensitized to green, red, and infrared, which detect solar radiation at the near end of the electromagnetic spectrum. The distinctive "false" colors on the film reflect different reflections from cultural and natural features. For instance, vigorous grass growth in floodplain areas sometimes shows up bright red. Systematic and continuous aerial photography is a powerful research tool in the studies of the Avebury and Stonehenge areas, where low-altitude shots are often revealing.

Fig. 6.2. Aerial view of Avebury. Photograph from English Heritage.

dozens of long and round barrows, causewayed camps, and stone circles representing different episodes in a long history. Unfortunately, almost none of these archaeological sites contains stratified occupation layers with characteristic pottery styles that allow stratigraphic analysis of individual sites. For generations, archaeologists have puzzled over the Avebury region; after years of sporadic research, we are still unsure as to whether it was a self-contained area.

Since 1987, a group of archaeologists headed by Alisdair Whittle of the University of South Wales has embarked on an ambitious, long-term field project to investigate the cultural sequence, changing natural environment, settlement patterns, and sacred sites of the region. Whittle's problem-oriented research has yielded rich dividends. For example, the excavators of the 1920s and 1930s did not have the benefit of radiocarbon dating; and until the 1980s, only a handful of radiocarbon dates was obtained from the Avebury region. A systematic dating campaign using highly accurate AMS radiocarbon dates calibrated with tree rings now allows Whittle's group to look at blocks of 3 centuries or so in the Avebury region, which equals about 12 generations.

By combining blocks of carbon 14 dates with environmental data from soil profiles and under earthworks, Whittle and his colleagues are chronicling a constantly changing ancient landscape, quite different from the fully domesticated lands of today. We now know that 15,000 years ago the chalklands were bitterly cold, open tundra. As sea levels rose and glaciers retreated, birch forest, then dense woodland colonized both valley and uplands. Ash, oak, and other deciduous species covered the now-open landscape. These trees, especially the oaks, were vitally important to Stone Age farmers, as building material for dwellings, fences, and revetted banks and ditches. Whittle found evidence of the use of as much as 20,000 linear meters of oak timbers in two palisaded Stone Age enclosures near West Kennet; some posts were up to an estimated 8 meters long.

The evidence for this rapidly changing, and increasingly artificial, landscape comes from the samples obtained from original ground turf and ancient turf lines, which were preserved under long barrows and earthworks. These thin chalk soils are little more than 10 centimeters thick, capping the subsoil, where such features as tree hollows and plow marks sometimes survive. Interpreting such exiguous stratigraphic indicators requires meticulous excavation and a detailed knowledge of how soils and earthworks decay over short and long periods of time.

Nearly 40 years ago, local archaeologists decided that they needed accurate control data for studying earthwork decay and soil weathering. In 1960, the British Association for the Advancement of Science erected an experimental "prehistoric" earthwork at Overton Down near Avebury to measure such processes in a scientific manner. The supervising archaeologists built the earthwork on chalk subsoil, with profiles approximating those of ancient equivalents. They buried samples of organic and inorganic materials in the ditches and earthwork, erecting it partly with modern tools and partly with the red deer antler picks and ox shoulder blade shovels employed by the ancients. Overton Down was then abandoned, but small and very precise excavations of the ditch and bank are being carried out at regular intervals: 2, 4, 8, 16, 32, 64, and 128 years. These limited probes check the decay and attrition of the earthwork and the silting of ditch and bank over a lengthening period, yielding priceless information that can be used for interpreting equivalent ancient sites on chalk soils.

Whittle used the data from Overton Down and other environmental data when he excavated the Easton Down long barrow at Bishops Canning southwest of Avebury. One trench cut a cross section across the southern ditch, the other traversed the northern ditch and exposed the original land surface, as well as the core of stacked turves, chalk, and topsoil. The excavators called in experts to take molluscan, soil, and pollen samples from the original land surface dated to 3350 to 3100 B.C.

Eleven mollusk samples taken from a well-sealed section of premound soil showed dramatic changes in snail species from woodland to open grassland forms. A tree hollow under the mound yielded woodland shells. An increase in open-country mollusca followed, and was so sudden that human clearance of the land seems the only logical explanation. Significantly, soil scientists found signs of lateral movement of the soil, which could result only from cultivation a short time before the mound was built.

Small amounts of pollen grain were found in the Easton Down molluscan samples—enough to show that no woodland grew close to the barrow site (see information in box below). No domesticated grain cereals or cultivation weeds were found in the samples, therefore indicating open-country grassland, wasteland, bracken, and hazel characteristic of soil disturbance, and possible burning and manuring of cultivated land.

PALYNOLOGY

Swedish botanist Lennart van Post developed palynology—the science of pollen and spore analysis—in 1916, as a way of studying vegetational change in Scandinavia after the Ice Age. By studying pollen grains from deep bogs, he reconstructed the complex botanical changes in northern Europe over 15,000 years. Subsequently, pollen analysis has revolutionized our understanding of both Ice Age and more recent climates throughout the world. Once confined only to swampy deposits and northern latitudes, pollen analysis has now yielded information about ancient vegetational cover and early agriculture in both tropical and temperate latitudes. It has also helped to chronicle human intervention in the natural environment through forest clearance.

The principle is simple. Large quantities of pollen grains are distributed in the atmosphere and preserve well if deposited in an unaerated geological horizon. Pollen experts take column samples from a wide variety of organic deposits such as swamps or lake beds. Back in the laboratory, grains of each genus or species are examined, for minute pollens can be identified microscopically with great precision, thereby providing a portrait of vegetation close to the spot where they were found. Palynologists have developed long sequences of changing vegetational cover, that determine such developments as the spread of forests over Europe after the Ice Age. Pollen analysis is priceless to archaeologists, and is used for modeling climate change and for identifying human modification of the landscape, such as was the case at Easton Down, where forest clearance had taken place. So precise is modern palynology that botanists can even identify cultivation weeds associated with the growth and harvesting of different crops. Pollen analysis can also identify grains stuck to mollusks, allowing the scientists to correlate two forms of evidence directly.

There are, of course, limitations to the approach: Pollen blows over long distances, and trees growing nearby can be overrepresented in the samples. At a general level, however, palynology provides a unique window into the past—and the basis for identifying the siting of burial mounds in reasonably precise, and perhaps intended, locations in ancient landscapes, especially when pollen is combined with other environmental data.

The precise dimensions of cleared and uncleared woodland, and the intricate patterns of cultivation, and abandonment—from villages moving from one location to another as land became exhausted—are difficult to determine. Occasional samples, however, give insights into an exacting, often harsh, and always changing ancient environment. For example, mollusca and soil profiles under the great henge (ceremonial enclosure) tell us that Avebury rose on long-standing, noncalcareous, and perhaps underutilized grassland, close to secondary forest. These brief samplings provide a limited environmental background for changing ritual activity and landscapes over a period of more than 2,000 years.

HOUSES OF ANCESTORS

Avebury lay at the heart of an ever-changing landscape comprised of earthworks, avenues, and burial mounds. The sequence of structures evolved over many centuries as customs and observances changed, beginning with the construction of communal long barrows (as opposed to conical burial mounds) and causewayed enclosures.

The story begins in a landscape of primordial woodland, inhabited by a few bands of Stone Age hunter-gatherers as early as 6000 B.C. Only a few thousand people lived throughout the British Isles at this time, subsisting off deer, small game, and wild plant foods. The same woodlands covered Avebury's chalklands, but only a handful of foragers exploited the forests and small clearings.

Some 1,800 years later, cattle herders and farmers spread thoughout southern Britain, clearing patches of woodland for cultivation and pasturage. Plant pollens from farming communities between 4350 and 4000 B.C. show dramatic reductions in tree cover. Cultivation weed pollen from open fields appears from this same time period. Avebury was still a quiet backwater until 4000 B.C. A few communities settled on the chalkland over the next 5 centuries.

The number of inhabitants rose gradually over the centuries. They still dwelled in dispersed communities, but over the years, began to pay close attention to the burial of the dead, and to the veneration of ancestors. They buried their dead in communal long barrows, the earliest burial places on the chalkland.

West Kennet

The Avebury long barrows form a dense group, some with stone internal burial chambers, others with now-decayed timber compartments. A few are little more than piles of chalk and sarsen stones, with flimsy rows of hurdles (determined from soil discolorations) subdividing the interior. The first sepulchres were modest structures, but the builders soon became more ambitious, as is seen at West Kennet, the most famous of Avebury's ancient burial places, built after 3600 B.C.

The West Kennet long barrow lies on a ridge 2 kilometers southeast of Avebury, a much-eroded earthen mound 100 meters long and about 2 meters high. The tumulus lies low against the skyline, its contours softened by 5 millennia of plowing and natural weathering. Only a single monolith at the entrance marks the sepulchre from afar.

The intangible relationship to the natural environment and the spiritual world can be felt in seeing burial mounds for the first time. As I climbed the path to the ridge, the tomb suddenly rose before me, as if ascending from the underworld. More than 5,000 years later, West Kennet still has an effect on the imagination. In its heyday, the fresh chalk of the steep-sided mound glittered brilliant white in the sun. Even on a rainy day, the wet chalk gleamed like a beacon in the gloom. Marked irrevocably on the landscape, the illusion of light, color, and ritual endures.

West Kennet began as a fairly unambitious burial place, marked out on an east-west line atop a low ridge. The builders laid out a mound of rough sarsen stones and boulders, which they heaped into a ridge about 2 meters high. They piled chalk rubble over the sarsen core, using 7-by-2-meter deep ditches on either side as the quarry. No one has yet excavated the western end of the barrow, and therefore we do not as yet know whether the imposing stone chambers at the eastern end were part of the original plan or added at a later date (figure 6.3).

At some point, the builders laid out a passage with four side rooms and a single end chamber. Great sarsen stones (blocks of natural sandstone) from nearby valleys formed the chamber walls, which were set in shallow holes. Gaps between them were then filled with oolite limestone from about 11 kilometers away. Large sarsen stones rested on the upright boulders to form a massive roof, with stone courses lapping one another to form corbeled, beehivelike roofs for the chambers. The passage opened into a crescent-shaped forecourt facing east and blocked with massive upright stones.

West Kennet is a classic megalithic tomb (from the Greek "megalithos," meaning big stone), a form of the "transepted gallery grave," which may have first developed in western France, then spread to England. Megaliths are widely distributed in western Europe, and are, in essence, large stone monuments—burial structures built over a long period from about 4800 to 2500 B.C. They vary greatly in design, from simple, boxlike chambers under circular burial mounds, to enormous tumuli with multiple passages and chambers, such as Knowth in Ireland (not on the map in figure 6.1). The earliest and most widespread of megaliths are the "passage graves," in which a narrow corridor led to the burial chamber, thus allowing repeated access long after the mound was completed. Numerous local variations flourished, among them "gallery graves" consisting of an elongated burial chamber reached by a short vestibule, and the transcepted version, which included side chambers. Megalithic tombs, notably those in Ireland, sometimes bear elaborate decoration—mainly abstract designs such as circles, meanders, and spirals, although more naturalistic representations also occur. Recently, some authors have suggested that some motifs resulted from trances and

Fig. 6.3. A rendering of West Kennet long barrow by archaeologist Stuart Piggott. Photograph from English Heritage.

altered states of consciousness, arguing for elaborate rituals conducted outside of or in the mounds.

Megaliths were usually communal sepulchres. Several archaeologists dug the West Kennet burial chambers, most notably, Victorian digger Dr. John Thurnam in 1859, and twentieth-century excavators Stuart Piggott and Richard Atkinson in 1955–56. Together, they recovered the fragmentary remains of at least forty-six people, many of them children or infants. The large chamber at the west end contained mainly adult male burials, whereas the southwest and northwest chambers held both male and female remains. The southeast chamber housed mainly children and young people; the northeast room, older individuals. Piggott and Atkinson believed that the placement of young and old in spatial opposition had symbolic significance. The West Kennet skeletons lay in groups that suggested that some adult males may have been buried separately. There were clear signs of geometric arrangement of opposing groups by age and sex. The West Kennet skeletons also provided valuable information on ancient disease. The bones displayed an unusually high incidence of spina bifida, suggesting some form of genetic relationship between some of the burials. Spondyloschisis of the spine occurred in an additional three skeletons, perhaps also linked to the individuals with spina bifida. The population as a whole had numerous medical problems, among them spinal arthritis caused by the cold, damp conditions. Many male skeletons bore healed bone fractures.

The mourners deposited the original dead with clay pots, stone artifacts, or food offerings on the original turf floor of the burial chamber. They sealed the tomb while the bodies decomposed. Judging from several incomplete skeletons, it appears that at intervals, the living entered the sepulchre and removed some of the now-defleshed skulls and limb bones, perhaps for ceremonies elsewhere. Each generation laid its dead in the same chambers, sometimes moving older bodies aside and piling up the bones in confusion.

West Kennet remained in use for more than 5 centuries. The last burial was a man who perished from an arrow wound in his throat. His skeleton rested on specially laid stone pavings in the northeast chamber. Soon afterward, his descendants filled the now-dilapidated burial chambers with chalk rubble and domestic rubbish. The entrance was sealed, and a line of large sarsen stones was placed across the crescent-shaped forecourt.

We can only guess at the reasons why West Kennet or other long barrows lie where they do. Analysis of pollen and snails shows that most of the barrows lie on once-cultivated land. Additional clues come from ancient ditch fills that are observed in the sides of cross-section trenches that cut across earthworks.

The archaeology of ditch silting is complex, even under ideal preservation conditions. "Primary silting" results from erosion when the ditch is newly excavated, causing a thin layer of silt at the base of the ditch profile. "Secondary silting" is a longer process, which occurs as the edges of the ditch settle to a natural angle of rest, often while the site is still in use. Secondary fills are invaluable sources of information on site abandonment and vegetational changes. For instance, the people living near Easton Down cleared and farmed the woodland for only a short time before they built the barrow. The mollusca in the later (secondary) fills of the Easton ditches show how regenerated woodland covered the tumulus within a very short time.

Whittle has calculated the thickness of the ditch fill between the base and the point where woodland regenerated at this site. The mollusk counts tell him how complete the regeneration was. The more complete the coverage, the less intense the human activity and the closer the woodland to the site. The mollusca and pollens from Easton Down place the long barrow close to formerly cultivated land. The original land surface under the South Street barrow near Avebury bears the criss-cross scratch marks from a simple ard (plow) drawn by oxen, perhaps at the edge of a field (figure 6.4). Other long barrows yield signs of cultivation and clearance. Since many such burial sites lay close to, but not on, cultivated land, we assume that they were significant to individual communities or settlement areas, leading us to believe that the burial sites may have served as territorial markers.

Communal burial chambers used over many generations, tombs that straddled the landscape like conspicuous markers, and sepulchres presumably reserved for people linked by birth and other ties of kin provided a vital bond between the farmers, their ancestors, and the land.

Windmill Hill

Built by local farmers, Windmill Hill lies nearly 4 kilometers across a valley northwest of West Kennet—a causewayed enclosure atop a strategic hill with a commanding view. Windmill Hill is situated close to the Kennet and Thames rivers, at a natural junction between well-defined high downs and river valley lands. The ancient Ridgeway Path—a communication route between the Thames, the chalklands, and areas to the west for thousands of years—passes nearby. Three concentric ditches and banks enclose about 9 hectares, the outer, irregular 2.13-meter ditch varying from 305 to 387 meters in diameter. Unexcavated "causeways" break up the ditches, which are progressively shallower from the outer ring. Low banks revetted with wooden posts lie on the outer sides. When newly built, Windmill Hill, with its chalk-white banks, was visible for kilometers. Thus, it may have served as a place of refuge in unsettled times.

People lived on, and cultivated, Windmill Hill as early as 3600 B.C. Storage pits and hut foundations predate the construction of the

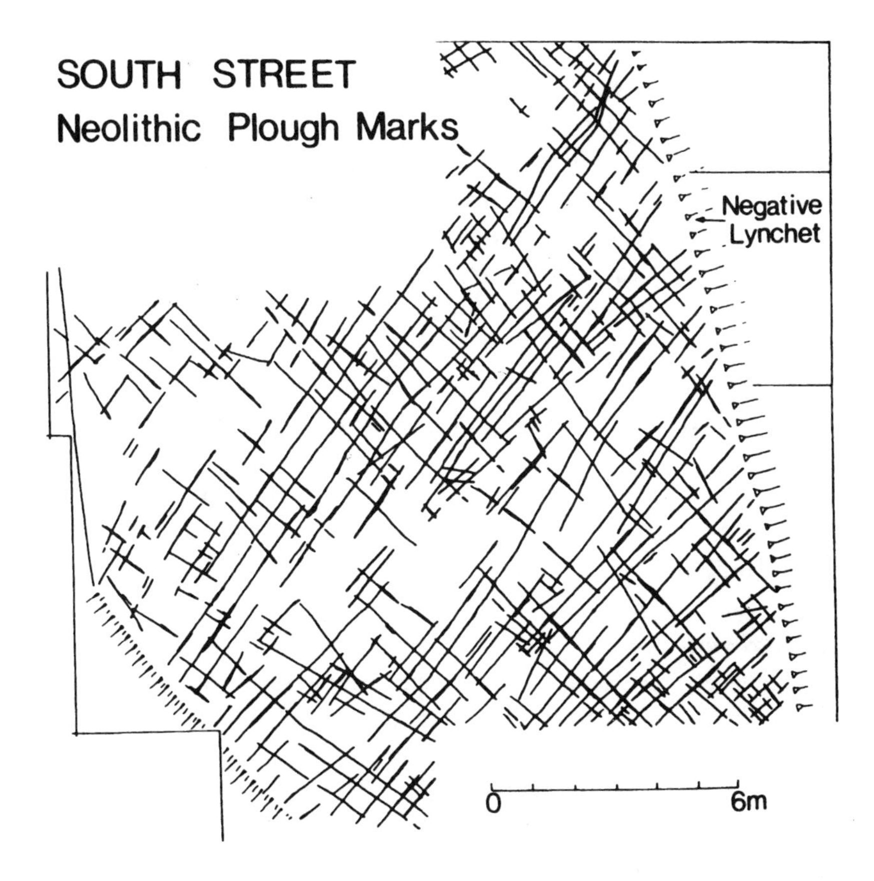

Fig. 6.4. Characteristic scratch marks of a Stone Age plow preserved on cultivated land under the South Street long barrow. Courtesy Prehistoric Society.

ditches. The actual digging of ditches and banks began in about 3500 B.C. Judging from the interrupted earthworks, individual groups dug their sections at different times, perhaps during slack months of the farming year. Some 62,000 to 64,000 people hours went into the digging of the earthworks, most likely over the course of several years.

Windmill Hill remains a fascinating enigma. The concentric ditches and earthworks enclose an open, and apparently unoccupied, space. Ditch pollens reveal an uncultivated scrub environment, common to similar enclosures elsewhere. With its protected open space, the 9-hectare enclosure was a logical gathering place at certain times of the year. People from afar may have visited the hilltop to trade animals, exotic stone axes, pottery, and other commodities, perhaps within a symbolic setting. But Windmill Hill served both the living and the dead. A square, wattle-and-daub building once lay outside the eastern main entrance of Windmill Hill. The excavators believe that this served as a mortuary house, where the bodies of the dead lay until the flesh had rotted from their bones. Then the defleshed bones of the ancestors joined those of their predecessors in nearby long barrows, or in the ditches of the great enclosure. Remains of sacrificial feasts of cattle, sheep, and goats lie in packed caches and clusters alongside human remains in the bottom of these ditches.

Eventually, Windmill Hill fell into disuse. The great henge at Avebury, less than 2 kilometers away, became the focus of a new sacred landscape.

Avebury

Avebury rests in a natural amphitheater ideal for large stone circles. In its final form in 2550 B.C., the great henge encompassed 11.5 hectares (figure 6.2), and measured approximately 350 meters in diameter. As John Aubrey pointed out over three centuries ago, the great bank and ditch do not form a perfect circle. It is as if the builders paced out the enclosure on the ground, then started digging. Aubrey surmised that most likely, teams of diggers, perhaps from different villages, worked independently, excavating the ditch and piling basketloads of earth to form the external bank. Eventually, their individual trenches met, to form a far-from-precise circle.

Four causewayed entrances divide the henge into 4 unequal arcs. As I stood on the much-eroded bank and looked down the steep side of the half-filled ditch on the northeast arc, I thought of the original ditch. Between 7 and 10 meters deep, its square-bottomed base had averaged nearly 4 meters across. The bank had been just as impressive. Between 23 and 30 meters across at the base, it had stood nearly 17 meters above the bottom of the ditch. In places, the builders erected crude chalk walls to prevent the bank from slipping across the narrow flat berm into the deep gully. At each entrance, they built timber revetments to keep the slopes in place, and scraped the surface of the entranceway to give an impression of great height on either side. Archaeologists guess that wooden barriers once blocked the defiles, restricting access to the enclosed area. When freshly dug, both bank and ditch of white chalk stood out for kilometers. The symbolic significance of the brilliant whiteness to their culture is unknown, but it certainly made a conspicuous landmark that extended beyond the mere desire to create a visible monument.

While the diggers trenched, they also hauled in and set up the 98 large standing stones, some up to 14 meters high, on the inner perimeter of the ditch. The largest stones stood at the northern and southern entrances. The builders erected both long, columnlike monoliths and massive, roughly triangular-shaped boulders that stood on relatively narrow bases. Avebury's architects followed a long-standing tradition of using local sarsen stones, weighing about 26 tonnes, for the uprights (figure 6.5).

Two smaller inner circles, laid out with almost perfect precision, stand inside the perimeter uprights, oriented roughly north to south. A 3-sided arrangement of 3 huge stones, open to the north, lies inside the center of the 49-meter, 27-stone northern circle; and a single high boulder stands inside the 51-meter, 29-stone southern ring.

Archaeologist Caroline Malone observes that Stonehenge, in its entirety (located 50 kilometers to the south), would fit comfortably within either of Avebury's inner circles.

By any standard, Avebury was a staggering construction project, even more so for a sparse population of subsistence farmers with no wheeled carts and only the simplest stone, antler, and wooden tools. Avebury's builders used simple technology, such as rollers, levers, fiber ropes, and ladders, which scientists can reconstruct only by experiment. Controlled tests conducted by archaeologist Richard Atkinson showed that stones weighing up to 4 tonnes could be dragged readily on a wooden sledge by 32 schoolboys. By adding wooden rollers, he was able to reduce the number of his haulers to 14. When waterborne on a simple raft, only 4 people were needed to transport the stone in calm water.

Careful excavation reveals that the same basic methods were used at both Avebury and Stonehenge to erect boulders weighing up to 25 tonnes each. The builders prepared the stone holes with a ramp

Fig. 6.5. Reerecting an Avebury upright in the 1930s. Photograph from English Heritage.

on one side. Then they rolled the foot of the stone over the hole, and inserted a long lever supported on wooden rollers under the end. Slowly, they levered it upward, placing additional timbers under the rising stone until they could prop it up, and then pulled it upright with long fiber or hide ropes attached to the top. Eventually, the stone slid into the waiting hole, at which point it was wedged into place with stone packing.

There is nothing mysterious about the building of either stone circle. The builders were farmers, who were expert with simple levers, fiber and hide lashings, and rollers. They could move heavy weights and shift vast quantities of earth. They could also call on large numbers of people to help them move large weights. The teams of people who excavated the ditch removed 0.12 million cubic meters of heavy chalk and mounded it into the bank. This enormous task took between 650,000 and 1,540,000 hours to complete.

Stone circles abound in Britain, but none approaches the scale of Avebury, with its vast bank, ditch, and elaborate standing monoliths. Why was this particular location so important? What compelled a small population of village farmers to erect such a stupendous center? What rituals unfolded within Avebury's stone circles, attracting people from distant communities? Avebury holds its secrets well. Generations of people kept the interior scrupulously clean, and, thus, almost no domestic or sacrificial rubbish survived the millennia. Only the fortifications themselves have provided a few revealing clues.

When archaeologist Harold St. George Gray trenched small cuttings across the bank and ditch between 1908 and 1922, he unearthed fragmentary human remains from different individuals, perhaps brought to Avebury from other communal graves. The original builders buried these skulls, jaws, or limb bones in the ditch with considerable ceremony, often with pottery or other burial offerings. Gray also unearthed the skeleton of a female dwarf surrounded by small sarsen stones and animal bones in the ditch on the east side of the southern causewayed entrance. Three human jaws and numerous antler picks lay close by. Apparently, the dwarf was a dedicatory burial placed in the ditch on its completion. All these finds point to a close tie between ancestral burials and the henge, as if the revered ancestors played an important role in ceremonial life within the great enclosure. This place, and the landscape of which it was a part, were sanctified by the ancestors.

I have visited Avebury and walked among its stones many times over the years, sometimes on fine summer evenings, and in September when the great stones cast long shadows. The high bank and ditch exclude the outside world. On crisp days, fast-moving clouds lead your eyes to the far horizon, placing the stones at the center of a much larger universe. I remember one fierce winter storm with rain lashing horizontally across the chalkland, and the earthworks mantled in a dank mist. Avebury then became a place of menace; of hidden forces. Haunting strands of white fog wisped around the wet monoliths, and I imagined figures among the stones. Perhaps Avebury, in its changing moods, served as a likely place of ritual. Certainly, these ghostly ancestors exude a powerful presence.

The Sanctuary and West Kennet Avenue

Three kilometers across the valley from Windmill Hill, another ceremonial structure stood on prominent Overton Hill, close to the Ridgeway. The Sanctuary, so named by eighteenth-century antiquarian William Stukeley, comprised a series of roofed, circular buildings, originally built of wood, and later of timber and sarsen stones.

The most recent investigators of the Sanctuary used proton magnetometers to investigate the site (see information in box, p. 135). The Sanctuary's surveyors laid out a grid of squares, then took systematic measurements with a staff to which were attached two small bottles filled with water or alcohol enclosed in electrical coils. They measured the magnetic intensity by recording the behavior of the protons in the hydrogen atoms from the electrical coils. The magnetometer itself amplified the weak signals from the coils. A computer recorded the data and converted them into a printout.

The magnetometer survey and radiocarbon dates showed that the first simple, round hut rose in 3000 B.C. (figure 6.6). A century or more later, a second much larger structure, detected from only 2 rings of postholes with a diameter of 11.2 meters, enclosed the original house. A third circular building, almost twice the size of its prede-

SUBSURFACE SURVEY METHODS

Magnetometers are an invaluable nonintrusive archaeological tool, especially when combined with pulse radar and other forms of remote sensing, such as restitivity survey. Magnetometer surveys measure the remanent magnetism of the soil. When the remanent magnetism of fired clay, or of other materials in a pit or similar feature, is measured, it will give a reading different from that of the intensity of the earth's magnetic field normally obtained from undisturbed soils.

Pulse radar uses a pulse induction meter to apply pulses of magnetic field to the soil from a transmitter coil. The instrument is very sensitive to metals and can be used to find pottery, metal objects, and graves containing such artifacts. Pulse radar was used successfully at the Mayan village at Cerén, San Salvador, to locate collapsed huts buried under 5 meters of volcanic ash.

Restitivity survey measures the electrical resistance of subsurface features. Rocks and minerals conduct electricity, mainly because the deposits have moisture containing mineral salts in solution. A restitivity survey meter measures the variations in resistance of the ground to an electrical current. Stone walls or hard pavements obviously retain less dampness than a deep pit filled with soft earth or a large silt-filled ditch. These differences can be measured accurately so that disturbed ground, stone walls, and other subsurface features can be detected by systematic survey. A restitivity meter attached to probes gives readings across a grid, which appear as contour lines on a plot, showing the presence of features such as walls or ditches.

cessor, with a conical thatched roof and possibly a central raised portion, replaced the second structure some time later. Two massive posts marked the entrance on the northwest side of the house, facing toward Avebury. At some point, the builders incorporated a circle of sarsen stones into the middle ring of posts, making a near-continuous internal wall. A dedicatory offering of a young man lay under a stone on the east side of this wall.

Like nearby Windmill Hill, the Sanctuary is an enigma. Early antiquarians and 1930s excavators recovered large numbers of human

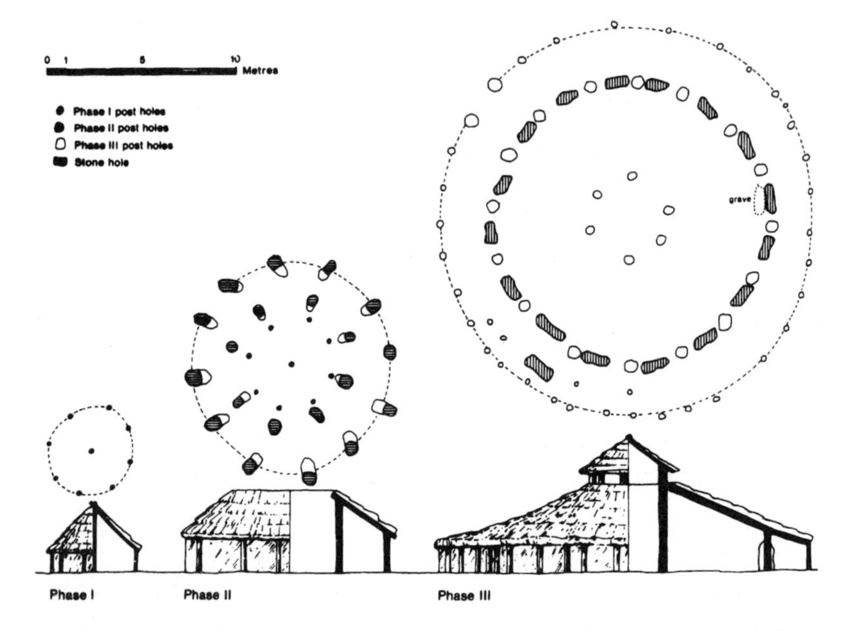

Fig. 6.6. The stages of the Sanctuary, as reconstructed by archaeological excavation. Courtesy Caroline Malone.

bones, and fragmentary animal remains, as if funerary rites and feasting occurred within the precincts. In its earlier stages, the Sanctuary may have been a charnel house where the corpse of a dead person lay (perhaps on a platform) until decomposed, or until animals and birds had defleshed the bones. Conceivably, the building performed a similar function to Windmill Hill: a place where the newly deceased turned into ancestors (a process that took months), before their bones entered a long barrow or the soil.

By 2500 B.C., the Sanctuary lay inside a sarsen-stone circle of 42 stones, 40 meters in diameter. Two stones set radially on its circumference, more or less in line with the door posts of the central building, linked the entire complex to the nearby ceremonial West Kennet Avenue leading to Avebury. This avenue connected the northwest side of the Sanctuary to the great henge at Avebury via a winding defile of about 100 pairs of standing stones. About 15 meters wide, the avenue extended along a sinuous 2.4-kilometer route, delineated by sarsen stones placed at about 25-meter intervals. The builders positioned long, cylindrical and triangular-shaped sarsens opposite one another. Similar stone shapes appear in opposition at Avebury itself. Alexander Keiller suggested many years ago that they were symbolic representations of males and females.

William Stukeley was fascinated with West Kennet Avenue and the incomplete and little-known Beckhampton Avenue, which passes from Avebury to the west. He recorded dozens of standing stones (figure 6.1), and noted that modern-day farmers were using fires to break them up for building material. Ever-fanciful, Stukeley interpreted the winding avenues as a great serpent, its head at the Sanctuary, the tail at the end of the Beckhampton Avenue. We dismiss the serpent as a figment of Stukeley's ever-lively imagination, but his notes and drawings are of priceless value. By 1908, only four West Kennet Avenue stones still stood upright and only thirteen were still visible; Beckhampton Avenue is now marked by only two standing stones.

Alexander Keiller's great wealth enabled him to purchase many original Stukeley papers describing Avebury and the West Kennet Avenue. While excavating Windmill Hill, he became obsessed with finding the original course of the standing stones. In 1934, Keiller and Stuart Piggott started a long-term program of selective excavation to solve the problem. They combined antiquarian observation and scientific digging, opening a parallel trench 6 meters wide on each side of the avenue. Whenever they came on a buried or fragmented stone, they levered the intact sarsen back into an upright position and set it in concrete, pinning it together with steel rods where necessary. The excavations also revealed the positions of 30 stones that had vanished without trace. In these instances, Keiller erected small concrete plinths to mark the spot, turfing and fencing the restored avenue for visitors.

The West Kennet Avenue follows a sinuous course, with little regard for the natural contours of the land and no seeming logic to the route. Perhaps the curves and kinks avoided existing settlements. Some of the stones were also grave markers for people buried at the foot of uprights in shallow pits. Keiller excavated four graves, all of them containing males. Scatters of human bone were found elsewhere along the avenue, the remains of burials disturbed long ago by subsequent cultivation.

Both the West Kennet Avenue and the main earthworks at Avebury date to about 2550 B.C., when the landscape surrounding the great henge reached a brief maturity.

Silbury Hill

By 2500 B.C. the West Kennet Avenue linking the Sanctuary and the Avebury earthworks was completed. The farmers then built Silbury Hill, 1 kilometer south of Avebury (figure 6.7). It is a stupendous earthen mound that rivals in size the Pyramid of Khufu at Giza by the Nile River.

Fig. 6.7. Silbury Hill from the air. Photograph from English Heritage.

Measuring 40 meters high and covering an area of 2.2 hectares, Silbury Hill stands out like a benign carbuncle on the rolling chalkland. Stone Age farmers built the great mound in 3 stages, piling up about 0.35 million cubic meters of chalk and soil and surrounding it with a ditch nearly 5 meters deep and 21 meters wide. Many archaeologists, some with the help of mining engineers, have dug into Silbury, with the vain hope that they would find a spectacular burial chamber in the center. Geologists from Cardiff University used echo-sounding equipment to establish the depth of the eroded ditch. Few, however, worked with the precision exercised by Richard Atkinson in 1968, but he was not looking for burials. The objectives of his excavations were to date the mound and study the history of its construction. Atkinson tunneled 70 meters into the mound, exposing a long section of the base. He identified 3 construction phases and acquired vital environmental data.

From Atkinson's samples, we know initial construction of Silbury Hill began in the late summer of about 2200 B.C., most likely after the harvest, when people had time for construction work. The farmers laid out a 37-meter circle on a small projection on the slope of a hill, heaping a thick layer of gravel on the original ground surface. Next, they laid sods, so well preserved that ants and anthills survived. The sods show that the ants were beginning to grow wings and fly away from their anthills, as they do in late summer. Four layers of soil, clay, chalk, and gravel followed, forming a drumlike mound about 4.5 meters high.

A second mound of chalk blocks laid in sloping steps and held together with interconnecting walls was built above the original tumulus. The builders began to dig a 6-meter-deep ditch around the conical structure, but they never completed it, as more grandiose plans came to the fore. By this time, the hill was nearly 17 meters high, as the villagers had dug into the neighboring chalk hillside for new raw material.

The third phase of Silbury enveloped the 2 earlier mounds. The earlier, incomplete ditch was filled in, as a new even larger one was dug to surround the new structure. Six concentric steps, each about 5 meters high and formed from chalk blocks rose one above the other. The mound now angled at 30 degrees. Chalk rubble and silt held together the honeycomb of chalk walls, forming a remarkably stable structure. By this time, the builders had extracted a hectare of chalk from the nearby hillside and additional chalk from other sources. More than 3 million hours, the equivalent of 700 workers laboring for 10 years, went into this extraordinary engineering feat.

Why did a small population of village farmers exert such extraordinary effort, spread over so many years, to create Silbury Hill? This question has baffled archaeologists since Stukeley's day. No great chief's burial lies within the mound. No sacrificial victims or shrines came to light in the excavations. The site location baffles conventional logic. You cannot see Silbury from all directions, nor can you locate it from within the Avebury earthworks. Had the builders wanted it to be visible from afar, they would not have built it where they did, in the lowest area of the Upper Kennet Valley. No avenue links Silbury to Avebury—it is as if the mound lay outside the sacred geography of the great henge. We can only assume that the chosen site had overwhelming symbolic importance, but whether it was a place of an important event, or a territorial boundary, or even a gate to the spiritual world, we have no idea. To this day, Silbury Hill remains a conundrum.

THE MYSTERY OF AVEBURY

Thousands of years ago, Avebury, with its avenues and associated stone circles, rose on the chalkland. The same physical elements appear around Stonehenge, near the town of Dorchester close to the south coast, and in the nearby Thames Valley. All of these landscapes are variations on a common theme—an association between burial places, enclosures, and henges—and all are incredibly difficult to decipher, in regard to their spiritual meaning.

What compelling spiritual associations made the Avebury chalklands a focus of religious activity to people living nearby as well as dozens of kilometers away? How did the religious beliefs behind the monuments on the landscape change over the centuries? Science tells us that the people lived in the midst of an ever-changing natural environment of woodland and open country, increasingly domesticated by their own labors.

Radiocarbon dates and meticulous excavations tell a story of a 2,000-year-long dialogue between the living and the dead, expressed in a complex archaeological record of burial mounds, earthworks, and stone circles. The interaction began with simple burial rituals, and continued with the erection of long barrows and causewayed enclosures that commemorated the ancestors and served as settings for communal rituals.

Around 2550 B.C., the farmers built Avebury, an enclosure in the old tradition of stone circles, ditches, and earthworks, but with a new formality. Avebury was larger, with carefully separated spaces and restricted entrances. The interior was clean, a formal setting for performances that enshrined ancient traditions and commemorated revered ancestors. The sacred tradition was powerful enough to enlist the labor of people from villages near and far. But the ritualistic performances did not last long. A new cycle of ritual activity saw the abandonment of the stone circles, as Silbury Hill rose 2 kilometers to the east. By 2000 B.C., the Avebury monuments had fallen into complete disuse.

Avebury's sacred monuments evolved over 2,000 years. They tell a story of changing religious beliefs and rituals, but also of spiritual continuity. A profound dialog between living people and honored ancestors continued over the centuries, with a changing backdrop of barrows, earthworks, and henges—as signposts to a world where barriers between the domains of the living and those of the dead and the forces of the spiritual world did not exist. Archaeologists have unearthed a long-forgotten ritual narrative, acted out in stone circles, in the forecourts of burial mounds and outside charnel houses, and within the serenity of the family and household. The intangible chants, dances, prayers, and incantations that passed an ancient, but ever-changing sacred narrative from one generation to the next, however, remain buried in the past.

When I stood atop the Avebury bank looking over the rolling hills and watching the cloud shadows chasing across the chalk, I realized I could not divorce the secular landscape from the sacred. Avebury is far more than a spectacular stone circle. It forms part of a once-enormous symbolic environment of carefully placed avenues, earthworks, and aligned stone circles, which linked communities and kin groups to the sepulchres of the dead, and both the living and the ancestors to places of communal worship. The excavation and study of this landscape has shown that the vast expenditure of time and labor reflected in the circles and earthworks was devoted to imprinting a spiritual landscape inhabited by ancestors on the living world in material form. The builders of West Kennet and Avebury had moved far from the household shrines familiar at Jericho and Çatalhöyük. They revered their ancestors on a humanly made stage that encompassed their entire landscape, using their earthworks to celebrate their close bond with the otherworld around them.

chapter seven

STONEHENGE AND THE IDEA OF TIME

"For everything there is a season, and a time for every matter under heaven." The anonymous author of Ecclesiastes knew full well that the passage of time gives an ordered sequence to human life. I cannot remember the day or the minute I was born, but I can remember being very young: a London air raid in 1940, fragments of antiaircraft shell in our garden, my third birthday party on a warm summer day—these memories of my youth defined a rhythm to my life long before I was conscious of the measured passage of time. Summer passed into fall and winter. The long, cold, gray months gave way to the budding leaves of spring. Each season brought its own activities: sailing in summer, picking apples in autumn, skating in winter, brisk walks in bracing winds as the leaves turned green in spring. As the years pass and the tyranny of time impinges ever more into my life, I become increasingly aware of how time ebbs and flows in quicker and slower beats. Today I live by book deadlines and scheduled classes. And yet, as time goes on, I find myself paying more attention to the movements of the heavenly bodies, astrology, and the idea of time.

STREAMS OF TIME

More than a century ago, the great German statesman Otto von Bismarck remarked that all humans float on an ever-moving stream of time. We do our best to navigate its waters; to control our individual and collective destinies. But none of us, whether hunter-gatherer, village farmer, or twentieth-century industrialist, can stop the waters. We Westerners have chosen to embrace the notion of linear time. Our history of human existence is cast in a precise chronology, extending back nearly 5 million years: it reflects sequences of time, marked off in divisions of millions of human actions. We experience life's moments in decades. The landmarks of thirtieth, fortieth, and fiftieth birthdays assume significance and become irrevocable.

We live in the grip of the inevitable. The certainty that the stream of time will continue to flow after we are gone lies at the center of our thinking about ourselves, and how we relate to one another, to the dead, and to those soon to be born. Time defines the essence of the spiritual elements of human life, and, thus, defines the unconscious life of humankind. Astroarchaeologist Anthony Aveni rightly calls time "the music, the rhythm, the road . . . the framework in which we recollect the past and anticipate the future" (1989, 17).

Long before the first words appeared on a clay tablet or papyrus, humans developed intricate, and often highly sophisticated, ways to measure time that passed from one generation to the next through myth and oral tradition, and through the power of human memory. I remember in high school that one of my teachers knew all 27,000 lines of Homer's *Iliad* and *Odyssey* by heart, in the original Greek. He mastered the rhythmic patterns of words, the formulae and memnonics, which the ancient bards repeated again and again, accompanying themselves with lyres or other musical instruments. To hear my teacher recite Helen's lament over the dead Hector was a spellbinding experience. He took me back to a darkened banqueting hall illuminated by flickering firelight, where men and women wept at the poet's words.

As students we learned that day what it was like to live in a nonliterate society, where all human knowledge—practical and spiritual—passed orally from one generation to the next. We came to appreciate how staggering amounts of legend, myth, and spiritual teaching were passed on by men and women who learned by rote from reciting chants; and who practiced intricate rituals that were used to share stories, which would, ultimately, be transposed into spiritual awareness, social conformity, and power. Ancient life unfolded against a backdrop of ever-changing seasons, of predictable movements of heavenly bodies in a cosmos inhabited by ancestors and spirit beings. Human calendars were used long before the Romans developed a form of the linear calendar we use today. These calendars of the past survive only in the form of material remains, in the artifacts, buildings, and landscapes that were the settings for the rituals that marked the seasons. They hint at very different concepts of time from those embraced by Western scientists.

In seventh century B.C., a northern Greek farmer named Hesiod set down the realities of a harsh farming life in a curious self-help book, *Works and Days* (Athanassakis 1983). Like Homer, he wrote familiar ballads about the correct season and time for completing farming chores, which dated back for centuries. His readers were familiar with his ideas and metaphors, for they, too, lived in a harsh and unpredictable farming environment. Hesiod wrote *Works and Days* as a realistic calendar, in the context of how to run a farm. He began with a broad chronological framework of ages, which began with a Golden Age, where everyone lived a blessed life. Then followed Silver, Brass, and Iron Ages, the latter being an era when humans found goodness through labor. The farmers worked hard, to shore up a degenerating world of human suffering.

Hesiod's farming almanac based the agricultural year on the movements of the heavenly bodies, where events that marked the passage of time coincided with the changing constellations. He adjures: "Start reaping when the Pleiades rise, daughters of Atlas/ and begin to plow when they set" (ibid., 76). Forty days from late March to early May separated setting and reappearance. His advice to serious wine growers: "When Orion and the dog star rise to the middle of the sky/and rosy-fingered dawn looks on Arcturus,/then, Perses, gather your grapes and bring them home" (ibid., 82). These astronomical events occurred in early September. After harvest, he adjured them to expose their grapes to the sun for 10 days and 10 nights, cover them for 5 (simple units of finger measurement), and press the wine into jars for aging on the sixth, at the fall equinox. Local Greek village winemakers follow the same procedure to this day.

Work and Days creates a telling portrait of the tension-filled world of the village farmer, always at the mercy of the elements. Time unfolds. Will the rains come on time? Will they drown freshly planted seed? Or will a sudden frost decimate a bountiful harvest? The relief at a successful harvest is palpable. The cycles of the farming year wax and wane in a seesaw of tension and relaxation, measured by the signpost of the heavens. The farmers curse nature, yet must follow her, and attempt to control and manipulate her vagaries.

Hesiod and his farming contemporaries thought of time not as an abstract formulation measured by clocks, but as an ordered cycle of natural events. The idea of time was a continual dialogue between the natural world and human society. Nature sent messages to the farmer through a repetitive, chain of interlinked events: the appearance of heavenly bodies, the migrations of birds, the flowering of specific plants, and even harsh winter northers that kept ships ashore. Time was a cyclical stream that helped create a harmonious order in society and in the repetitive events of human life.

Those who cultivated the soil maintained a dialogue with the forces of the spiritual world. In areas where humans lived and farmed, the constant conversation with the dead—inhabitants of the spiritual world—was usually part of a larger interaction between people and the natural world. Cycles of life and death revolved endlessly around the passing seasons, measured by warmth and cold, dry and wet periods, the flowering of plants and ripening of nuts, and the regular passage of the heavenly bodies. Hesiod measured time with reference to constellations and the changing night sky, as Greek farmers had done for millennia. The heavens were an integral part of the ancient farming world, of enduring lore and myth, and of the relationship between the living and spiritual worlds.

Crossing the Atlantic Ocean in a small sailing vessel some years ago, I learned much about the stars and the changing phases of the moon. Far from land, the heavens shimmer with brilliant constellations, planets, and the white tapestry of the Milky Way. I used a star atlas to identify bright stars, and to select heavenly bodies for fixing my position 1,000 kilometers from land, using a sextant. As the days passed, the stars became my friends, a part of my world, where sea and sky merged into one and the familiar landmarks of Europe and America were out of mind. For the first time I realized that the heavens were a clock—an endless calendar that required but simple observation to tell the season and the passage of days.

To this day, I still look up and recognize old friends as they appear overhead in May and June, the time of my passage. Ancient farmers had as close a relationship to the heavenly bodies as I did for those brief 24 days. They observed the changing seasons from land, using simple alignments and the positions of sunrise and sunset on the horizon to do so. Ancient astronomy was not a science, however; it was based on a knowledge of the heavens passed from one generation to the next.

Astronomy was a serious preoccupation in many ancient societies. Outrigger canoe navigators in Polynesia used zenith lines and other phenomena to make long passages over open ocean and to return safely. They served long, arduous apprenticeships with master pilots, where they learned practical observations and a vast body of ritual (see information in following box). Mayan priests and scribes

---

TRADITIONAL PACIFIC NAVIGATION

The analogy is a little far-fetched, but an excellent example of how ancient peoples used the heavens comes from the Pacific. For generations, archaeologists have argued over the first settlement of Polynesia. Did the first human settlers arrive in battered canoes, accidently stranded on hitherto-unknown islands after being blown far off course? Or did Polynesian seafarers deliberately colonize isolated landmasses deep in the Pacific, with the expectation that they could return to their homelands? The eighteenth-century British navigator Captain James Cook, himself a navigational genius, had no doubt as to the Polynesians' seagoing abilities in outrigger canoes. He theorized that they had sailed from mainland southeast Asia to their island homelands. While in Tahiti in 1769, he met a local pilot named Tupaia, who knew sailing directions for thousands of square miles of the Pacific by memory. Cook and Tupaia sailed together on *Endeavor* from Tahiti to New Zealand, but, unfortunately, the Tahitian died of fever on the way to England before passing on his traditional lore. For 2 centuries afterward, ancient navigational skills passed into a seeming limbo as scholars argued about deliberate or accidental migrations.

The controversy ended in the 1960s, when British yachtsman David Lewis met one of the last Micronesian navigators and was astounded by his uncanny ability to navigate accurately out of sight of land. He apprenticed himself to the pilot and learned much of the ancient ocean-going lore, navigating his yacht to New Zealand without the aid of a sextant. Lewis learned some of the subtle signs of sea and sky that had been passed from one generation to the next by the navigators: recitations that listed the appearance of zenith stars in the heavens for different courses and latitudes, key landmarks and sighting lines, and the messages given by cross-swells or by refracting waves from invisible islands over the horizon. Traditional Pacific navigators were as familiar with the sea and heavenly bodies as farmers were with the heavens and their land ashore. The ocean environment was replete with meaningful landmarks and signposts, provided that these navigators knew how to identify them. Their skills were developed during long apprenticeships; they learned sailing directions through ocean-going experience and memnonic recitation. Since Lewis's apprenticeship, a major revival of traditional Polynesian canoe-building and passage-making knowledge has begun, culminating in the passages of the Hawaiian double canoe *Hokulae'a* from Hawaii to Tahiti and back, and on a triumphant circuit of Polynesia. Unfortunately, archaeologists cannot apprentice themselves to Stone Age farmers of 5,000 years ago, but Pacific seafaring shows us the amount and depth of environmental and astronomical information that existed at one time.

---

acquired their skill through many years of training in written scripts and through memnonic recitations. Like religious performances, astronomical knowledge itself was intangible. Only the settings—the ancient observatories—have survived for archaeologists to study; many of these yield their secrets unwillingly. Stonehenge, in the heart of Wessex in southern Britain (figure 7.1), is the most famous and provocative of these ancient observatories devoted to the idea of cyclical time.

Fig. 7.1. Stonehenge from the air. Photograph from English Heritage.

STONEHENGE

Nineteenth-century antiquarian Richard Colt-Hoare wrote of Stonehenge in 1812: "How Grand! How Wonderful! How Incomprehensible!" (Chippindale 1994, 123). Many modern-day archaeologists echo his sentiments. Stonehenge is one of Europe's most famous monuments, revered as an ancient sacred place since Roman times. Scholars have puzzled over its trilithons (2 uprights with a horizontal lintel) and stone circles for more than 800 years. A twelfth-century cleric, Henry of Huntingdon wrote: "No one can conceive how such great stones have been so raised aloft, or why they were built there" (ibid., 20).

Six hundred years later, the charming and eccentric William Stukeley spent the summers of 1721 through 1724 putting together "a most accurate description" of Stonehenge with "nice plans and perspectives." He and his company even dined atop one of the trilithons, where he found space enough "for a steady head and nimble heels to dance a minuet" (Stukeley, as cited in ibid., 71–79).

Stukeley realized Stonehenge, like Avebury, flourished amidst a landscape of burial mounds and other ancient earthworks. His natural curiosity took him riding into the surrounding countryside, where he reveled in the chalk downs and cursed the numerous visitors hammering at the trilithons to obtain souvenirs. He was the first to identify an avenue of two parallel ditches running from Stonehenge's entrance to "where abouts the sun rises, when the days are longest" (ibid., 73). Stukeley and his patron, Lord Pembroke, investigated a strange, cursuslike (Latin: roadway) monument and dug into several barrows, the "artificial monuments of this vast and open plain." In one barrow, he found "bits of red and blue marble, chippings of the stones of the temple. So that probably the interr'd was one of the builders" (ibid., 75).

Astronomer Edmond Halley and Stukeley estimated that Stonehenge was of an "extraordinary antiquity," and discovered that it had a greater exactness." Stukeley observed a regular "variation from the cardinal points" in the alignment of the monument. The axis of the stone circles lay 6 to 7 degrees clockwise of northeast, facing toward the point where the sun rose at the summer solstice. "What would be more probable . . . than that unlettered man in his first worship and reverence, would direct his attention to that glorious luminary the Sun" (ibid., 75). All subsequent astronomical research on Stonehenge stems from Stukeley's commonsense observation that the midsummer solstice was of cardinal importance to those who built the stone circles.

Unfortunately, Stukeley, obsessed with the Druids, considered them to be the builders of Stonehenge and the direct descendants of Abraham. Modern-day Stonehenge expert Christopher Chippindale remarks: "Stonehenge has never fully recovered from the Reverend

Stukeley's vision" (1994, 86). To this day, modern-day Druids act out their bizarre rituals in the heart of Stonehenge every summer solstice. English Heritage, the government organization responsible for ancient monuments throughout England, keeps the stone circles under 24-hour guard to preserve the "grand and incomprehensible" against cultists and vandals.

The Stonehenge we see today evolved over more than 1,500 years. Its story began around 4000 B.C., when farmers with herds of cattle and sheep settled on the chalk hills and valleys of this part of Wessex. As at Avebury, pollens and mollusca reconstruct the changing ancient environment. Land snails and pollen grains from excavations in the Stonehenge car park tell us that open pine and hazel woodland flourished in the area before farming began. Soon, the ancient woodland gave way to a complicated mosaic of primordial forest cover: less-dense oak and hazel woodland, and areas with shrub and grassland that had been cleared by a scattered population of Stone Age farmers. We know little of these people, but they may have moved considerable distances over the landscape, leaving exhausted gardens fallow for long periods of time. Some places may have acquired particular importance within this landscape, among them the Stonehenge location. These people's dialogue with the supernatural world began, as it did at Avebury, with the building of communal sepulchres—long barrows set on ridges, perhaps as territorial markers—and causewayed camps, one only 3 kilometers from Stonehenge.

About 2950 B.C., the farmers built a simple earthwork enclosure with a ditch and internal bank (and a small external one), and 3 entrances on a plot of open, well-established downs turf at Stonehenge (figure 7.2). We do not know when the land was first cleared, or why the builders chose that particular location close to a dry valley system. Occasional signs of earlier occupation in the vicinity show that the farmers erected their new enclosure in an already inhabited landscape. A circle of wooden posts set in the so-called "Aubrey Holes," named after John Aubrey of Avebury fame, soon lay just inside the ditch. The original ditch and bank gleamed white against the surrounding landscape, thus making the new enclosure visible from a considerable distance. Cattle bones

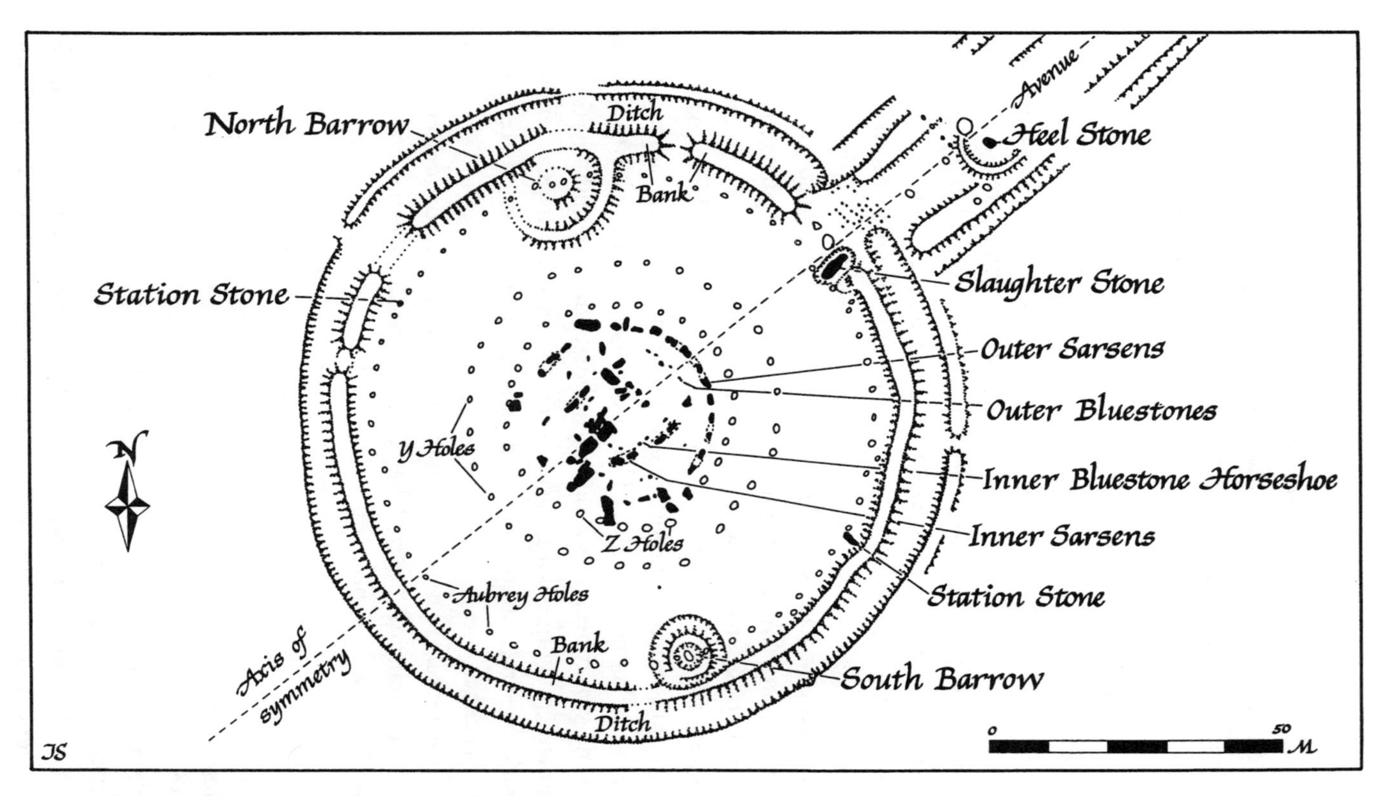

Fig. 7.2. Plan of Stonehenge showing major features.

(including a skull as much as 300 years older than the earthworks) were placed in the bottom of the ditch, especially near the entrances. A long, narrow enclosure with chalky bank and ditch extending over nearly 3 kilometers ran along to the north of Stonehenge. The new earthwork was quite unlike Windmill Hill or other causewayed camps; it may actually have been an early form of henge monument, with a timber circle.

Over the next 300 years, the original enclosure changed from a monument in the ancient tradition of causewayed camps into one with a concern for human remains. The posts in the Aubrey Holes vanished, and some cremations were placed in the filled holes. The ditch fell into disuse, while an elaborate, but undecipherable timber structure rose in the center of the monument. Even using computer-aided design (CAD) mapping systems, archaeologists have been unable to decipher the various timber circles inside the enclosure. Narrow corridors formed the northeastern and southern entrances, as if someone at this time had controlled access to Stonehenge.

In 2550 B.C., Stonehenge stood in the midst of open pastureland. People still lived in small settlements, moving around frequently with their herds. Within a few centuries, the environment had changed. The farming population increased and grew larger areas of cereal crops. Soon formal field systems appeared, subdividing the land into small, intensively farmed blocks. The farmers settled down in larger, more permanent villages, for their fields required intensive ground preparation, sowing, weeding, and harvesting. With larger village populations, the leaders of individual communities could deploy labor for other activities at slack times in the farming year.

Over the next 1,000 years, Wessex achieved great prosperity, with its rich agricultural land, abundant sources of flint, and strategic position astride long-established trade routes. Wealthy chieftains ruled over the region. Their grieving followers buried them under conical burial mounds within sight of a transformed, and spectacular, Stonehenge.

This new Stonehenge appeared soon after 2550 B.C.; a monument with the same overall function and plan, but executed in stone, instead of timber. For 10 centuries, successive generations of builders modified and tinkered with the concentric circles of

Stonehenge, and with the immediate periphery of the henge. The most important change came at the beginning, in 2550 B.C., when the builders changed the axis of the original enclosure. Using CAD software, researchers have shown how the axis changed from a centerline through the northeast entry to a new orientation through the main entrance, which matches the position of sunrise at the summer solstice. The earlier Stonehenge designs had no astronomical orientation. The new monument, aligned on the solstice, now provided a connection between earth and the heavens.

Stonehenge in its various later forms comprised a series of concentric stone circles. It is difficult for archaeologists to study the relationships between circles, since their arcs never touch (figure 7.3). Nor is it easy to associate the changing circles inside Stonehenge with isolated features at the edge. The sequence of architectural events at Stonehenge laid out in table 7.1 summarizes current knowledge, based on over a century of stratigraphic excavation on major features of the monument, and forty-nine radiocarbon dates, twenty-nine of these being AMS readings.

After 2550 B.C., the builders worked in stone, using bluestones from the Preseli Hills of South Wales and sarsens from the Marlborough Downs, some 30 kilometers to the north. The bluestones at Stonehenge encompass a number of different rock types, all of which source to the Preseli Mountains in southwest Wales, nearly 200 kilometers from the site. The exotic slabs came to southern England by raft and then over land, a hazardous journey across the strong-running tidal waters of the Bristol Channel. Some geologists have argued that the Stonehenge builders used local glacial erratics—that is, that the bluestones had been transported to near the site by Ice Age glaciers—but there are no other large examples of these stones within the whole of Wessex, except for those found at archaeological sites, to support this theory.

Once the stones were transported, the architects erected a still-undeciphered bluestone arrangement in the northern and eastern segments of the monument, then tore the stones out and erected the famous sarsen circle of thirty uprights with lintels, seventeen of which still stand, most on the northeast side. A smaller circle of upright bluestones was built inside the sarsen circle, and inside this,

Fig. 7.3. Stonehenge trilithons from the air. Photograph from English Heritage.

five more sarsen trilithons were set in a horseshoe, the tallest standing opposite the entrance.

By this time, the builders trimmed even large stones carefully. Large numbers of worn sarsen stone hammers ("mauls") from the Stonehenge excavations show that the stone masons battered and flaked the surfaces of the boulders to produce smooth and even slightly curved surfaces for the lintels and tapered uprights. The masons also pecked out mortise-and-tenon joints between uprights and lintels using the same simple artifacts. Most experts believe that

TABLE 7.1: THE ARCHITECTURAL SEQUENCE AT STONEHENGE*

Stonehenge I: ca., 2950–2900 B.C.
A circular ditch and outside bank with northeast entrance was built. Fifty-six Aubrey Holes were dug and filled in inside the ditch. Two stone uprights and a wooden structure marked the entrance.

Stonehenge II: ca., 2900–2400 B.C.
The avenue was begun, and the entrance was shifted slightly to the east, oriented to the rising sun at the midsummer solstice. A double circle of bluestones was set up in the center, but soon removed.

Stonehenge III: ca., 2550–1600 B.C.
Stonehenge III comprised 3 phases. During the first, the 10 lintelled sarsen structures were erected in a horseshoe formation. Thirty smaller uprights formed an outer stone circle, with 30 slightly smaller lintels making a continuous ring 4.8 meters from the ground. Two uprights were erected at the entrance. This is the stereotypical Stonehenge and the apogee of the monument.

During the second phase, around 2290 B.C., about twenty bluestones were set up in an oval inside the horseshoe of trilithons. Somewhat later, two circles of holes were dug outside the horseshoe, but never filled with uprights.

During the third phase, the bluestones were reset in two groups, with a horseshoe inside the trilithon horseshoe, and a circle inside the existing outer circle.

Stonehenge IV: ca., 1600 B.C.
No changes were made at Stonehenge itself, but the avenue was extended 2.4 kilometers, to the Avon River. By this time an extensive field system surrounded the site, which eventually fell into disuse.

*Note: The architectural sequence at Stonehenge is provisional at best, and subject to continuous refinement.

the trilithon lintels were erected atop their uprights by encasing the latter in timbers, erected layer by layer as the lintel was levered up level with the top, then slid into place. However, British amateur archaeologist Cliff Osenton believes the builders worked the other way around, raising the lintel first, by using a pole as a fulcrum and three wooden blocks arranged in a triangle as levers. First one point of the lintel was raised and supported with wedges, then the next and the next. This technique requires only two people and simple methods that involve rotating the lintel around the center of gravity. Using this method, Osenton has been able to raise stones weighing several tonnes, in quarries with only one helper.

The tinkering continued, as following generations removed an arc of bluestones, leaving just a horseshoe. Even later, new leaders dug two circles of holes outside the main circles, but never filled them with stones. The builders also constructed an avenue leading to Stonehenge from the River Avon, delineated by twin, parallel banks, with quarry ditches on their outer sides. The Stonehenge Avenue is the longest in Britain, followed by the West Kennet Avenue at Avebury. Archaeologists have established the route of the avenue through judicious excavation, field walking, and aerial photography. The ceremonial way runs for a kilometer in a more-or-less straight line from the River Avon toward a group of burial mounds, before it curves gently toward the west. From this point, you can see Stonehenge on the horizon at the skyline to the left as you walk along the defile, before it curves into a shallow dip, and then runs straight toward the entrance of the henge about 500 meters away. As I walked toward the distant trilithons, I felt drawn along a set path toward the monument. At the same time, I almost felt excluded, as if the avenue controlled access to the interior. Archaeologist Richard Atkinson believed the bluestones arrived at Stonehenge along this processional way.

Over a period of 50 to 70 generations, Stonehenge was preeminent on the landscape, a focus of activity for kilometers around. After 1600 B.C., Stonehenge lay in the midst of an intensively farmed landscape. Wessex's leaders devoted their energies to commemorating individual leaders with burial mounds. The stone circles fell into disuse, a symbol of authority that had endured for more than 1,500 years.

STONEHENGE AS AN ASTRONOMICAL OBSERVATORY

What was Stonehenge? What prevailed on Stone Age and Bronze Age farmers, with the simplest of technology and only relatively limited human resources, to construct such an imposing set of stone circles? Was Stonehenge the center of some long-forgotten religious cult, or was it an observatory, a sophisticated place of dialogue with the sun and stars?

In 1906, Britain's Astronomer Royal, Sir Joseph Norman Lockyer, announced that Stonehenge was an astronomical observatory, its great stones aligned with the movements of the sun and other stars. Although Stukeley had drawn attention to the orientation of the axis of the stones to the midsummer sunrise, Lockyer took this idea further. He studied the orientations of ancient Egyptian temples before turning his attention to Stonehenge. He used a straight sighting line down the center of the avenue as his primary axis for observing astronomical alignments, as it passed close to an earthwork on Sidbury Hill, 13 kilometers away. Lockyer assumed Sidbury Hill (which has no relationship to Silbury Hill near Avebury) was contemporary with Stonehenge. In fact, it was built many centuries after Stonehenge, rendering his main sighting line meaningless. At best, Lockyer's research was naïve. He adjusted azimuth lines, moved stars, and relied heavily on coincidence, to the point that critics laughed at his work.

Most experts dismissed the observatory theory outright until the 1960s, when amateur astronomer Peter Newham found new alignments for the equinoxes and the moon. At about the same time, Boston astronomer Gerald Hawkins used an IBM mainframe computer to plot the positions of 165 key points of Stonehenge: stones, stone holes, earthworks, and other fixed points. He found "total sun [and moon] correlation" with a network of 13 solar and 11 lunar alignments, all of them based on features of early, rather than later, Stonehenge, where the alignments were less precise. In his bestselling book, *Stonehenge Decoded* (1965), Hawkins called Stonehenge a "Neolithic computer" used for predicting lunar eclipses.

Under Hawkins's scheme, concocted with the aid of an early computer known at the time as an "electronic brain," the Aubrey Holes had lain open for generations, as they served as tally pits for counting the passage of time. Archaeologists, however, remained unconvinced, especially since their excavations showed that the Aubrey Holes were filled soon after they were dug. Furthermore, the two Station Stones (figure 7.2), cover some filled holes. Stonehenge could not have been an observatory and a "computer" at the same time, for the very features allegedly used as timekeepers were invisible and under the ground during the very centuries when the "computer" was purportedly in use. From the archaeologists' point of view, the fatal flaw in Hawkins's reasoning was his assumption that any alignments he saw, as a twentieth-century astronomer, were also known to the original builders. Hawkins was familiar with abstruse astronomical data. How could one assume Stone Age or Bronze Age farmers had the same expertise, especially since they had to cope with the cloudy and unpredictable sighting conditions of the British heavens? Therefore, Hawkins's astronomy was viewed by archaeologists as little more than an anecdotal way of explaining Stonehenge.

Retired engineering professor Alexander Thom cast a wider astronomical net. Since Stonehenge is only one of many stone circles in Britain, he wondered if they were all astronomical alignments. Did their builders have advanced mathematical and engineering skills? Thom developed a theory of "megalithic astronomy," measuring dozens of stone circles throughout the British Isles. His precise figures convinced him that the builders had used a standardized measurement of 0.82 meters, which he called a "megalithic yard." He believed that they had built circles in standardized shapes and aligned them with stars or the directions of the solar and lunar cycles.

Thom left his study of Stonehenge for last, waiting until he had an unrivaled experience with alignments, stone circles, and their settings. He arrived in 1973 and began by surveying the entire monument from scratch with absolute precision. He also studied the visible distant horizon, where artificial or natural sighting points may have lain. Thom came up with two startling conclusions. First, at the time of the main sarsen stones, Stonehenge was oriented on the half-risen solstice sun. Second, Stonehenge was a highly sophisticated astronomical observatory. Thom considered the stone circles as a central "backsight," used with no less than eight "foresights," mostly earthworks identified on the visible horizon.

Unfortunately, Thom was an engineer, not an archaeologist. He failed to reconcile the archaeological and astronomical evidence. Nearly all his "foresight" earthworks are of a later date than the stone circles of the "backsight." A quarter century after his scientific tour de force, Thom's theory lies in archaeological ruins, despite recent attempts to identify Stonehenge as a cyberspace computer of unimagined sophistication.

Was Stonehenge an observatory? One cannot speak of it in the same breath as sophisticated Mayan observatories; or of its builders as astronomers on the same par as Babylonian priests. No one doubts Stukeley's observation that Stonehenge was aligned on the axis of the midsummer sun. But I believe that none of the many astronomical calculations of recent years has proved that the stone circles were an elaborate device for predicting eclipses or measuring the sky. Rather, I think that Stonehenge, with its straightforward solar alignment, reflects a distinctive *idea* of time, which revolved around the cyclical movements of sun, moon, and stars across the heavens, as indicators of the passing seasons.

The farmers who built the later Stonehenge lived in a demanding environment where the passage of the seasons governed their lives. I believe they set their existence within a symbolic environment, a variation of that constructed earlier over the course of many centuries, at Avebury, to the north. Every year, the eternal cycles of planting, growth, and harvest, of symbolic life and death, repeated themselves in endless successions of good or bad harvests, drought or excess rainfall, and famine or plenty. The archaeological evidence shows that the people placed great store in death rituals, and on the reverence of the ancestors—the guardians of the land. They devoted enormous resources to the great stone circles in the midst of their sacred landscape, where their priests and shamans used the mighty trilithons and simple stone alignments to observe the passage of the seasons. We can hypothesize that at midsummer sunrise, and, perhaps on the shortest day of the year (around December 21) the priests stood at the open side of the horsehoes to observe sunrise or sunset. In winter, the setting of the winter sun at the solstice signaled the beginning of lengthening days, "the certainty . . . that the seasons were going to follow their natural order, spring would come after winter, crops would grow and life would go on as before" (Aveni 1989,145).

STONEHENGE AS A SACRED PLACE

Stonehenge symbolized an idea of time, and the continuity of farming life from season to season, year to year. However, I believe that the stone circles were also a powerful political statement, a symbol of traditional religious beliefs at a time of ongoing political and social change. This statement, which came from Stonehenge and its setting, was both tangible and intangible, designed to play on human emotions.

We have great difficulty envisaging Stonehenge in its heyday, sitting as it does in the midst of the late twentieth-century landscape. Today's visitors view Stonehenge from a carefully monitored distance. They walk around the exterior of the site along a tarmac pathway and gaze at the stone circles and great trilithons from a respectful distance. Without being able to step inside the circles, the Stonehenge experience is effectively meaningless, for you cannot look along the solar alignment or gaze up at the weathered uprights or the lintel where William Stukeley enjoyed his dinner. Stonehenge's powerful ambiance escapes the casual visitor.

Our only impressions of this political statement can come from our physical perceptions of what it was like to move around both outside and inside the monument. These perceptions must be fundamentally similar to those of the ancients, for the trilithons still tower overhead and restrict the view, as they did 4,500 years ago. With the exercise of ritual power, setting is everything. Stonehenge was such a setting.

To visit the interior, I had to obtain a special permit from English Heritage in London, and enter early in the morning before busloads of tourists descended on the surrounding path. As I walked into the great circle, I felt a powerful sense of enclosure, as if I were entering a room delineated by vast stone uprights. Nineteenth-century diarist Francis Kilvert likened the experience to entering a great cathedral: "As I entered the charmed circle of the somber Stones I instinctively uncovered my head" (Cleal, Walker, and Montague 1995, 333). I was alone in the great henge, a chill wind blew between the trilithons; a bright, cool light contrasted blue and green, sky and earth. The interior, even in a ruined state, was mesmerizing in its isolation from the outside world. I believe this may have been part of the message of Stonehenge. Much of the surrounding landscape is invisible from the innermost part of the stone settings. For a thousand years, the worshippers inside the stone circles used a sacred place cut off symbolically from its surroundings. When I moved to the area between the Bluestone Circle and the Sarsen Circle, I could see vistas of countryside between the ancient uprights. Perhaps these segments of the outside world bore symbolic relationship to the interior.

Most people who dwelled in the area probably never ventured inside Stonehenge. They always saw the stone circles from outside, from beyond the encircling ditch and bank, perhaps congregating there for major ceremonies. Stonehenge stood amidst a much larger landscape, one that, like Avebury, was endowed with deep spiritual meaning. I believe the spatial relationships between Stonehenge and other, still little-understood components of this landscape, were of vital importance over many centuries.

The ancient landscape boasted many human-built monuments: long barrows, avenues, cursuses, round burial mounds, and other henges (figure 7.4). They have come down to us as a palimpsest of earthworks and archaeological sites, little excavated, and still mysterious. We do not know exactly what sites appeared when, or what

Fig. 7.4. Barrow group near Stonehenge. Courtesy Audrey Foxx.

conspicuous monuments would have stood out in different centuries, during a period of major cultural changes. During these transitions, henge monuments went out of use, more complex village societies appeared, burial customs changed from formal inhumations to cremations, metalwork and new pottery forms appeared, and more intensive farming practices altered field systems. Stonehenge itself changed over the centuries, but the circles always symbolized the close, and unchanging, relationship between the farming year and the ever-changing world of the heavens.

Some major changes of intense significance can be discerned, apart from some important places losing their importance, and shifting village settlement patterns. As at Avebury, they are reflected in changing burial customs, and especially in the deliberate placement of burial mounds only a short distance from Stonehenge. I believe that these settings had profound importance in validating political and spiritual authority by associating the dead with an ancient shrine that stood in the heart of a densely farmed landscape.

At around 3000 B.C., round burial mounds came into use on the chalk uplands of Wessex (see information in box, p. 164). These distinctive burial mounds cluster around major monuments like Stonehenge, sometimes arranged in formal cemeteries. The earliest burials in them were inhumations associated with decorated pottery; exotic ornaments such as pendants made of Baltic Sea amber; and gold, bronze, and copper metalwork. Later, people were cremated, as though more emphasis were placed on the funerary monument itself. Their ashes sometimes lay with ornaments and weapons that recall those made in central Europe, even as far away as Mycenae in Greece. The rich grave furniture buried in these barrows is eloquent testimony to chiefly power and prestige.

The settings of some important burial mound groups close to Stonehenge leave no doubt as to the importance of their owners. For example, barrows on the summit of Stonehenge Down, only about 200 meters from the monument, could be seen from the interior. The Normanton Down barrow cemetery directly to the south (including the famous Bush Barrow, containing one of the richest graves) contains elaborately furnished burials. This tumulus was set slightly higher than the other mounds and in clear view from Stonehenge.

BURIAL MOUND EXCAVATION

Burial mound excavation is a specialized genre of archaeological digging, developed to a fine art in Europe. The Stonehenge area was a hotbed of intense burial mound excavation in the late eighteenth and nineteenth centuries, when landowners vied with each other in regard to their treasure hunting. Barrow diggers like wool merchant William Cunnington and antiquarian Sir Samuel Colt-Hoare thought nothing of opening two or three mounds a day. Come evening, the two friends would dine together surrounded by the "rude relicks of 2000 years." Modern-day archaeologists spend much of their time trying to reconstruct vital burial information destroyed by their hasty predecessors.

Early excavators were content to sink trenches across a mound, or simply to dig a pit into the center to locate the burial as soon as possible. Such rough-and-ready digs gave way to thorough dissections, often based on total excavation of the mound to the original ground surface. Under classic modern practice, commonplace until recently, the excavator would dig the mound in quadrants, leaving intact cross sections across the mound. This approach allowed the study not only of primary and later burials, but also of the methods in which the mound was constructed, as well as the original land surface, where pollen samples, mollusca, and other environmental data could be collected. Few burial mounds are now excavated completely, because archaeologists are aware that future research methods may refine their data recovery methods, and also because of the need to preserve the finite archaeological record. Limited excavations are now the order of the day, unless total destruction of a site from industrial activity is imminent and other ways of mitigating the destruction are impracticable. As we saw at Easton Down, each cutting has a highly specific objective, such as the collection of environmental or chronological information, or the solving of a narrow question, and minimal amounts of earth moving and intrusion are now the ethic.

Many barrows must have had a wealth of association for those who used and maintained Stonehenge. The identity of the men and women who lay in these tumuli survived in oral tradition, and later, in myth, for many generations, in the form of revered kin leaders and ancestors. Those who led the age-old rituals within the stone circles lived in the heart of the sacred landscape and were buried so that their ancestral tumuli could be seen from the monument that symbolized the continuity of human existence.

Avebury and Stonehenge both lay at the center of intensely farmed landscapes with powerful spiritual associations. At Avebury, the relationship between the living and the dead formed a fundamental part of human life. The relationship changed continually over many centuries, culminating in the great monument itself, which remained in use for only a few centuries. Stonehenge was more permanent and long lasting. Although the stone circles also changed, they remained in use, and relevant to changing farming societies, for 1,500 years—longer than any Norman cathedral. During many centuries of profound permutation, villages prospered and more people crowded the densely farmed landscape. At the center of this busy world lay a set of ancient values and beliefs epitomized by the weathered stone circles of Stonehenge. Even though the dead were buried in sight of Stonehenge, one has the impression that this monument was more about the idea of time, of fundamental continuity, than of ancestors. The seasons came and went, and so did human life itself, while Stonehenge remained a place where the cyclical passing of the seasons and of human existence fostered the continuities of farming life.

chapter eight

TWO LIVINGS: AGRICULTURE AND RELIGION

I stood in the bottom of the eleventh-century great kiva (ceremonial room) at Casa Rinconada in Chaco Canyon, New Mexico (figure 8.1). The late afternoon sun bathed Chaco's cliffs in yellow-red light, as shadows lengthened over the canyon floor. I looked up into the darkening heavens, imagining the ancient Pueblo Indian cosmos of layered worlds. The *sipapuni*, the gateway between these worlds, lay close to my feet. I imagined the four great wooden pillars supporting the heavy beams of a now-vanished roof, wood smoke clinging to the rafters. A chill wind breathed across the kiva floor. I shivered momentarily, touched by the world of centuries ago. An ancient Hopi origin tale came to my lips: "In the beginning there was only Tokpella, endless space . . . Only the Tawa, the Sun spirit, existed, along with some lesser gods. There were no people then, merely insect-like creatures who lived in a dark cave deep in the earth" (Swan 1994, 657). The storyteller recalls how the Tawa led the creatures through two levels of the world. Eventually they climbed up a bamboo stalk through the *sipapuni*, the doorway in the sky, into the upper world. There the War Twins hardened the earth by freezing it. The gods gave the people corn and told them to place a small *sipapuni* in the floor of each of their kivas.

Casa Rinconada's main doorway faces toward celestial north, a fixed point in the nighttime sky. All stars seem to revolve around this fixed point. The four kiva roof supports once defined the cardinal directions, symbolizing the four trees that earth people once climbed

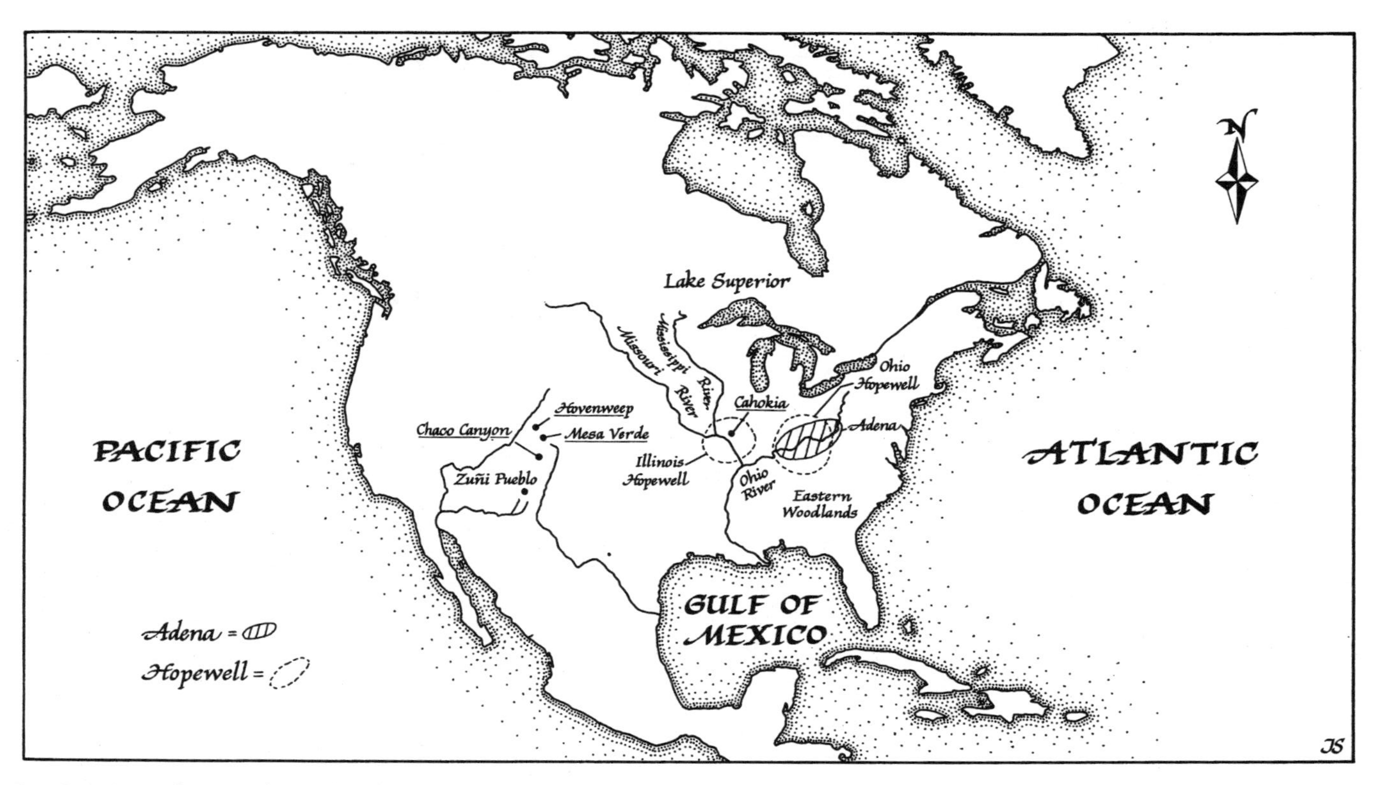

Fig. 8.1. Map showing the archaeological sites (underlined) mentioned in chapter 8.

to reach their homeland. At winter solstice sunrise, the sun's rays enter to the right of the main doorway. They shine into a niche in the northwest wall, which marks the "house" where the sun stays at the northernmost point of its cyclical journey.

A Hopi elder once told an archaeologist: "There are two livings, agriculture and religion" (Zeilik 1985, S2). The Pueblo Indians link the cycles of planting and harvest to the changing panoply of the heavens. Like the builders of Stonehenge's stone circles and the Greek farmers of Hesiod's day, they anchor their activities in an orbit of cyclical time, based on ecological realities of uncertain rainfall and changing seasons. Farming and the measurement of time go hand in hand in Pueblo society.

The native American universe in all its rich variety links humans and the vast forces of nature into a single continuum. The relationship between the cosmos and humanity is the foundation of the very fabric of life and the afterlife. The heavenly bodies, like humans, act out their roles in a layered world floating in a great abyss. Some native Americans are still expert astronomers. Countless observations of the sun and the night sky form a repository of solid information for establishing precedents, for explaining natural phenomena like eclipses, and for ensuring the continuity of human life.

Native American astronomers use rich heritages of traditional myths, songs, and poems passed down orally from generation to generation to interpret human life and the cosmos. This indigenous knowledge is recited by Pueblo storytellers from memory, as they sit face-to-face with their audiences, often for as long as an hour at a time, holding their listeners spellbound with tales that bind living people to the world of the sun and moon. The plots twist and turn, with heroes and terrifying hazards, tests of skill and wisdom. Almost invariably, the tale involves deeply felt religious beliefs. Only a fraction of these tales survive, lovingly transcribed, edited, and translated as a permanent record of a vanishing world. Many tell of the sun, whose powers of warmth and light sustained life itself. One Hopi tale recounts how a young man was conceived by the sun and journeyed to visit his father. After many adventures, including a journey across the heavens, he returns happy: "I saw for myself how he attends to our needs every single day of our lives. Therefore we must live out our lives in a good manner, and he will never forsake us" (Swan 1994, 678).

Many Indian astronomers are also shamans, who use hallucinogens and trances to pass effortlessly from the living into the spiritual world. They believe that supernatural beings (animals, plants, and natural forces) in the other world possess the same characteristics as humans—rational thought and behavior, a wide range of emotions, and spiritual powers. Like their ancestors, modern astronomers strive to keep the forces of the cosmos in balance. They use their ritual knowledge and special powers to influence the living forces of nature, timing important ceremonies to coincide with moments of change, such as the solstices. Evidence of this close relationship between what we may call ecological time and human life appears at many ancient North American archaeological sites.

PUEBLO SUN PRIESTS AND ECOLOGICAL TIME

Frank Cushing (1857–1900) was a dashing and flamboyant character, so much so that sober-minded modern-day anthropologists find him puzzling. Largely unschooled and the son of an unconventional physician, Cushing spent much of his youth alone in the woods and countryside around his family's upper New York state farm. He learned as much from nature as from books. By age ten, he was fascinated by Indian artifacts and lifeways; he claimed, in later years, that his life had become centered on recreating their life for himself. In his youth, Cushing experimented with the replication of projectile points and other artifacts and developed his powers of acute observation that later made him famous. As a young adult, Cushing became an anthropologist for the Smithsonian Institution, and in 1879, arrived at Zuñi Pueblo on a mule (figure 8.2). A pall of wood smoke covered the town as the sun set behind the settlement. The mud-brick walls melted into the landscape: "It seemed still a little island of mesas, one upon the other, reared from a sea of sand, in mock rivalry of the surrounding grander mesas of Nature's rearing" (Green 1979, 48).

At first the Zuñi were hostile, but Cushing's cool demeanor and quiet determination made a deep impression on the people. He

Fig. 8.2. Zuñi women making pottery. A drawing by Frank Cushing (redrawn by Farny).

found all his preconceptions about scientific observation of the Zuñi swept aside. Effectively abandoned by his white colleagues, he soon found himself dependent on the Indians and unable to work on his own terms. The people accepted him on *their* terms, vowing to make him a Zuñi. Wisely, Cushing decided to let the Indians do what they wished with him, although he often created tensions due not only to his insistence that he sketch and take notes, but also to his breaking of taboos.

A pioneer of the anthropological method known as participant observation, Cushing lived among the Zuñi for 4½ years, learning their language and recording their traditional life in great detail. After 3 years, the Zuñi initiated him into the secret Priesthood of the Bow. Cushing now dressed in Indian clothing. He spent many hours sitting in kivas watching: "the blazes of the splinter-lit fire on the stone altar, sometimes licking the very ladder-poles in their flight upward toward the skyhole, which served at once as door-way, chimney, and window." He listened to: "the shrill calls of the rapidly coming and departing dancers, their wild songs, and the din of the great drum, which fairly jarred the ancient, smoke-blackened rafters" (ibid., 112). His membership in the Priesthood of the Bow involved years of devotion; fasting; motionless, quiet contemplation; and dancing to exhaustion. The psychological and spiritual investment was deeper than he realized. When Cushing resumed to Washington and abandoned his religious observances, he was troubled by dreams about assuming his true self.

Frank Cushing enjoyed the trust of the Zuñi. But the doors of their towns soon shut to outsiders as the Indians learned to distrust visiting scholars and their observations. His legacy to anthropological science is unique because of his insistence on participant observation and his deep identification with people still little involved with the Western world. Unfortunately, he never published a full account of his Zuñi research, and many of his field notes are lost. The remaining notes and writings are a priceless source of information on pueblo life before large-scale European colonization.

Cushing was the first to identify the two "livings"—agriculture and religion—of the Pueblo peoples, where the realities of an arid, harsh environment formed the framework of ecological time measured by the movements of heavenly bodies. One day, he watched a Zuñi Sun Priest approach a square, open tower. The man sat just inside the structure before a sculpted pillar, praying as he awaited the rising sun. Cushing wrote: "Nor may the Sun Priest err in his watch of Time's flight; for many are the houses in Zuñi with cores on their walls or ancient plates imbedded therein, while opposite, a convenient window or small port-hole lets in the light of the rising sun, which shines but two mornings in the three hundred and sixty-five on the same place" (ibid., 116).

In 1893, another early anthropologist, Alexander Stephen, visited Tewa Pueblo. He talked to village elders, who described the "houses" where the sun stopped on its way across the horizon. Stephen compiled a horizon map and visited shrines where Sun Priests would place prayer sticks to welcome the sun and encourage its journey. He found that the Pueblo people tied their world to the horizon and the heavens by making a calendar out of the environment around them. "The Hopi orientation bears no relation to

North or South, but to the points on the horizon that mark the places of sunrise and sunset at the summer and winter solstices" (as cited in Zeilik 1985, S5).

Sun Priests still observe the summer and winter solstices. The Pueblo people anchor time, and their ritual cycles, to these events, especially to the winter solstice, when the sun is at its southernmost point. They believe that if the sun does not turn around, then it will fall off into the underworld. Some groups observe a period of "staying still," to keep the sun in its winter house. Winter solstice ceremonies guide the sun in the correct direction. The summer solstice, after the planting season, is of lesser importance. The Sun Priests set the days for these two ceremonies—and the other lesser celebrations in the annual calendar—starting their prayers and observations about 28 days before the winter solstice and 29 days before the summer solstice. They use a chosen spot in the village for their work, tracking the sun's seasonal position with the aid of horizon markers, which show up clearly at sunrise. Sometimes they employ windows in buildings to manipulate light and shadow.

Sun watching requires much more than observation of solstices and other events. Sun Priests have to be able to make anticipatory observations about 2 weeks before an event such as a solstice celebration, to allow for preparations to be made. They also need to be able to predict events during times of bad weather. Thus, Sun Priests go through lengthy training to learn how to gauge the position of sunrise relative to horizon markers several weeks *before* the event. At the solstice, the sun stands still on the horizon for 4 days, making the observation of the actual solstice day impossible beforehand. So Sun Priests have to make their observations at a time when the sun is still moving perceptibly—approximately 10 minutes of arc (arcmins)—every day. They predict the day of the solstice by making these observations and then keeping a tally of days on a notched stick. This approach solves the problem of cloud cover and bad weather. By using any clear day well ahead of time, Sun Priests can use their notched sticks to calculate the correct day even if the sky is overcast.

Predictions of the summer solstice are remarkably accurate, almost invariably to within a day and a half. Such accuracy is essential. A major disaster would transpire if the ceremonies took place and the sun had already turned or still moved toward its turning point, as if it were about to fall off the earth. Accurate predictions reinforce priestly power and strengthen bonds within the community, as well as validate the world view.

Ecological time served the Pueblo people well. They lived a well-regulated life attuned to the solstices and to the realities of their arid environment. The yearly cycle repeated itself endlessly. As one year ended, another began, measured in the passage of moons and days, and the seasons of planting and harvest.

THE ANASAZI: CHACO CANYON

Fortunately for science, pre-Columbian astronomers used buildings, pictographs, and other humanly manufactured objects for their observations, enabling us to trace Pueblo astronomy back to the Anasazi, "the ancient ones," whose primordial roots extend back as far as 2,500 years ago. The first farmers of the Southwest dwelled in small communities of pithouses. By A.D. 900, southwestern farming populations rose considerably. Many Anasazi communities moved into large, well-constructed towns, epitomized by the great pueblos of Chaco Canyon. For 2½ centuries the Chaco Canyon pueblos flourished, during a time of constant climatic change. By A.D. 1050, the Chaco Phenomenon (an archaeological term) was in full swing, and expanded from its canyon homeland to encompass an area of more than 65,000 square kilometers of the surrounding San Juan Basin and adjacent uplands. Roads and visual communication systems linked outlying communities with the canyon. Great towns like semicircular Pueblo Bonito (figure 8.3) housed hundreds of people. The population of Chaco Canyon rose from a few hundred to at least 5,500.

Chaco Canyon comes as a complete surprise. You drive in from the main road along a dirt road, which winds down a steep switchback and through a narrow defile to the canyon floor, lined with stark cliffs. Suddenly you have entered an enclosed world, protected by precipitous cliffs; a dry valley paved with sage and occasional cottonwood trees. The human imprint is inconspicuous on the land. The brown mud walls of eight-hundred-room Pueblo Bonito lie close by, under a high escarpment, melting imperceptibly into the

Fig. 8.3. Pueblo Bonito. Photograph from Photo Researchers.

arid landscape. I remember my astonishment at seeing the size of the carefully planned town the first time I visited. How could so many people live and thrive in such a dry, unpredictable environment? The question has intrigued archaeologists for generations.

The study of tree rings (dendrochronology) plays a vital role in southwestern archaeology, both to date wooden beams found in ancient pueblos and also to document climatic changes over thousands of years, such as the drought that seems to have caused at least partial abandonment of Chaco Canyon in the twelfth century (see information in box, p. 175). The tree rings from the beams of Chaco Canyon's pueblos form a calendar of unpredictable rainfall. In some years, it rained at one end of the canyon, but not at the other, in a constant cycle of dry and wet years, depicted in thick and thin tree rings.

The Chacoans responded to their uncertain environment by building three large, semicircular pueblos, "Great Houses," at the junctions of major drainages, so that they could maximize crop yields from floodwater agriculture. The Chaco Great Houses ranged in size from Pueblo Bonito with 900 rooms to Wijiji, with just over a

DENDROCHRONOLOGY

Short of historical records, dendrochronology (tree-ring dating) is the most accurate chronological method available to archaeology. Astronomer A. E. Douglass developed the first tree-ring time scales in the Southwest in 1913, using sequoias and bristlecone pines, two of the longest-living trees in the world. Douglass also counted tree rings from ancient lintel and roof beams in Anasazi pueblos. At first, his pueblo dendrochronologies "floated," as he was unable to anchor them to sequences based on living trees. However, in 1929, a log from Show Low, Arizona, provided the critical link in the first master tree-ring sequence that could be used to date beams from pueblos throughout the Southwest.

Dendrochronology is based on the concentric annual growth rings visible in the cross sections of felled tree trunks. Growth rings are formed in the cambium—or growth layer—between the wood and the bark. As new cells are produced during the growing season, a distinct line forms between the wood of the previous year and the new growth. Tree-ring scientists collect full cross or V sections from beams or tree trunks, or use a specially designed core borer to penetrate the heart of the tree. In the laboratory, the sample is leveled to a precise plane, allowing the researcher to record individual ring series. Comparisons are made by eye, by plotting rings on a uniform scale, which allows them to be computer matched with the master tree-ring chronology for the area.

Few regions enjoy the precise chronologies of the Southwest, but tree-ring dating has now been applied to oaks and other species in Europe, the Aegean area of the Mediterranean, and western Asia. North American sequences based on the bristlecone pine tree extend back to more than 8,200 years before present; those based on German oaks, to 10,021 years before present. Scientists have even used tree rings to date the oak boards behind Dutch old master paintings, and to date firewood aboard the late Bronze Age Uluburun shipwreck off southern Turkey (see chapter 11).

Climatic variations within circumscribed areas, such as southern California, coastal Peru, or the Aegean, tend to run in cycles of wetter and drier conditions, reflected in patterns of thicker and thinner tree rings. Pueblo beams record drought cycles that may have caused the abandonment of Chaco Canyon and other population shifts. Jeffrey Dean of the University of Arizona's Laboratory of Tree-Ring Research is extending his research far beyond dating and climate change to study the detailed architectural history of individual pueblos. For example, he collected beams from as many rooms as possible in Kiet Siel Pueblo in the Kayenta region of northwestern Arizona, dated them, and showed how this 150-room settlement, first founded in the late 1240s, had added 12 new room clusters by 1271. A major influx of newcomers between 1275 and 1286 saw the construction of several new room sets and additional kivas. Construction and remodeling ceased in 1286, and Kiet Siel was abandoned by 1300.

Dendrochronology also provides accurate calibration for radiocarbon dates extending back to before 6000 B.C., thus allowing for much more precise chronologies of such well-known sites as Avebury and Stonehenge than even a few years ago.

---

hundred. Most were multistory buildings made of beautifully coursed sandstone walls with rubble cores up to a meter thick. They were erected in a series of components, often over quite short periods of time. Much larger than other pueblos, Great Houses are also remarkable for their formal layout and many kivas, one or more being of considerable size. The builders built one story at a time, using the completed level as the platform for constructing the next. At Pueblo Bonito, the earliest room blocks—made of rough stone masonry—formed crescents of irregularly shaped rooms. After A.D. 1030, the architecture became more standardized, based on stone veneer masking rubble-core walls. The construction of each Great House consumed many people hours and vast quantities of sandstone quarried from nearby cliffs. Over 200,000 trees, sometimes from forests as far away as 80 kilometers, went into the roof beams and floors of the Chaco pueblos.

During the eleventh century, the Chaco people also built at least 9 Great Kivas, each requiring at least 40 wooden beams. The Great Houses were important ritual and storage centers. They supported considerable numbers of kivas over and above the conspicuous Great Kivas, some of which needed nearly 30,000 people hours to build. No one knows how many people lived permanently in Pueblo Bonito and its neighbors, but estimates range from a few hundred to several thousand. Chaco Canyon's Great Houses display such variation that they may have served many functions, both ritual and residential. Without question, many of their rooms were empty most, if not all, of the time. Some were designed to restrict lateral access, as if they were owned by different kin groups.

A team of researchers known collectively as the Solstice Project has spent nearly 20 years studying the lunar and solar cosmology of Chaco Canyon with the latest astronomical, engineering, and survey technology. In research to be published after the completion of this book, they have studied the orientation of the Great Houses and shown that the extremes of the lunar and solar cycles played a significant role in the construction of 12 of the 14 major Chaco pueblos.

The Chaco Great Houses may have been conspicuous structures erected on the landscape that legitimized political authority. Above all, they reinforced shared beliefs in the nature of the Anasazi cosmos. Significantly, many Chaco Great Houses boasted a low wall on their entrance side, perhaps separating the sacred and secular worlds architecturally. At Pueblo Bonito, a pair of earthen mounds, originally thought to be trash heaps at the entrance, were found to be defined by masonry walls and formal stairways that led to plastered horizontal floors. The space between the two mounds formed a visual link and a symbolic avenue between the Great Kiva at Pueblo Bonito and the isolated Great Kiva at nearby Casa Rinconada. Outside Chaco, some Great Houses are surrounded by sunken avenues and delimited by raised berms, tangible barriers, perhaps, between visiting pilgrims and sacred space.

Even in the best years, the soils in the canyon could support only about 2,000 people, yet there are rooms in these Great Houses for at least 6,500. Only one explanation fits this fact: that hundreds of visitors descended on the canyon at certain times of the year. Chaco was a hub of the Anasazi world and bright blue turquoise was the Anasazi's gold. The stone flowed into the canyon from sources near Santa Fe, New Mexico, about 160 kilometers to the east. Important festivals, especially at the solstices, may have been occasions for large gatherings, when hundreds of people converged on the great pueblos, bringing turquoise, brightly painted pots, and a variety of foodstuffs. Major ceremonies brought outlying communities together, and allowed for intensive trading and important religious activity.

THE MYSTERY OF THE CHACO ROADS

Early excavators focused on large, spectacular sites like Pueblo Bonito and Casa Rinconada (figure 8.4). They noticed a few tracks converging on Chaco Canyon from outside, but did not attach any importance to them. Back in the 1930s, early aerial photographs revealed faint traces of what appeared to be canals. During the 1970s and 1980s, the Chaco system was plotted using aerial photographs and side-scan radar, which senses the terrain to either side of an aircraft's track by sending out long pulses of electromagnetic radiation. The radar allows the observer to track pulse lines in the form of images, no matter what cloud cover obscures the ground. The images are normally interpreted visually, using radar mosaics or stereo images, as well as digital image processors.

Side-scan radar placed the canyon at the center of a vast ancient landscape. A web of over 650 kilometers of unpaved ancient roadways links Chaco with over 30 outlying settlements. The Anasazi had no carts or draft animals, but they built shallow trackways up to 12 meters wide, which were cut a few centimeters into the underlying soil or demarcated by low banks or stone walls. Each highway runs straight for long distances, some as far as 95 kilometers. The people approached the canyon along straight walkways, then descended down stonecut steps in the cliffs to the pueblos. The road system defied environmental logic, and, thus, must have had some deeper meaning. Something compelled the Anasazi to travel over 65,000 square kilometers of the San Juan Basin to converge in a single location at specific times of the year.

The Chacoan "roads" remained a mystery. Were they used for travel or transportation of vital commodities? For years, archaeologists have proposed the theory of an integrated Chacoan cultural system

Fig. 8.4. Casa Rinconada. Photograph from Photo Researchers.

that unified a large area of the Southwest a thousand years ago. One authority, archaeologist James Judge, believes that the San Juan Basin's harsh and unpredictable climate, with its frequent droughts, caused isolated Anasazi pueblos to form loosely structured alliances for the exchange of food and other vital commodities. Judge states that Chaco lay at the hub of the exchange system and also served as the ritual center for major rainmaking ceremonies and festivals. The Canyon's Great Houses were the homes of privileged families who were able to predict the movements of heavenly bodies and controlled ritual activity. Under Judge's scenario, the roads were pilgrimage and trading walkways.

Other archaeologists, such as David Wilcox of the Museum of Northern Arizona, disagree with Judge. Wilcox believes that the powerful families of Chaco obtained food and labor from outlying communities, using their ritual powers and military force to control tributary communities; and that the roads carried tribute and military detachments. However, archaeologist John Roney of the Bureau of Land Management points out that there are no signs of domestic rubbish or encampments along the roads. He has walked along many of the "fuzzy lines" depicted on air photographs, verifying more than 60 road segments, many of them short and without specific destination. Roney is certain that major north and south tracks radiated from Chaco, but he is cautious about connecting segments into long lines to join distant places on the map. He is sure of the existence of only a mere 250 kilometers of roads, and believes that the Chacoans constructed the walkways as monuments—as a ritual gesture—and not as roads of use.

Do roads have to lead to a destination, as we Westerners always believe? The answer to the Chaco road mystery may lie not in the archaeological record, but in Pueblo Indian cosmology. The so-called Great North Road is a case in point. Several roadways from Pueblo Bonito and Chetro Ketl ascend Chaco's north wall, to converge on Pueblo Alto. From there, the road travels 13 degrees east of north (347 degrees) for about 3 kilometers, before heading due north for nearly 50 kilometers across open country to Kutz Canyon, where it vanishes. Michael Marshall of Cibola Research Consultants emphasizes that north is the primary direction in the mythology of Keresan-speaking Pueblo peoples, who may have ancestry among Chaco communities. North led to the place of origin, the place where the spirits of the dead went. Marshall thinks that the Great North Road may have been an "umbilical cord" to the underworld—a conduit of spiritual power. Another Pueblo concept, that of the Middle Place, was the point where the four cardinal directions converged. Pueblo Bonito, with its cardinal layout, may have been the Middle Place. The Great Houses and trackways of Chaco Canyon may have formed a sacred landscape, a symbolic stage where the Anasazi acted out their religious beliefs and commemorated the passage of seasons.

Fortunately, new scientific technologies such as GIS are combining multispectrum imagery, color infrared photographs, and 1930s halftone images. By enhancing different light, heat, and vegetational conditions, fieldworkers can go into the field with information on hitherto invisible features, and obtain their position on the ground within a few meters using the satellite-driven Global Positioning System. A new generation of survey work will establish the. true extent of Chacoan roads and place them in a precise topographical context.

Another archaeologist, Steve Lekson, has taken a more ambitious and controversial look at the southwestern landscape. He points out that Aztec Ruins 88 kilometers north of Chaco Canyon, Chaco itself, and Casas Grandes in northern Mexico, 630 kilometers to the south, lie along a 725-kilometer meridian line. Tree-ring dating tells us that the 3 sites were occupied successively: Chaco from A.D. 850 to 1125, Aztec from A.D. 1110 to 1275, and Casas Grandes from A.D. 1250 to 1500. As Chaco lost its political power and was abandoned, Aztec was founded on the same meridian, seemingly connected to the former by the Great North Road (although the final 30 kilometers of the road are still uncertain). No road appears to connect Chaco and Casas Grandes, which came into prominence as Anasazi power finally dispersed to the north. Lekson believes a vast symbolic landscape covered the Southwest, organized on a north-south axis. He is now searching for an extension of the "Great South Road" toward Casas Grandes, using remote sensing and ground survey.

ASTRONOMY AT CHACO AND HOVENWEEP

The careful orientation of Casa Rinconada and other kivas ties the world of Chaco Canyon to the Pueblo people's cosmos, where cardinal directions were of fundamental importance. Some arcane and long-forgotten symbolism lies behind this vast human landscape, where dozens of kilometers of straight track converge on pueblos set in a stark, sometimes foreboding, setting of bluffs and canyon walls. Walking along one such roadway, I marveled at the ability of Anasazi astronomers to draw people into the canyon on specific days. Only a few traces of their observations have survived in the archaeological record. Fajada Butte is one example. Long ago, 3 sandstone slabs, each about 2 meters high, fell from the cliffs at Fajada Butte, a dramatic break in the landscape at the eastern end of Chaco Canyon. By chance, they landed in such a way that they deflect sunlight against the vertical wall of the mesa at certain times of the year. Someone long ago had pecked one large and one small spiral on the wall behind the boulders, that they could use the intersecting points of light and spirals to measure the passage of the sun. The summer solstice sees a single dagger of light cutting across the large spiral. Two vertical stripes of sunlight bracket either side of the same spiral 6 months later, at the winter solstice. The spring and autumn equinoxes cast sunlight across the small spiral and to the right or left of the center of the large circle.

Astroarchaeologist Michael Zeilik points out that the Fajada Butte slabs are a natural rockfall, at least three-quarters of a mile from any large pueblo. Access is difficult when winter ice and snow make the steep approach slick. Thus, it is impossible to use the site to make accurate (to within a day) predictions regarding the solstice. When access finally becomes possible, the movements of the sun before the solstice are simply too small for the human eye to detect. Zeilik, therefore, believes that Fajada Butte was not an observation post, but rather a Sun Shrine: a pile of rocks marked by petroglyphs some distance from a settlement.

Zeilik has also searched Pueblo Bonito for corner doorways and windows that could have served as observation points. He notes that the horizon seen from the eastern side of the pueblo at sunrise provides adequate sighting landmarks until October. The sun then moves across a flatter horizon with no obvious features. At this moment, however, sunlight first appears in one corner window opening in the Great House, as if the astronomers built their own marks when the horizon failed them.

The best archaeological evidence for Pueblo astronomy comes from Hovenweep Pueblo in Colorado, erected by Anasazi people related to nearby Mesa Verde communities between the late twelfth and mid-thirteenth centuries. The pueblo includes round, square, and D-shaped towers. At least one, Hovenweep Castle, has special sun-sighting ports that align with the summer and winter solstices. Nearby Holly House contains petroglyph panels with symbols that may represent the sun and other heavenly bodies.

■ ■ ■

It is one thing to observe an impending solstice, and quite another to announce a ceremony over 65,000 square kilometers. The Chacoan Great Houses lie in the canyon itself, linked to outliers by the road system. Outlying structures are often well-planned pueblos, most of them lying close to one of the roads. Their features include kivas, some multistory residential blocks, and occasional tower kivas several stories in height. Other outliers straddle high mesas and promontories. Some experts believe that the Anasazi signaled one another over considerable, line-of-sight distances from tower to tower, or from tower to high pueblo. However, we have no proof of such activities. Most likely, word of impending ceremonies came from runners who were dispatched from the Great Houses in the canyon. Long-distance running has a long history among Pueblo people, both as a method of communication and as a ritual activity. Anthropologist Peter Nabokov has studied Pueblo Indian running and shown how mythic races helped to order the world, by giving a model and mandate for human activity, and creating the cardinal directions.

The Great Houses of Chaco Canyon influenced the lives of thousands of people living over a vast area. Like powerful computer servers, they processed and distributed information, and helped make decisions for the benefit of farmers scattered over thousands of square kilometers. Chaco's sacred places could not function without an Anasazi version of telecommunications, even in a world timed by astronomical observations. The Pueblo world was one of close interdependence, between farming and religious observance, and isolated communities and pueblos large and small. Reciprocal obligations tied village to village, family to family, and Great House to outlying hamlets. The major ceremonies of the solstices brought people together in implicit recognition of this interdependence, in communal rituals that turned the sun around in its winter and summer houses.

Agriculture and religion: the two lives of the Anasazi intersected throughout the year. The farmers occupied a tenuous place in the natural order. People depended on maize, maize depended on skillful cultivators. No one could farm effectively without paying attention to the cycles of the heavens, and to the heavy rain clouds building on the horizon as planting time came near. Both the timing of planting and harvest—vital to survival—relied on irregular rainfall, and an environment dictated by the sun and the procession of the seasons—as predicted by the astronomical events in the heavens. Thus, the most important Pueblo rituals revolved around the celebration of important astronomical events.

chapter nine

THE MOUNDBUILDERS OF EASTERN NORTH AMERICA

Some 15,000 years ago—the date is much debated—tiny groups of late Ice Age hunter-gatherers crossed the Bering Land Bridge from Siberia into Alaska, initiating the first human settlement of the Americas. Within a remarkably few millennia, their descendants had settled throughout North, Central, and South America, exploiting every kind of natural environment imaginable.

All native societies of America come from the same biological and cultural roots, which extend beyond the memory of oral tradition or written history. Science and archaeology offer the only chance to study "the trunk and branches" of this great genealogical tree. Radiocarbon dates and tree-ring analysis, described in chapters 2 and 8, respectively, allow us to develop a linear chronology for early native American society, and to probe the spectacular mortuary cults and sacred places in eastern North America, inhabited by people known to generations of archaeologists as "the Moundbuilders."

After the Anglo-French war ended in 1756, thousands of settlers moved westward from the Colonies. Wagon trains fanned out over the midwest, across the Alleghenies, and into the Mississippi Valley. As farmers cleared hectares of dense woodland, they discovered that they were not the first inhabitants of this apparent virgin landscape. From the mantle of forest and undergrowth, silent earthen mounds stood in solitary splendor. Others lay in impressive clusters, as if constructed over many centuries. Hundreds of mounds and other earthworks dotted the Ohio Valley alone. The largest tumulus, at Cahokia in the Mississippi Valley, opposite modern-day downtown St. Louis, covered over 6 hectares (figure 9.1). East of the Alleghenies, the mounds stretched out from western New York along the southern shores of the Great Lakes and into Wisconsin and Nebraska. They also extended deep into the south and east, from Texas to Florida.

Philadelphia's intellectual community, which included Thomas Jefferson, puzzled over the mysterious earthworks. Who had built these "graves and towers like pyramids of earth"? (Silverberg 1968,

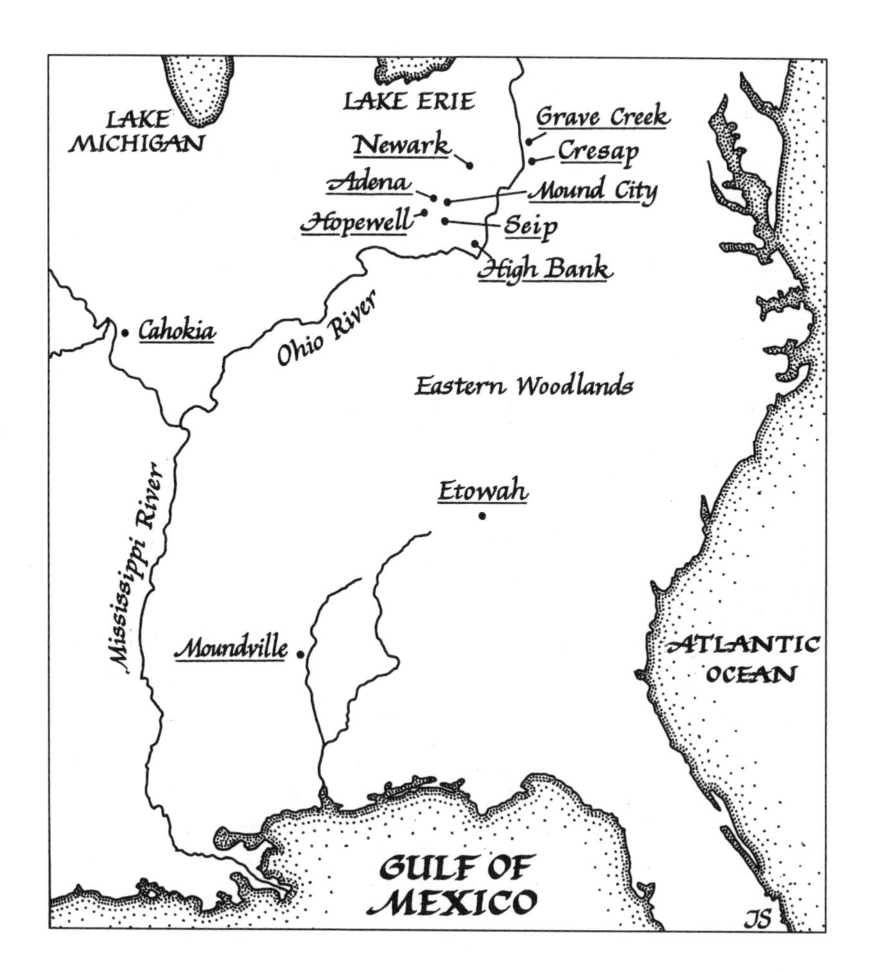

Fig. 9.1. Map showing the archaeological sites (underlined) mentioned in chapter 9.

222). Unlike many of his contemporaries, Jefferson believed in firsthand observation. He occupies a small, but honored niche in archaeological history for his stratigraphic excavation of an Indian burial mound on his Virginia estate in 1784. In his *Notes on the State of Virginia,* he described layers of human bones and argued that the mound was built by ancestors of local Indians, who still visited the location in his day. Jefferson's sober observations were ignored by his scholarly contemporaries, despite being based on one of the first stratigraphic excavations in the world. Was Jefferson correct when he argued that native Americans constructed the tumuli? Or were Canaanites, the Ten Lost Tribes of Israel, or some other long-vanished alien civilization responsible for them?

The Moundbuilder controversy continued throughout the nineteenth century, and even today, at times, reverberates through fringe, pseudoscientific literature. Popular writers seized on the earthworks with joy. They wrote stirring epics of vast armies and warring civilizations, who buried their dead by the thousands in high burial mounds. Cornelius Matthews's *Behemoth: A Legend of the Moundbuilders* (1839) told how a huge, mammothlike beast ravaged towns and villages and decimated entire armies. Then a heroic Moundbuilder warrior named Bokulla defeated Behemoth with cunning wiles and saved civilization. *Behemoth* is epic material, but absolute historical nonsense, born, like its many imitators, out of once-commonplace racist beliefs that native Americans were incapable of constructing large earthworks, or for that matter, anything other than simple foraging and farming.

As so often happens, speculation outpaced scientific fieldwork. Only a few antiquarians, like Caleb Atwater, postmaster of Circleville, Ohio, measured and surveyed the earthworks. Atwater believed that shepherds and farmers from India, China, and the Crimea had built the mounds after a long journey across the Bering Strait after the Biblical Flood. The postmaster was a sober observer by the standards of the day.

As the Moundbuilder debate raged on, the earthworks themselves vanished in the face of expanding agriculture and industrial civilization. Only a few scientists followed in Atwater's footsteps. Newspaper editor Ephraim Squier and physician Edwin Davis explored more than two hundred mounds and a hundred earthen enclosures between 1845 and 1847. Their accurate surveys and occasional excavations were exemplary for the time. In many cases, they represent the only record of now-destroyed archaeological sites (see figure 9.2). Squier and Davis's *Ancient Monuments of the Mississippi Valley* appeared in 1848, the first volume on the natural sciences published by the newly established Smithsonian Institution. Prophetically, Squier and Davis referred to some of the enclosures as "sacred earthworks," placed on lowlying, indefensible ground. They also remarked on the many skeletons found in mounds both large and small. For all their sober observations, however, Squier and Davis also considered native Americans incapable of building such imposing structures.

In 1881, John Wesley Powell of Colorado fame, director of the U.S. Geological Survey and the Smithsonian's Bureau of Ethnology, appointed a scientific jack-of-all-trades named Cyrus Thomas to resolve the Moundbuilder issue once and for all. Thomas used a team of assistants to investigate 2,000 earthworks in 24 states, over the remarkably short time of 4 years. His excavations produced 40,000 artifacts, none of which were of foreign manufacture. Once a believer in the existence of a "race of Moundbuilders," Thomas's research changed his mind. In his definitive 730-page final report published in 1894, he showed how native Americans built very different kinds of earthworks over a long period of time—and continued to construct them even after European contact. He wrote: "The links directly connecting the Indians as the mound builders are so numerous and well established, that there can be no longer any hesitancy in accepting the theory they are one and the same people" (1894, 35). The Smithsonian survey, combined with other research, exploded the myth of a separate race of Moundbuilders, but left challenging questions unanswered. Which native American societies built the earthworks of the Eastern Woodlands and why did they do so?

A century of research since Cyrus Thomas's day has transformed our knowledge of the Moundbuilders. We now know that their cultural roots go back at least 4,000 years, and that the beginnings of moundbuilding were closely connected to a new social complexity in society and to the increasing importance of worshipping prominent individuals as revered ancestors. Archaeology tells us that

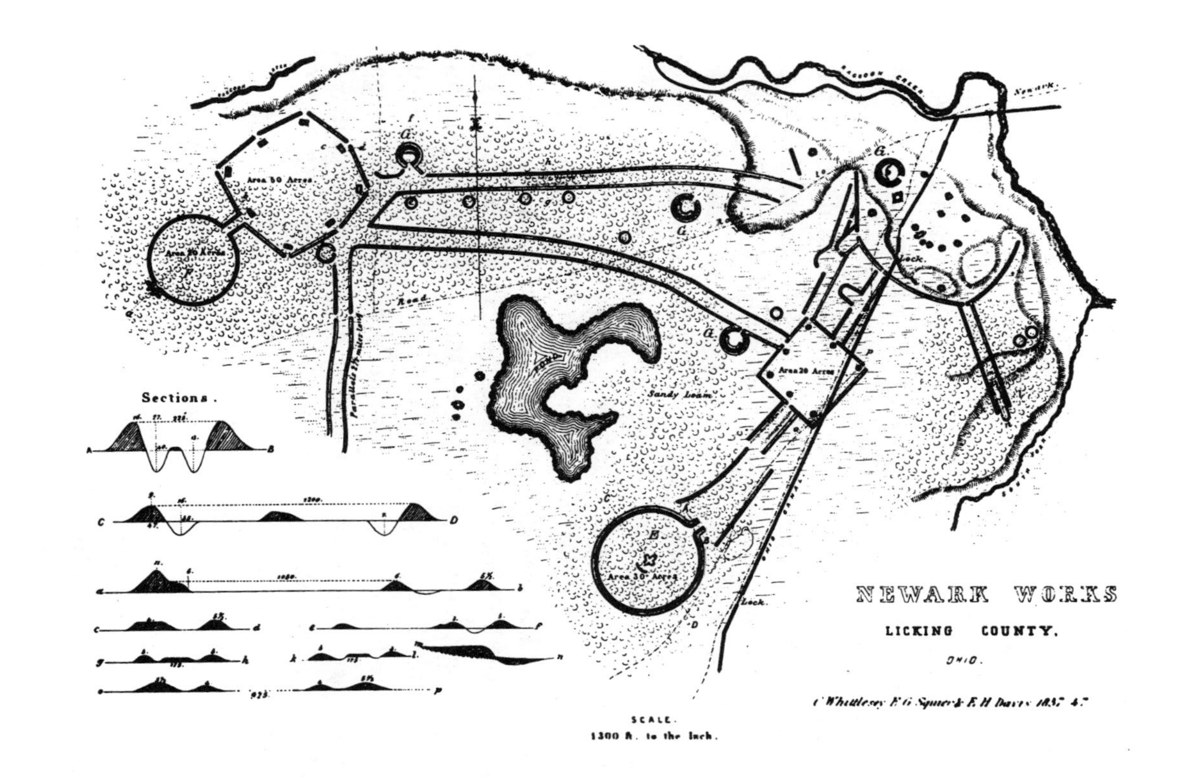

Fig. 9.2. Squier and Davis's mid-nineteenth century plan of the Newark earthworks, drawn before extensive disturbance and the building of a golf course.

in 2000 B.C., thousands of small foraging groups inhabited the diverse environments and river valleys of the Eastern Woodlands of North America. The densest populations flourished in fertile river valleys, along seacoasts and around estuaries, where there were abundant food resources. Every society adapted to local conditions: Some relied heavily on fall nut harvests, others on freshwater fish and migrating waterfowl, as well as deer hunting in winter. As time went on, favored areas attracted larger human populations, and people settled in extensive, semipermanent villages. The landscape filled up gradually with foraging groups, to the point that mobility became restricted and territorial boundaries better defined.

People's ties to the land became closer, when ancestral roots in known territory assumed overwhelming importance. By 1500 B.C., the increase in burials and cemeteries testified to a much greater concern with death and funerary rituals in eastern North American life. Ancestors, and ancestral lineages, tied people to the land. Many groups now buried their dead in small burial mounds. Some communities deposited the deceased in low, natural ridges overlooking river valleys, or in other strategic locations. Others used low, artificial mounds, intended as territorial markers. Neighbors respected kin-based burial grounds. The variety of grave goods deposited with the dead reflected subtle differences in social ranking for the first time. Sometimes the old wore more shell bead necklaces than the young, a reflection of greater status for elders. Sometimes a single individual lay in a richly adorned grave, whereas neighboring burials had few possessions. Perhaps this person was the leader of a senior lineage—the one from which social and ritual rulers, or "chiefs," traditionally came.

What kinds of individuals were these newly important people? Archaeologists writing about emerging social and ritual complexity have been strongly influenced by the research of anthropologist Marshall Sahlins on chieftainship in south Pacific Islands. Sahlins was concerned about definitions of "chieftains," and saw them, originally, as what he called "Big Men," individuals with unusual charismatic, entrepreneurial, or supernatural powers, who exercised leadership by attracting followers to them. (A few scholars call such entrepreneurial individuals "aggrandizers.") Some of these people were also expert acquisitors and traders, or skilled arbitrators, people who attracted the loyalty of others. These followers perceived their leader as someone with beneficial associations, but his death left them without a chief, and the need to find another in his place. In some places, like Tahiti, chieftainship eventually became hereditary. The interpretation of social ranking given here relies on this anthropological perspective. In time, the descendants of the first "Big Men" became the leaders of more complex human societies, which placed profound importance on the burial of the dead.

ADENA: CARE FOR THE ANCESTORS

By 500 B.C., the Adena moundbuilding culture (named after a burial mound on the Adena estate near Chillicothe, Ohio, excavated in 1901) appeared in the Ohio Valley. Adena is a somewhat shadowy cultural entity, virtually indistinguishable from dozens of contemporary foraging societies throughout the Eastern Woodlands, except for an extraordinary preoccupation with burial and mortuary rites, reflected in funerary earthworks—especially burial mounds. Three to five hundred Adena tumuli lie within a 150-mile radius of Chillicothe, with outliers extending into neighboring regions. The Adena people cremated most of their dead. However, a few individuals received special treatment. Their bodies lay on platforms or in temporary graves until their flesh decayed. Then their kin gathered together the bones and deposited them in a burial mound. Like British long barrows such as West Kennet, many Adena mounds remained in use for generations.

Numerous Adena mounds were excavated generations ago, when digging methods were crude at best.* A few noteworthy exca-

*Until recently, American archaeologists excavated native American burials with little regard to the feelings of living groups, or the ancestry of the people lying in a site. However, the passage of the Native American Grave and Repatriation Act (NAGPRA) in 1990 mandated the cataloging of native American human remains and ceremonial objects in all museums receiving federal funds, the repatriation of certain of these remains, and the protection of all graves and grave furniture found on federal and tribal land. Most of the excavations described in this chapter were carried out before NAGPRA went into effect.

vations reveal complex interments. The Cresap Mound, 10 kilometers south of Moundsville, West Virginia, excavated in 1958, contained at least 54 interments dating to between 200 B.C. and A.D. 50, lay in 3 major concentrations within a tumulus 4.5 meters high and 21 meters across at the base. Archaeologist Don Dragoo discovered that the mound covered the postholes and hearth of a dismantled house. The dead lay in every kind of grave, from simple pits to elaborate, roofed log structures. At first, the funerary offerings consisted of utilitarian artifacts like stone tools. In later centuries, exotic objects like stone pendants, copper bracelets, and silhouettes cut from mica sheets predominated.

One skeleton from the Cresap upper levels held a lightly polished human skull in its lap, as if cradling a revered ancestor or a prized war trophy. The Ayers Mound in Kentucky yielded the skull of a man, whose front upper incisor teeth were missing. Next to the cranium lay a wolf's palate with the incisors still in position. The excavators believe that the man was a shaman, who impersonated a wolf, an animal that was thought to have special powers in more recent southeastern Indian religion because it can move around in the dark. Adena burials sometimes contained smoking pipes and masks made from the skulls of elk and other animals, reflecting a close relationship between people and their prey.

The Adena mortuary cults flourished at a time of profound change in eastern North American society. As people settled down in more permanent communities for much of the year, membership in social groups assumed a greater importance. Membership in lineages and clans gave individuals access to a much wider range of contacts, and the ability to acquire goods from afar. The major players in these exchanges were lineage leaders, individuals with unusual spiritual powers honored at death with special artifacts and elaborate funeral ceremonies. After death, they entered the supernatural world, and became the ancestors and guardians of Adena society.

HOPEWELL

Adena ceremonialism reached its height at a time of slow change in eastern North America. Hitherto isolated and nomadic foraging groups now lived in well-delineated territories and were in regular contact with neighbors near and far. By 200 B.C., the entire Eastern Woodlands of North America formed a gigantic exchange zone, where local societies participated in long-distance trade of every kind. Winding intervillage paths linked the copper outcrops of Lake Superior with the Mississippi Valley, and brought Gulf Coast shells into the heart of the Midwest. Flint traveled all the way from the Yellowstone region of the Rocky Mountains into the Eastern Woodlands. Such trading activity required well-developed diplomatic skills and much closer ties with individuals and communities living not just in the next valley, but also great distances away. Inevitably, powerful chiefs in distant villages developed individual ties with their social equivalents elsewhere, cherishing personal contacts over many years, and cementing them with carefully orchestrated exchanges of gifts. Such presents developed complex ceremonial associations. Masks, copper artifacts and effigies, mica ornaments, finely carved pipes, and other fine objects became standardized over large areas. When their owners died, the same items accompanied the dead to the otherworld, as burial ceremonials assumed much greater elaboration and significance.

In addition to artifacts, expanding trade routes carried ideas and new spiritual beliefs that developed from earlier and simpler religious practices. About 200 B.C., ancient Adena mortuary ceremonialism in Illinois and the Ohio Valley fused with some fresh religious ideas, perhaps from southern Indiana and the Midwest. These new notions revolved in part around the use of the heavenly bodies to measure the passage of the seasons and the use of symbolic geography to reproduce the cosmos on the landscape. They also encompassed a world view that thought of human existence and the cosmos in terms of a powerful duality of opposites, such as fire and water, and the upper- and underworlds. Archaeologists identify these new beliefs from the distinctive artifacts found in the graves of their owners, and group them under the generic name "Hopewell," named after a complex of mounds and other earthworks on a farm of that name in Ross County, Ohio. Like the Adena people, the Hopewell folk were foragers and perhaps simple farmers, who lived in small villages. Only their mortuary customs and religious beliefs distinguish them from their neighbors.

Fig. 9.3. Hopewell Mound group, Hopewell, Ohio. Photograph from Photo Researchers.

The new mortuary rites achieved dazzling complexity, set in the midst of carefully aligned groups of earthworks. Hopewell burial mound complexes covered areas as large as several New York City blocks. The Seip Mound in the Ohio Valley is 76 meters long and 9 meters high, and as wide as an American football field. The mounds at the Hopewell earthwork complex (figure 9.3) itself are about 30 meters in diameter with a height of 9 meters, and a volume of about 14,000 cubic meters. Each represents over 200,000 hours of earth-moving with stone-bladed tools and baskets. Unlike the Adena people, who occasionally fashioned simple, circular embankments near their burial places, Hopewell communities built ceremonial compounds in varied and precise geometric shapes, combining some of them with open plazas and avenues that sometimes extended for several kilometers. At Mound City, Ohio, 24 burial mounds lie within an enclosure covering 5.2 hectares, an area equivalent to the base of the Great Pyramid of Khufu at Giza, Egypt.

The Hopewell people cremated most of the dead, reserving lavish treatment for prominent families, to the point that some observers call the Hopewell people "the Egyptians of North America." In 1892, Warren K. Moorehead dug into 15 of the 30 or more burial mounds in the ceremonial enclosure at the Hopewell mound complex itself to acquire exhibits for the 1893 World Columbian Exposition in Chicago. He dug 5 trenches into the largest mound, which was 153 meters long, 55 meters wide, and 10 meters high. Moorehead uncovered a group of 48 burials accompanied by a remarkable collection of copper artifacts, among them a deer antler headdress fashioned in wood and covered with copper sheet, metal breastplates, and 67 axes, one weighing 17 kilograms. Thirty years later another archaeologist, Henry Shetrone, dug into the same tumulus, where he found the skeletons of a young man and woman lying side by side. Both wore artificial noses fashioned of copper sheet, and necklaces of grizzly bear teeth and copper ear ornaments. Thousands of freshwater pearl beads and buttons of copper-covered wood and stone swathed the woman's body from head to knees, the remains of an elaborately decorated garment. Another Hopewell mound yielded a skeleton buried with a cache of several hundred pounds of obsidian, which was determined through sourcing methods to have originated in Idaho and Wyoming (see information in box, p. 195).

Archaeologists and treasure hunters have unearthed more than 1,150 prominent individuals from Ohio Hopewell burial mounds alone. Sometimes the dead lie in large burial crypts about 3 meters square, which were often built on isolated high ground clear of the settlement, and then sunk into the soil and covered with heavy roofs. These simple boxes were communal graves, used to house fellow kin or members of a single community. The Hopewell people also used charnel houses, apparently reserved for more important individuals. Both corpses and cremated remains were placed inside a substantial thatch-roofed house. Once the houses were full or the bodies had decayed, their descendants burnt down the structure and erected an earthen mound atop the ruins. Occasional caches of animal bones lie nearby, perhaps the remains of meals.

Charnel houses, commonplace in more recent times, were also used by the Choctaw people, who lived in Hopewell country centuries later. They employed them at the time of European contact.

SOURCING COPPER ORE

Sourcing copper ores can be achieved through X-ray fluorescence spectrometry. A sample is irradiated by an X-ray beam, which excites the electrons on the surface of the material. The beam is turned off, at which point the copper produces secondary X-rays that fluoresce in wavelengths characteristic of different elements, with energies proportional to the concentrations of elements in the metal. This quick and inexpensive technique is useful for determining basic ore composition. Atomic absorption spectrometry involves dissolving a small ore sample in diluted acid, then spraying it onto a flame, where a single element of interest can be measured. Neutron activation involves bombarding an ore sample with neutrons in a nuclear reactor, transforming the atomic nuclei of the elements into unstable radioactive isotopes. These isotopes release gamma rays characteristic of each element present, their varying intensities reflecting the amount of each present. Fast and nondestructive, neutron activation is accurate, but lead isotope analysis, which measures lead trace elements in a thermal ionization spectrometer, is ideal for accurate sourcing, since the composition of the ore remains unchanged during smelting and working.

Some southeastern native American societies confined their use to the elite; however, the Choctaw placed everyone in such structures. The chiefs lay in a special house, while everyone else was placed in kin-based charnels. The dead were placed on covered platforms in the open with special food offerings. Their flesh decayed over a period of several months. Then the deceased's relatives called in clan members—priestly specialists who defleshed the bones and put them in a chest. A communal feast followed, during which the bones were placed in the charnel house. They remained in the house for a considerable time, honored at regular feasts and ceremonies until the charnel house became full. Then the community removed the dead and interred them.

Superficially, Choctaw burial customs appear similar to those of the Hopewell people: the dead were exposed, the bones defleshed and sometimes cremated, then the remains were placed in charnel houses, and finally reinterred. Upon closer inspection, however, important differences emerge. The Hopewell placed great significance on exotic artifacts and raw materials. They also destroyed their charnel houses when they became full. Hopewell expert Mark Seeman argues that the charnel houses were important social and ritualistic catalysts, where local chieftains reinforced their secular and spiritual authority. In ceremonial feasts and rituals, they may have redistributed food and gifts to the dead throughout the community, while conducting the rituals that displayed their spiritual authority and close relationship to the animal and spirit worlds. Seeman's arguments were based in part on Sahlin's "Big Man" research, and on the belief (still held by some) that redistribution of goods through society was a primary function of chiefs and other popular leaders.

The Hopewell people were expert artisans. Native eastern North Americans never smelted copper, for they lacked the specialized equipment to melt metals systematically. However, they cut and hammered native (raw) copper and mica into simple weapons, tools, and ornaments, reserving their finest work for thin sheet metal silhouettes of people, birds, and clan animals, often hammered against a wooden prototype to give a repoussé effect (figure 9.4). In addition to copper and mica objects, Hopewell craftspeople also created fine clay vessels, stone tools, pipes, and human figurines. Most of these standardized ceremonial artifacts display no wear and were, thus, most likely prized objects of high prestige that were destined only for display, trade, or burial with their owners.

Hopewell ritual linked village communities living over a wide area. Common religious beliefs, including mound burial, elaborate ceremonies, and special reverence for important individuals who were buried with exotic artifacts, fueled an insatiable demand for both finished ceremonial artifacts and raw materials on the part of a minority of the farming population. Thousands of kilometers of winding paths and rivers formed the tentacles that linked leaders in different areas. These individuals created economic and ritual bonds with partners, cementing these links with gifts of ceremonial objects, such as axes. Through the exchange of gifts, they fashioned lasting ties of reciprocal obligation and trade alliance.

Fig. 9.4. A profile of a male head in hammered copper, from the Spiro site, Oklahoma. Photograph from the Ohio Historical Society.

Astronomy and Hopewell Earthworks

Circles and squares, octagons and causeways: the Hopewell earthworks bewilder the eye, especially when seen from the air. Despite more than a century of research, we still know little of the chronology of Hopewell earthwork construction, which makes the task of deciphering their meaning doubly hard. However, the process of decipherment begins logically with astroarchaeology, which provides evidence for purposeful alignment of earthworks with heavenly bodies.

The Newark, Ohio, constructions, dating to about A.D. 250, cover 10.4 square kilometers of countryside (figure 9.2). Today, a private golf course winds through the octagon-circle combination that encloses some 25 hectares with large earthen embankments 12 meters across at the base and 1.7 meters high, an area twice the size of Çatalhöyük in Turkey (see chapter 4). Eight oval, flat-topped mounds lie inside the octagon opposite the vertices. A large flat-topped mound overlooks the entire complex. A large square enclosure and another circle lie southeast of the octagon. Physicist Ray Hively and philosopher Robert Horn used data from surveys by Atwater, Squier and Davis, and other Victorians, combined with their own transit and steel tape survey to produce an accurate map of the much-disturbed site. The earthworks display astonishing precision, with exact corners and precise orientation. The Newark octagon covers 24 hectares, with openings that lead to the center at each corner, built with the aid of the diameter of the nearby circle. The builders of the circle and octagon at Newark constructed them with great care. A narrow defile leads to a perfect 321.3-meter diameter circle nearby. Hively and Horn studied the diameters of the circles, the sides of octagons, and diagonals, in addition to the sides of earth squares and calculated that the builders used an exact unit of measurement 321.3 meters long.

The 2 men went on to study the astronomical alignments of the Newark earthworks, a daunting task for such a large, and much-obliterated complex. They considered only alignments along linear embankments, axes of symmetry, and special points like centers of circles separated by an integral number of circle diameters. Using astronomical tables, they calculated the azimuths for the rising and setting of the sun and moon in A.D. 250—the estimated building date. They compared these azimuths with earthwork features and found no convincing evidence for solar alignments. However, the Newark octagon fits the northern and southern extremes of the rising point of the moon along the horizon. These extreme rising and setting points oscillate between 2 fixed azimuth points every 18.61 years. The architects even distorted the western side of the octagon slightly to accommodate additional alignments. The long 1.7-meter high walls of the octagon would have allowed an observer to define precise azimuths within 0.25 degrees. The Hively and Horn tables show the axis of the avenue between circle and octagon and 4 sides of the octagon marked 5 of the 8 extreme lunar set points with a mean accuracy of 0.5 degrees. Furthermore, the observation points for these alignments lay at 4 vertices of the octagon. Hively and Horn proposed no less than 17 independent lunar alignments, 8 of which were, most likely, deliberate. These alignments were probably accurate enough to allow for the prediction of years when lunar eclipses occurred near the winter or summer solstices. They certainly allowed the Hopewell people to monitor the monthly and 18.6-year lunar cycles. The precision of the earthworks and sheer number of alignments strongly suggest the Newark complex provided a means of observing cyclical, lunar time.

Hively and Horn also studied the only other Hopewell earthwork complex known to include an octagon, at High Bank Works near Chillicothe, Ohio. High Bank is 113 kilometers away from Newark, but the layout is basically the same as Newark. Both have a large circle joined by an avenue to a large equilateral octagon, associated with an adjacent array of avenues and smaller circles. By comparing surveys from both sites, the 2 scholars established that the circles were of identical diameter (321.3 meters) from bank summit to bank summit. Furthermore, the same circle diameter formed the basis for each octagon, even though there were minor differences between the High Bank and Newark earthworks. The builders had laid out the portion of the High Bank octagon that lies east of the avenue alignment with great care, while the western segment is more haphazard. One could argue that this was due to poor workmanship, but, given the amount of work involved, Hively and Horn believe it was deliberate.

Using astronomical tables for the High Bank location from A.D. 250—again, the estimated building date—they calculated the rising and setting points for the sun and moon. They estimated the altitude of the sun using topographic maps, assuming that the builders cut down any trees that had interfered with the horizon. Four alignments marked the extreme north and south rising points of both the sun and moon, the walls being aligned within 0.5 degrees in one case, 0.1 degrees in the other. These determined the main features of the shape and orientation of the High Bank octagon. When Hively and Horn examined their calculations carefully, they found that the 16 percent deviation in one western wall allowed the builders to produce a lunar alignment between a southwestern and eastern entrance. The octagon design defines 4 lunar alignments: 2 rising points and the corresponding setting points for the northern maximum and southern minimum of the moon's passage in the heavens: the points where it pauses for several days before changing direction at the summer and winter solstice. These alignments would have been sufficient to monitor the 18.6-year lunar cycle. The axis of the High Bank avenue through the circle and octagon allows an observer standing on the west side of the circle to observe the winter solstice over the northwestern corner of the octagon. A convenient gap in the circle earthwork permits the watcher to sight the solstice line through the left edge of the gap.

Unfortunately, most Hopewell earthworks have vanished under the plow, so the meticulous surveys needed to establish alignments are often impracticable. However, Hively and Horn have made a strong case that at least two earthworks allowed Hopewell priests to observe the extreme rising and setting points of the sun and moon. And, like Stonehenge, prominent ancestors lay in burial mounds within sight of the sacred observatories.

Adena and Hopewell communities lived by hunting and plant gathering, sometimes cultivating some native grasses, and eventually maize, to supplement their diet. Over the centuries, the larger villages became more permanent, as different groups settled in increasingly well-defined territories. As was the case in western Asia, the relationship between living people and their land, and between the living and their ancestors, became stronger. We see this in flamboyant Hopewell mortuary rituals, in great elaboration of grave furniture, and in the burials of prominent members of society, respected for their entrepreneurial and ritual skills. Adena earthworks were simple circles and burial mounds, which have no orientation toward the movements of sun or moon. On the other hand, Hopewell enclosures display precise alignment to solstices, and a concern as much with the burial of ancestors as with the observance of seasonal rituals governed by astronomical phenomena, as if the Hopewell people arranged their earthly environment to mirror the sky itself.

In some respects, Adena mounds and Hopewell earthworks are similar to Avebury, with its communal barrows, causewayed enclosures, and stone circles, where the dialogue between the living and the dead continued for centuries. Newark and other astronomical monuments recall Stonehenge. The last time I walked the Newark ruin field, I admired the precision of the layout and the exactness of the entrances to the octagon. I could not help but compare it to Stonehenge, where the stone circles welcome the sun's rays at the solstice. There, worshippers greet the midsummer and midwinter sunrise, as early morning sunlight flashes into the interior of the ancient temple. Newark has the same intent.

Symbolic Geography

The Hopewell dead at Newark and elsewhere lay in the midst of sacred landscapes, where the living dug enclosures, circles, and avenues as a setting for public ceremonies regulated by the passage of seasons measured in the heavens. There, as elsewhere, as the settlement became more permanent and the population grew, the rituals that marked the environmental calendar became increasingly central to human existence—to the welfare of society. Planting and harvest ceremonies moved to the center of the ritual stage, as people attempted to steer a sober and moderate course between the serene order of the upperworld in the heavens and the chaos and violence of the underworld beneath the earth.

Ancient Hopewell religious beliefs and the symbolic landscapes associated with them lie far in the past. The last burial mounds were erected about A.D. 400, more than a millennium before European contact. This chronological gap of over 10 centuries makes conventional ethnographic analogy from historic peoples impossible except in the most general of terms. Archaeologist Warren DeBoer has used a combination of archaeology, ethnographic observation in Ecuador, and data from historical peoples to attempt an interpretation of Hopewell symbolic geography.

DeBoer lived among the Chachi Indians of northern coastal Ecuador's Cayapas Basin in the late 1980s. He spent some time at Punta Venado, a Chachi ceremonial center laid out in a flattened U, which flanks a plaza with the church in the center. De Boer believes

Punta Venado is a map of Chachi country, a symbolic "Big House," with a settlement plan that reflected the cosmos itself. He points out that a broadly similar way of thinking and general settlement layout occurs among many eastern North American groups. DeBoer believes that this kind of notion was widespread in ancient native American societies, and that it offers a way of interpreting Hopewell earthworks.

Another method of interpretation is based on persistent notions of duality in human existence. The Cherokee, Creek, and other Eastern Woodlands peoples used a continuous form of seasonal duality in their domestic architecture. They built both circular winter ("hot") houses and adjacent or nearby rectangular summer dwellings. The same duality occurs in pre-Columbian Mississippian villages, and even in much earlier settlements, where rectangular and circular dwellings occur within the same site. Among historic groups, a rich geometric symbolic duality is associated with the placement of houses, which goes far beyond the seasons. Winter is to summer as circle is to rectangle. DeBoer extends this further to local and foreign, peace and war, upperworld and underworld. He says: "There is a historically deep-seated repertoire of polarities—a given structural field of play, if you will—that provides a lexicon for interpreting the form and meaning of both traditional and iconoclastic action performed by very real human actors" (1997, 3). We know the basic duality has historical roots, but is it also a part of the Hopewell earthworks?

Hopewell people laid out their geometrically precise earthworks over a course of about 5 centuries. We know they carried out mortuary rituals and other major ceremonies, as well as celebrated feasts, and organized causeway-directed foot races at these locations. Hopewell enclosures follow striking patterns of circles and squares, with the occasional octagon, but defy accurate chronological ordering. DeBoer's approximate seriation begins with Adena sacred circles, the square enclosure being added early in Hopewell times. At first the circles and squares are symmetrical, as if reflecting an orderly duality. Some time later, the earthworks became more irregular: shapes are blurred and fluctuate in size, and the squares always have openings. DeBoer speculates that these changes reflect the occurrence of a social crisis. For example, at the Newark, Ohio, earthworks, the largest square covering 20 hectares may reflect a local leader aspiring to prestige and power, as earthworks elsewhere shrank in response to power shifts. In time, the more standardized pattern, of a big circle of about 16 hectares in size and a smaller square about 11 hectares in size aligned obliquely and separated by a circular insert, again became predominant. It is as if powerful, and ancient, domestic orthodoxy reasserted itself in the face of new, reactionary ideas. Perhaps earthwork architecture served as a symbolic barometer of changing social conditions.

In addition to the duality of circles and squares, many Hopewell sites occur in pairs, sometimes on either bank of a river, or some kilometers distant from one another. In the most extreme example of this pairing, archaeologist Barry Lepper believes that 2 centers, High Bank and Newark, 113 kilometers northeast, were linked by a causeway he calls the "Great Hopewell Road," parts of which appear on aerial photographs.

All of this would be pure speculation from ethnography, historical records, and earthwork survey, but for some telling clues in the archaeological record itself. For example, Henry Shetrone unearthed northern and southern altars under Mound 17 at the Hopewell mound complex itself. White-colored mica and domestic artifacts lay nearby. The northern altar area yielded red-colored copper and stone projectile points. The artifact distributions hint at a powerful opposition. Two mounds at the Turner site reveal a striking polarity. Mound 4 to the northeast yielded offerings of white mica, and images of "horned serpents." The southwestern Mound 3 contained most of the copper offering and projectile points, as well as depictions of birds of prey. DeBoer theorizes that this duality reflects the upper- and underworld of the cosmos. However, the cappings of the mounds seem to reverse the duality. The summit of Mound 4 is covered with white mica sheets, while the mourners sprinkled black-colored cannel (a coallike substance) over Mound 3. DeBoer points to living societies where members of a clan opposite to that of the dead are major players in burial rites. He reads binary contrasts into the archaeological evidence: cannel as a foreign commodity, traded from afar long before Hopewell times; mica as new goods, an exotic material intimately, and distinctively, associated with the Hopewell exchange networks.

Societies like the Hopewell were poised between two extremes, those of egalitarian governance and tyranny. The native American anthropologist Alfonso Ortiz believes such duality presupposes a triadic intermediary, which existed between the two. Late Hopewellian earthworks are more tripartite and asymmetrical.

Stone platform pipes carved into animal effigies also provide some clues (figure 9.5). Ephraim Squier and Edwin Davis recovered a cache of more than 200 smashed pipes from Mound City in the Scioto Valley alone. The nearby Tremper Mound contained another 136 pipes. Almost all the animal motifs occurred in both mounds, and combined herons eating fish (sky-water), and otters with fish in their mouths (land-water), and so on, as if "a triangular superstructure with sky, earth, and water apices would appear to scaffold the choice of animal representations" (DeBoer 1997, 6). When DeBoer arranged the pipes on a sky-earth-water triangle, he found the earth-water axis to be the most prominent and otters to be the most common. He says: "The skyward wafting smoke from Hopewell pipes thus intimated a ternary cosmos mapped and wrapped in zoomorphic imagery" (ibid., 6).

DeBoer's research draws on data from anthropology, archaeology, and ethnohistory, and unashamedly uses analogy to bridge many centuries. Some archaeologists feel he has thrown caution to the wind, others that he has not gone far enough! Archaeologically, he relies on data collected over many years, and on an important unpublished doctoral dissertation by A. M. Byers on the earthwork enclosures of the Ohio Valley, which attempts an imaginative seriation of the major sites. DeBoer is concerned with general structures of Hopewell cosmology and earthwork layout. At no point does he claim that he has deciphered the meaning of the earthworks, but his analysis of duality gives a provocative insight into the ancient dualities of native American culture. He has shown convincingly that a polarity which was reflected in large earthworks existed in Hopewell society. The full meaning of Hopewell earthworks can, however, never be deciphered, for, as British archaeologist Richard Bradley

Fig. 9.5. Hopewell pipes in animal forms. A drawing from Squier and Davis's classic report on the Ohio earthworks.

reminds us, the interpretation of any monument made to last begins as soon as it is built, and is in the hands of all of us: builder, user, scientist, and visitor alike.

MISSISSIPPIAN

The date for the introduction of maize into eastern North America remains controversial, due to a lack of AMS radiocarbon dates for actual cobs or seeds; but corn seems to have arrived in the east during the beginning to middle part of the first millennium A.D. How much time elapsed before it became a staple is unknown, but it was probably around A.D. 900 to 1000. Beans came somewhat later, but added a vital protein source to the diet. In time, these two crops wrought major changes in Eastern Woodland life, fostering higher population densities, larger food surpluses, and great elaboration in human society. Maize and bean cultivation brought massive transformation to the natural environments of fertile river valleys, where Adena, Hopewell, and other foraging societies lived for many centuries. Between A.D. 800 and 1500, "Mississippian" societies developed throughout much of the Eastern Woodlands.

The term "Mississippian" covers a wide spectrum of farming societies, from minor chiefdoms to enormous ceremonial centers like Cahokia, in the Mississippi Valley of Illinois, or Moundville in Alabama. This multiplicity of chiefdoms lived in a constant state of political flux, as chiefs rose to prominence, then died; alliances were forged and fractured, and new people rose to power. Only strong, and still little-understood, religious beliefs linked Mississippian communities everywhere. They reached their apogee at Cahokia.

Cahokia

Cahokia was founded in the heart of a pocket of extremely fertile bottomland on the Mississippi floodplain near St. Louis known as the American Bottom. At that time, fish, game, and wild nuts and grasses abounded. Migrating waterfowl flew through the valley in spring and fall, and rich agricultural soils allowed bountiful crops of maize and beans. Cahokia's leaders presided over the highest population density anywhere north of Mexico. At the height of its power between A.D. 1050 and 1250, Cahokia covered an area of more than 13 square kilometers, about the size of the ancient city of Teotihuacán in Mexico. Some 30,000 people lived in pole and thatch houses covering 800 hectares (an area four times the extent of Sumerian Ur, Mesopotamia, in 2800 B.C.), clustered on either side of a central east-west ridge. More than 100 earthen mounds of various sizes, shapes, and functions dot the Cahokia landscape, most grouped around open plazas. The largest, Monk's Mound, dominates the site and the surrounding landscape (figure 9.6). Monk's Mound rises in 4 terraces to a height of 31 meters (more than twice the height of any other tumulus at Cahokia), and covers 6.4 hectares. The sides measure 316 to 240 meters in length; 614,500 cubic meters of earth form the mound.

Two rows of somewhat smaller mounds lie on either side, with a plaza area immediately to the south. Many were platform mounds for important public buildings or elite residences. A conical tumulus just south of the pyramid may have supported a charnel house where the elite lay until buried in the mound itself. One burial mound contained a male burial laid out on a platform of 20,000 shell beads. Three high-status men and women lay nearby; 800 arrowheads, copper and mica sheets, and other valuables accompanied these skeletons, perhaps those of close relatives sacrificed at the funeral.

Fig. 9.6. An artist's impression of Cahokia at the height of its power, looking across the Grand Plaza to Monk's Mound. By Lloyd Townsend. Courtesy of Cahokia Mounds State Historic Park.

Four decapitated men with their hands severed also served as sacrificial victims. The bodies of 50 young women, 18 to 23 years of age, lay in a pit close by. A stout wooden palisade with watchtowers and gates enclosed 81 hectares of the central precincts. Rebuilt at least 4 times, the palisade served to isolate the general population from what must have been a restricted, sacred area. In all probability, only a few hundred people lived inside the wooden fence.

Cahokia presided over a hierarchy of lesser ceremonial centers. Four sites nearby cover more than 50 hectares and boast several mounds. Five smaller settlements surround single tumuli.

Cahokia's Cosmos: Fertility and Duality. Cahokia is vast and unique, by far the largest ceremonial center in southeastern North America. Fortunately for science, the ancient cosmology and religious beliefs behind the site can be pieced together, at least partially, from a combination of archaeology in the American Bottom and ethnohistory derived from historic southeastern Indian groups.

The layout of Cahokia and other major centers reflects a traditional southeastern cosmos with four opposed sides, reflected in the layout of their platform mounds, great mounds, and imposing plazas. By A.D. 1050, the rectangular plaza surrounded by mounds reenacted the ancient quadripartite pattern of the cosmos, seen in much earlier settlements along the Mississippi. Four-sided Mississippian platform mounds may portray the cosmos as "earth-islands," just as modern-day Muskogean Indians thought of the world as flat topped and four sided. Archaeologist John Douglas used ethnographic and archaeological data to argue that the four-sided cosmos had a primary axis that ran northwest to southeast, with an opposite axis dividing the world into four diamond-shaped quarters. Cahokia is oriented along a slightly different north-south axis, but it certainly perpetuates the notion of spiritual links between opposites and a cosmos divided into quarters. Researchers believe that the orientation reflects observations of the sun rather than the moon. Astroarchaeologist Anthony Aveni thinks that Cahokia's rulers used the sun to schedule the annual rituals that commemorated the cycles of the agricultural year.

As we saw with the Hopewell, southeastern cosmology revolved around dualities. In the case of Cahokia, these included the upper-

and underworld, and a powerful and pervasive fertility cult linked to commoners and a warrior cult associated with the nobility. These dualities were carried through to the smallest ritual centers. Changing settlement layouts imply that, at first, local communities and kin groups controlled fertility rituals in villages divided into symbolic quarters, with ceremonial structures facing a central square. Later centers displayed more formal layouts, with central plazas, elaborate sacred buildings, and storage and ritual pits filled with pots and other ritual offerings made during fertility and world-renewal ceremonies. By this time, experts believe, power was passing from local kin leaders to a powerful elite based at Cahokia, a shift reflected in increasingly elaborate ceremonial architecture, residences for local nobles at local centers, and mortuary complexes controlled by full-time priests.

During the heyday of Cahokia, after A.D. 1050, some centers possessed temple structures, sacred fire enclosures, ceremonial courtyards, and grave houses and burial areas where the bodies of the dead from select lineage groups were processed. Archaeologist Thomas Emerson believes that such centers had close links with local kin groups, as well as political and religious leaders related to nobility at Cahokia itself. Their carefully laid out centers brought two central ritual themes together: the spiritual realm of fertility and life, and the validation of living rulers, who were intermediaries with the supernatural realm.

Mississippian religious beliefs are still elusive, but distinctive artifacts yield a few clues. Emerson investigated a small rural shrine at the so-called BBB Motor site near the great center, and carried out additional excavations at a larger rural ceremonial complex called the Sponemann site nearby. Both shrine excavations produced Cahokia ritual artifacts and paraphernalia, especially a series of Cahokia-style figurines fabricated from soft red stone, found in an undisturbed archaeological context. One figure shows a woman kneeling or squatting on a circular base, with a square pack on her back. Her head rests on the head of a feline-headed serpent and her right hand holds a hoe that is stroking or tilling the serpent's back. The serpent's back bifurcates on the woman's left side and changes into gourds growing on vines, which entangle the woman's body.

A second figure also depicts a woman with sloping forehead kneeling on a rectangular base formed of corn ears or reed bundles woven into a mat. She wears a skirt, her long hair hanging down her back. A rectangular basket lies in front of her. A now-broken maize stalk emerges from the base and passes through the woman's hand, sweeping back to an attachment above her ear. Emerson and others believe that the woman is an Earth Mother figure, historically depicted wearing a short skirt, and associated with the serpent monster and with the giving of plants to humankind. On her back she carries a sacred bundle, another gift to humanity. Ethnographic sources hint that the Earth Mother assumed many forms: goddess of death, and an Old Woman, to mention only two.

The feline-headed serpent motif occurs widely on Mississippian and later artifacts. In fact, the serpent monster theme was widespread in historic southeastern rituals, and has considerable antiquity. Serpents appear in temples and other public structures, as tattoos, and in dance themes. They possessed the attributes of several animals, and symbolized water monsters and other mythical beasts. Serpents were underworld creatures, associated with lightning, thunder, rain, water, and power over plants and other animals. The underworld was the source of water, fertility, and power against evil.

Validating Power with Artifacts

Cahokia and other Mississippian centers reflect an ancient cosmology in a symbolic language intelligible to noble and commoner alike. Like Avebury, Stonehenge, or Newark, these centers were powerful statements of the supernatural realm—the cosmos. However, Cahokia's rulers also used artifacts to meld the rituals of the commoners with the need to validate their autocratic rule.

Back in 1949, University of Michigan archaeologist James Griffin identified a series of sharp-shouldered jars with highly polished surfaces and distinctive, standardized trail designs on the upfacing shoulder. He named them Ramey-incised jars, after the site where they were first identified, noting their unusually fine finish. For years, Ramey jars were identified simply as a high-status or ceremonial ware, until Emerson carried out a detailed formal analysis of all known vessels of this type. He identified 9 basic categories of decoration, such as chevrons and scrolls, then calculated the frequencies of each one. He then examined the associations between different basic elements, discovered 2 major pairings, and established that the symbolism of the patternings was homogeneous, tightly focused, and capable of interpretation in symbolic terms. Emerson and his colleague Timothy Pauketat believe that the Ramey symbols represent a visual portrayal of the Cahokian cosmos. More than 90 percent of the Ramey symbols fall into 2 closely related groups. Judging from ethnographic analogies, the chevron arc, found on nearly 64 percent of the sample motifs, may relate to a continuum between the arch of the sky, birds, and the upperworld, as well as a continuum between the upper- and underworld. Another set of symbols that includes the scroll, circle, and chevron symbol, may refer to the underworld, serpents, and water. Emerson argues that the Ramey containers were important ceremonial containers, which were centrally produced and then distributed as part of Cahokian fertility rituals. He theorizes that they were manufactured and used as part of important religious ceremonies held during certain seasons of the year at rural centers, which were themselves linked to Cahokia, thereby integrating society with pervasive concepts of fertility and cosmos reflected in harvest and other ceremonies. Pauketat and Emerson call these rituals "rites of intensification"; they played a vital role in linking commoner and elite. The Ramey jars were an easily understood symbolic text in clay form that were disseminated throughout small communities as people dispersed after major ceremonies.

Moundville

Cahokia's vast size was a reflection of the political and religious authority of its builders, of the fertility of its floodplain hinterland, and its strategic position astride the Mississippi and trade routes with north and south. The great center was unique, and its religious beliefs, political institutions, and trade networks influenced other groups hundreds of miles away. In time, other important centers developed far from the Mississippi, many of them coming into prominence as Cahokia itself faded into insignificance during the thirteenth century, for reasons that are still little understood.

Another great Mississippian center, Moundville in Alabama, boasts of 20 large platform mounds and covers 122 hectares. The platforms surround a 32-hectare rectangular plaza sited near a river, and protected on 3 sides by a palisade. Some platforms supported shrines and other public buildings or residences. A few contained caches of what must be ancestral human skulls. A sweat lodge and charnel house lay just outside the southern side of the plaza. Over 3,000 burials have come from Moundville, many of them from small cemeteries near local centers. Archaeologist Chris Peebles studied the furniture from the Moundville graves. His social-ranking study was based on a statistical cluster analysis of more than 2,000 tombs combined with an elaborate chronological ordering (seriation) of the pottery found with the burials (see information in box, p. 213). He developed a social hierarchy for the site, based on the same grave analysis. Seven high-ranking males, all deposited with elaborate objects including copper artifacts, headed the social hierarchy and lay in or near earthen mounds. A second group of more than 110 people buried in mounds apart from the most important individuals possessed considerable numbers of copper earspools, bear teeth, shell beads, and other prestigious items. The lowest-ranking people had almost no grave furniture and were deposited at the edge of the site.

By pre-Columbian North American standards, Cahokia and Moundville are vast constructions. Even smaller Mississippian centers offer a dramatic contrast in size when compared with Adena and Hopewell sacred places. Larger Mississippian communities arranged platform mounds and tumuli around open plazas and, at times, fenced off the central area. Everyone, except for elite families, lived in communities clustered close by.

Etowah

Between 1954 and 1958, Lewis Larsen of West Georgia College excavated one of the three great mounds at Etowah, Georgia. He found more than 200 burials in the tumulus, deposited in layers as the mound grew higher in succeeding generations. Etowah reached the height of its powers between A.D. 1200 and 1400, a center built on fertile riverside soils, within easy reach of galena, graphite, greenstone, and marble. which were used to make artifacts and ritual objects.

SERIATION

Seriation, the ordering of artifacts by their morphology, is a fundamental part of archaeological research. It is based on the assumption that the popularity of any artifact or culture trait is transient. The miniskirt, whitewall tires, and the hula hoop, are examples of "cultural artifacts" that enjoyed a moment of popularity. Pottery styles, stone tool designs, metal artifacts, and other examples of ancient technology enjoyed similar, if longer, periods of popularity. When changes in artifact styles are plotted as percentage bar graphs against time, the period of greatest popularity forms the longest bar, the total effect being what archaeologists call a "battleship curve," the shape of a battleship when viewed from above, with the bow and stern marking the beginning and end of an artifact's life.

Seriation has been widely applied to undated archaeological sites, notably to hunter-gatherer camps and early farming settlements of unknown age in Mexico's Tehuacán Valley, where Richard MacNeish ordered three styles of local pottery in relative chronological order using the battleship curve principle. Seriation places sites or graves in their relative chronological order on the basis of changing artifact styles—sometimes even using changing designs within a single pottery class. Today's researchers use sophisticated statistical methods to produce seriations and to test the validity of their orderings. The most successful seriations are those combined with accurately dated cultural sequences. With a master chronology and seriation in place, the researcher can assign undated sites with similar, distinctive artifacts to an approximate relative position on the time scale. Thus, the routine process of artifact ordering lies at the very core of all archaeological science.

Early treasure hunters unearthed an astounding array of exotic objects from the burial mound: copper sheets embossed with mythical figures, engraved shell ornaments depicting elaborately costumed warriors, and necklaces of freshwater shells.

Larsen's systematic excavations of the burial mound recovered several undisturbed elite graves. "Burial 15" was a 3-meter long log crypt containing 4 dismembered bodies. Two painted 0.6-meter high marble figures of a man and a woman guarded the grave. The woman figure sat with her hands on her knees, her hair in a topknot, carrying a backpack-like object. The most important individual interred in "Burial 15," was a man, who wore a heavy necklace of shell beads the size of golf balls. A conch shell ornament, fashioned from a Gulf Coast shell, hung from the necklace. Bands of shell beadwork encircled his wrists and upper arms. He wore a dramatic headdress of feathers and copper-covered wooden coils fastened to a plate of sheet copper endorsed with the image of a bird's open wing. A magnificent greenstone ax lay across his chest, blade and handle fashioned from a single piece of stone. The man exuded power and ritual authority in his feather cape and lofty feathered headdress. Such leaders used compelling supernatural powers to justify their privilege and rule.

The "Southern Cult"

Of necessity, archaeologists study the material remains of human behavior, reflected in surviving artifacts of all kinds. Mississippian religious beliefs were complex and permeated all levels of society. These beliefs come down to us in a striking art tradition, fashioned in clay, shell, and stone, and presumably in more perishable mediums as well. Over half a century ago, Mississippian archaeologists noted the striking similarities between hundreds of artifacts buried with the elite throughout the southern and southeastern United States. Ceremonial stone axes, copper pendants, shell disks or gorgets, and engraved shell cups displayed common themes and motifs found over an enormous area, from Oklahoma to Florida. They believed that these flamboyant objects reflected a common religion, which they named the "Southern Cult." This generic term is still used as a convenient label today, although researchers now recognize great diversity in religious beliefs over thousands of square miles. The most distinctive "Southern Cult" motifs date to the thirteenth century: bi-lobed arrows, striped poles, and fringed aprons worn by shamans and warriors. Weeping eyes appear on embossed copper faces, shell gorgets, and clay vessels. Birds, serpents, human skulls, and human hands adorn clay vessels and ceremonial objects of all kinds. There are

Fig. 9.7. Shell gorget showing a flying shaman (diameter: 10 centimeters). Photograph from the American Museum of Natural History.

long-nosed gods depicted on copper masks, shamans flying through space carrying death's heads (figure 9.7), and ceremonial maces. "Southern Cult" art displays a preoccupation with themes of war and purity: wind, fire, sun, and human sacrifice.

Many years ago, experts pointed to the loose resemblances between "Southern Cult" art and Mesoamerican artifacts from hundreds of kilometers to the south. They theorized that the cult and its exotic objects came to the Mississippian homeland from the south, in the hands of professional Mexican merchants, known to the Aztecs as *pochteca*. Today, we know that many of the themes of the "Southern Cult" appeared centuries earlier. These ancient themes—denoted by weeping eyes, bird symbols, circles, and crosses—survived for many centuries, becoming more refined as Eastern Woodland societies evolved from foraging to maize and bean agriculture. Even the term "Southern Cult" is misleading. There was never a "state" religion that flourished throughout Mississippian country. The distinctive artifacts of the "Southern Cult" reflect a complex, highly variable set of religious beliefs and mechanisms that lay behind the authority of every chief, whether he presided over a small village or an enormous center like Cahokia or Moundville. Many of these objects, such as the copper sheets, mica silhouettes, and shell ornaments, served as badges of rank—of secular and spiritual status.

Mississippian Religious Beliefs

The question remains: What do we know of the actual religious beliefs of the Mississippians? Despite the passage of five centuries, oral traditions collected from Cherokee, Creek, and other southeastern groups are still able to provide some insights into ancient Eastern Woodland religious beliefs, which help interpret the Mississippian archaeological data.

The southeastern Indian cosmos, like that of other native American societies, envisaged the earth as a circular island resting on primordial waters. The sky above was a vault of stone in the form of an inverted bowl. At dawn and dusk, the bowl rose, so that the sun and moon, the two principal deities, could pass under it in their regular journeys across the heavens. The sky vault comprised an upperworld, peopled by the sun and moon, and by large versions of all the creatures that lived on earth. The underworld, beneath earth, was a chaotic realm of disorder and fertility. Earth came into being when the Creator brought up mud from beneath the waters to form the island where humans lived. Four cords supported the earth, each at a cardinal direction. Human behavior steered a middle course between the order of the heavens and the chaos of the underworld. Failure to follow strict moral and religious behavior would weaken the cords holding earth aloft. It would then sink into the dark waters again and humans would perish.

According to the Cherokee, and probably other southeastern peoples, living things fell into three broad categories: people, animals, and plants. Humans hunted animals, and when they did so carelessly, the animal world became angry and the spirits of animals would send disease as revenge. Many animals had deep symbolic meaning, such as the bald eagle (the bird of peace), and the perigrine eagle (the symbol of war). Owls and cougars traveled in darkness and had an edge over other animals. Plants were the friends of people, and the source of most medicines. Some plants, such as cedar, pine, and spruce, were considered sacred because they remained green in winter. The interrelationships between these three groups helped shape Indian life. Some living beings were more sacred than others, for they seemed able to resist everyday constraints. Some humans, such as priests and shamans, lived more spiritual lives than other people and possessed powers beyond those of ordinary folk.

The Cherokee spoke with awe of the *uktena*, a living beast with the body of a rattlesnake, but the girth of a tree trunk. The *uktena* resided on the borders of the living world, in deep water and near high mountain passes. Images of creatures like the *uktena* abound on Mississippian shell ornaments and pottery. Writhing serpent-like beings have wings, deer horns, and sometimes the teeth of carnivores. The *uktena* was a symbol of a human transformed into the spiritual world, wearing a blazing headdress in the form of a crystal, which reflected the future. Divination, shamanism, and altered states of consciousness lay behind the serpentine monster of the Southeast.

As we saw at Cahokia, the people lived in a world of opposition, between the upper- and underworld, between water and fire. Water was life-renewing, and fire never extinguished except when someone died, a symbol of the dissolution of life. Rites of passage marked every stage of life: from birth, when a baby was dipped in water, then rubbed in bear fat, to the elaborate ceremonies of death. Before European contact, Southeast groups celebrated a series of public rituals throughout the year, few of which survived past European settlement. The Green Corn Ceremony, held at a propitious moment between June and September, celebrated the filling out of the kernels of the late corn crop. The public ritual combined thanksgiving and a constant search for purity. Everything began with the eating of old food. Then the men cleaned public buildings, the women their family homes. Next, the men fasted for 1 day and 2 nights, while drinking vegetable emetics and vomiting all that remained of the previous year's food. While fasting, the men resolved disputes and forgave people for minor crimes. After the fasting ended, each family extinguished its hearth. The people fell silent, as if time had stopped. Then a priest used a fire-drill to kindle a new, pure fire. Everyone lit their fires from this first hearth, as the ceremony ended with feasting and dancing. The 4-day Green Corn Ceremony was the climax of the ecological calendar.

THE ROOTS OF EASTERN WOODLAND BELIEFS

The ultimate roots of Moundbuilders' beliefs go back deep into the remote past, to mortuary customs developed centuries earlier by Adena villagers in the Ohio Valley, and by Hopewell priests adorned with magnificent artifacts of rank. I thought of these long tentacles into the past when I examined Adena and Hopewell pipes in the American Museum of Natural History some years ago. Holding an Adena smoking pipe, my thoughts wandered far from the Ohio Valley. I thought of the Wiltshire countryside, where ancestral burial mounds stand out against rolling downland. The dialogue between the living and the dead is etched into this British landscape. The sacred environment of the Adena mounds has long vanished under modern fields. Nevertheless, just as it does at Avebury, archaeology reveals rites of reburial; symbols of a belief that death was a process of transformation from life to the spiritlike existence of the ancestors.

Hopewell religious beliefs evolved with even greater elaboration, as some aspects of historic southeastern cosmology first appear about 2,000 years ago. A new preoccupation with astronomical alignments and an ecological calendar developed, as cultivation assumed greater importance in daily life. Additionally, a basic duality governed symbolic geographic and human existence. A millennium ago, Mississippian society took these ancient beliefs one step further, and forged highly elaborate societies that lived in a world delineated by powerful oppositions, and encompassed the spiritual world: the domain of deities, spirits, and supernatural forces. A supreme being probably lived in the sky, closely connected to a pillar, which connected the layers of the cosmos. Many societies offered prayers to this being, often referred to as the Creator. The shamans and the spirit mediums have always enjoyed special status in human societies. Mississippian visionaries—astronomers, diviners, and others with a close relationship to the spiritual world—sometimes enjoyed vastly enhanced political powers, for they had special powers of social control at their disposal. The most important among them were, thus, able to control the destinies of thousands. Their knowledge of the ecological calendar and the spiritual cosmos, used in elaborate ceremonialism, maintained a balance between the realm of the living and that of the spiritual universe.

chapter ten

THE BULL BENEATH THE EARTH

I wandered through the deserted ruins of Knossos, Crete, on a windy spring afternoon. Bright wildflowers embroidered the weathered stone, bursting through cracks between ancient paving stones. Small dust puffs chased each other across courtyards and down quiet stairways. The sun settled in the west and the deep blue sky turned lighter, and then rose tinted. The northwest wind faltered and slowly died. Long shadows turned the standing walls into hard edges against the darkening heavens. The reconstructed columns of the great palace glowed a deep plum red in the setting sun's light. I sat and contemplated the panorama of buildings, losing myself in the Cretan past; imagining the buildings restored, and coming to life. A momentary rumbling and the earth shaking gently beneath my feet jerked me back to the world of the present. Birds twittered in alarm among nearby trees. Then the shaking ceased as abruptly as it began, and I remembered how, 50 years earlier, another archaeologist had felt the voice of the earth.

Englishman Arthur John Evans had devoted his career to Knossos. In June 1926 he was lying in bed reading when an earthquake shook his house. Church bells jangled, roofs crashed to the ground, and clouds of dust rose above the swaying trees. "A dull sound rose from the ground like the muffled roar of a bull," he wrote afterward. "It is something to have heard with one's own ears the bellowing of the bull beneath the earth" (1943, 299). He imagined the beast that tossed the earth with its horns, the great Minotaur, mythical bull-human appeased by the Minoans with constant sacrifice.

The Minoans of Crete—expert traders who sailed in black-hulled ships—are the stuff of legend. They appear occasionally in the epic stanzas of Homer's *Iliad* and *Odyssey*, shadowy figures on the fringes of known history. For more than 3,000 years they vanished into the oblivion of the remote past, their palaces and towns long forgotten. A century ago, only a handful of scholars thought that these shadowy Cretans had once actually existed.

Legend tells us that at one time the entire Greek world lived in fear of King Minos of Crete. His ships traveled throughout the eastern Mediterranean, striking fear into marauding pirates and keeping age-old sea lanes safe for everyone. Minos exacted a high price for his protection. Every nine years, his emissary would arrive at Athens and claim seven youths and seven virgins as tribute for the Minotaur, the terrible bull-human monster that dwelled in an impenetrable maze in the heart of the king's palace. The third time the Minoans arrived, Athenian King Aegeus's son, Theseus, was in the city after a long absence abroad. Theseus boldly joined the sacrificial victims, his objective being to kill the Minotaur. He promised to return in a ship with white sails if he succeeded. The Athenians were imprisoned in the Palace of Minos awaiting their turn to die at the sacrificial altar. King Minos's daughter, Ariadne, had fallen helplessly in love with the handsome young Athenian, smuggled a sword to him, and helped him escape from capitivity. As her lover made his way into the depths of the Cretan labyrinth to confront the Minotaur, she held one end of a ball of wool. A terrible battle between Theseus and the monster ended in the Minotaur's death. Theseus then made his way back to safety by winding up the wool. The two young people hastily set sail for Athens, but, in his excitement, Theseus forgot his promise of white sails. His father saw the approaching black sails and cast himself off a cliff in despair.*

*Everyone should read Mary Renault's fictional account of Theseus and the Minotaur, *The King Must Die*.

EXCAVATIONS AT KNOSSOS

German archaeologist Heinrich Schliemann, of Troy fame, was the first excavator to cast an eye on Crete in the 1880s. Schliemann believed in the literal truth of Homer's *Iliad*. Fresh from triumphant excavations at Mycenae and Troy, he set out to find the elusive King Minos on Crete. He tried to buy a hillside named Knossos near the town of Heraklion, where he suspected King Minos's sprawling palace had once stood (figure 10.1). While Schliemann negotiated in vain, Arthur John Evans was hot on the trail of the ancient Cretans.

Evans was a restless, birdlike man. In his early years, he traveled widely as a journalist in the Balkans, then, as now, a trouble spot. He had an insatiable appetite for political intrigue, which did not endear him to the Austrian authorities. They threw the irrepressible Evans in jail. In 1884, Evans was appointed keeper of the moribund Ashmolean Museum at Oxford University, which was, at that time, little more than a cabinet of exotic curiosities. Evans revitalized the Ashmolean with new collections and fresh displays. He spent time traveling for months on end in Mediterranean lands, where he collected antiquities in large numbers, leaving his assistant to answer all queries at the museum, by saying, "The keeper, sir, is somewhere in Bohemia."

In 1883, Evans heard Schliemann lecture on his excavations at Mycenae on the Greek mainland. Schliemann believed Mycenae, with its gold-laden warrior burials, was the citadel of the Homeric King Agamemnon. Evans disagreed. He felt Mycenae was a Bronze Age settlement, abandoned centuries before Homer. Later, browsing in the collections of antique dealers in Shoe Lane, Athens, Evans became interested in the engraved gems and seals so common in their stores. Evans was blessed with microscopic eyesight, and spent hours peering at the tiny inscriptions on the ancient sealstones (small stones carved with inscriptions that served as seals for various transactions; some may have had religious significance). All of the three- and four-sided stones bore symbols similar to scratchings he had seen on Mycenaean jars, some of them from Egyptian towns on the Nile. He bombarded the dealers with questions about the seals. "Crete," they all said, "they come from Crete." Utilizing his archaeological knowledge and strong perceptive powers, Evans speculated that

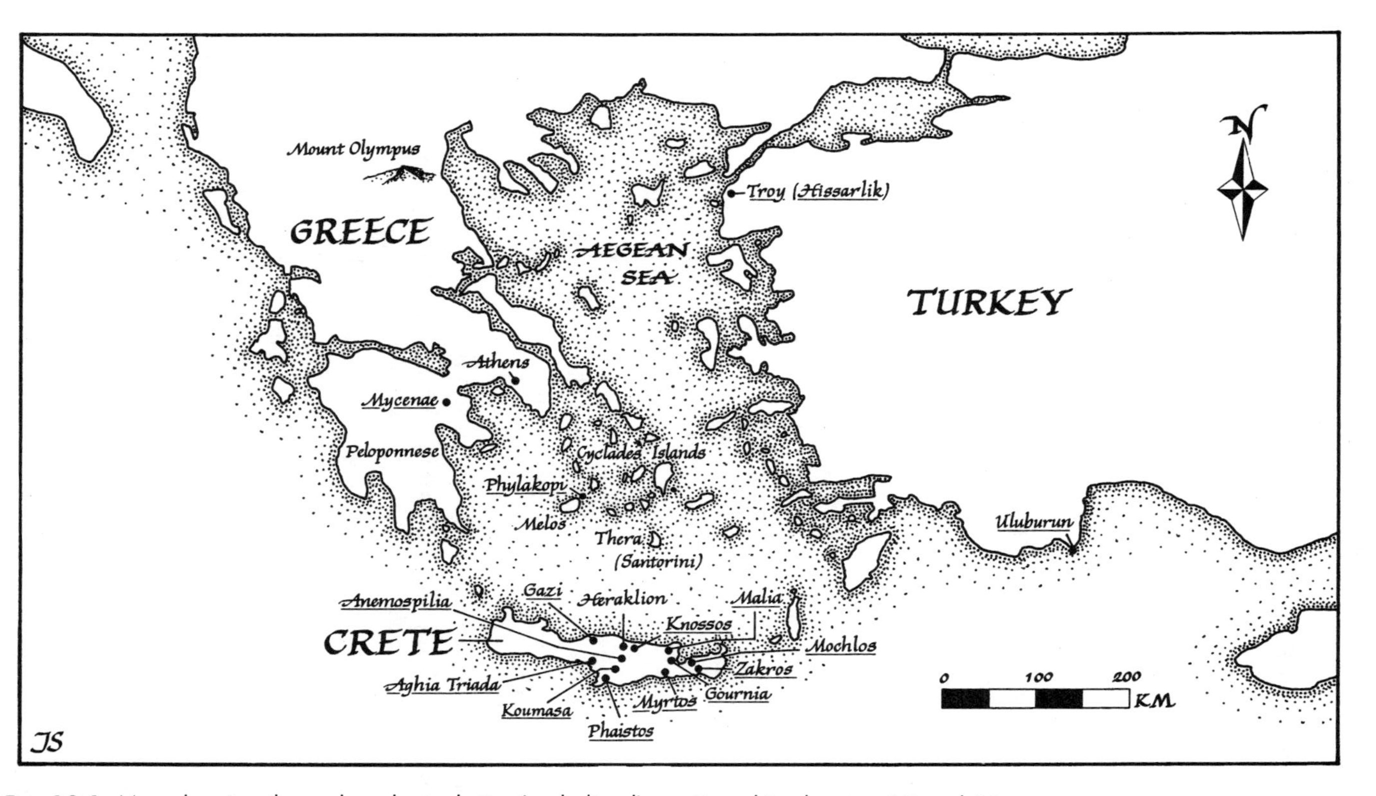

Fig. 10.1. Map showing the archaeological sites (underlined) mentioned in chapters 10 and 11.

perhaps Crete had been a staging post in the gradual spread of writing from Egypt to Mycenae.

In 1894, Evans landed in Crete for the first time, his mind full of seals and complex symbols. The very first day, he purchased twenty-two Cretan gems and a fine Mycenaean seal in a Heraklion market. Three days later, a guide led him to a flower-covered hillside at Knossos, where he found hundreds of potsherds and abundant traces of past human occupation. He returned to England in a rare state of excitement, convinced that Knossos would unlock the secrets of a long-forgotten Cretan civilization. Six years later, he embarked on the excavation of Knossos, the greatest archaeological and intellectual challenge of his career, and one that would span over 30 years.

Arthur Evans had never dug systematically in his life, except for some minor collecting forays. Fortunately, he had the sense to employ experienced assistants, among them architect Thomas Fyfe, and archaeologist Duncan Mackenzie (who had dug at Phylakopi). From the very beginning, Evans realized that Cretan civilization, and Knossos itself, were "an extraordinary phenomenon, nothing Greek, nothing Roman" (Evans 1943, 202).

The very first day of the dig yielded buildings and artifacts. On the second, Evans gazed at a newly unearthed house with faded wall paintings. A hundred men labored to clear additional rooms. Within a few days, the chambers became a labyrinthine palace of courtyards, passages, and storerooms. Dozens of incised clay tablets and gems came from the trenches. When a magnificent wall painting of a male cupbearer "with noble profile" emerged from the soil, Evans announced triumphantly the discovery of the Minoan civilization, named fittingly after the legendary King Minos. However, he refused to link the Minoans to the tale of Theseus, instead contenting himself by remarking: "The old traditions were true. We have before our eyes a wondrous spectacle—the resurgence of a civilization twice as old as that of Greece" (ibid., 206).

Arthur Evans, the epitome of a Victorian scholar and antiquarian—widely read and intellectually sophisticated—was a product of an industrialized nation at the height of its imperial powers. Like his contemporaries, Evans looked at the world through Western eyes. As his workers uncovered the maze-like buildings of Knossos, they cleared a room with painted walls and benches, where a stone throne still stood intact. Evans labeled this the Throne Room and referred to the entire complex as the Palace of Minos at Knossos—and a palace it has remained ever since (figures 10.2, 10.3).

Evans puzzled over the religion and cosmology of the Minoans for decades. His only clues were the paintings and sculptures from his excavations. Goddesses and bulls adorned Knossos's walls. After uncovering a painted stucco relief of a charging bull, Evans wrote: "What a part these creatures play here." In fact, bulls were everywhere—in frescoes, on vases, and on gems. The western entrance to the palace led through a paved courtyard past huge pictures of young men leaping over bulls. The Knossos frescoes moved the great Swedish classicist Martin P. Nilsson to describe Minoan religion as a picture book without a text. How, then, did one write the captions to the illustrations?

In the 1920s and 1930s, Evans and other early researchers had little data to work with. They considered Minoan religion a primitive

Fig. 10.2. General view of the Palace of Minos at Knossos. Photograph from Hirmir Verlag, Munich.

Fig. 10.3. The Knossos Throne Room with frescoes restored by Arthur Evans. Photograph from Hirmir Verlag, Munich.

configuration of classical Greek beliefs. During the past three-quarters of a century, Minoan archaeology has made gigantic strides. Meanwhile, new studies of Greek myth and ritual draw on what are often called "structural" perspectives, the notion that religion is a code, just as all societies have basic "codes" that govern the ways in which their cultures operate (see information in box, p. 227).

At the same time, archaeologists concentrate on meticulous excavation of putative shrines, using the principles of context, time, and space to prove ritual activity on-site without a shadow of scientific doubt. These powerful approaches have shown that both Minoan and Mycenaean religions were coherent, autonomous, and distinct systems, quite different from that of classical Greece.

BEGINNINGS: ANCESTORS AND THE FERTILITY GODDESS

The Minoans and their religion had deep roots in the indigenous farming societies of Crete. Small, self-sufficient communities cultivated fertile river valleys and coastal plains; their survival depended

PROCESSUAL ARCHAEOLOGY, FUNCTIONALISM, AND STRUCTURAL ARCHAEOLOGY

Archaeologists study material remains of past human activity, so, not surprisingly, much archaeological theory has revolved around the evolution of human culture and people's relationships to their environment. Such theory was the crux of the "new" processual archaeology of the 1960s, which stressed the study of the processes of cultural change and of cultures as complex systems interacting with, and being a part of, much larger environmental systems. Functionalism, the notion that a social institution within a society has a function in fulfilling all the needs of a social organism, has been integral to anthropology and archaeology since the nineteenth century. Under this approach, all aspects of society contribute to the whole. Thus functionalism is inextricable from processual archaeology's notions of systems and cultural systems. Processual archaeology has a strong scientific orientation, and is still the dominant theoretical paradigm in the discipline. However, many archaeologists feel that processual archaeology's explicitly scientific approach is too impersonal and de-emphasizes people at the expense of focusing on the processes of cultural change.

Several generations of "postprocessual" archaeology have developed in reaction to this perceived scientific impersonality. These theoretical approaches tend to stress the role of the individual and of groups in decision making, as well as the importance of social organization, religious beliefs, and world view. Structural archaeology is one such postprocessual approach (again derived from anthropology), which attempts to get at the active, social manipulation of symbols as they are perceived by their original owners. Archaeologists recover and observe ancient material culture. Structuralists believe all aspects of this material culture, including artifact styles and settlement patterns, can be understood in the context of a set of rules that perpetuated the culture from one generation to the next. Structural archaeologists attempt to go further than processual scholars. They believe that functional analysis can yield information on the underlying "codes," but that the codes themselves must be explained in terms of the logic behind them. Although the concept is great in theory, few archaeological projects (as opposed to studies of myth and religion) have yet provided convincing accounts of the intangible relationships between the "codes" and social and ecological organization.

---

entirely on good harvests and ample food surpluses. Like all early farming societies, the Cretans measured the passage of time by the rhythms of planting and harvest, birth, growth, and death; and they venerated their ancestors. They also worshipped a female fertility goddess, as did contemporary societies in Turkey, Syria, and Israel. A statuette of a goddess standing on a low bench, clasping a jug, was discovered in a small village shrine in the Phournou Koryphe on the island of Myrtos, dating to 2200 B.C. Nearby lay fragments of vessels used to make offerings of wine and other liquids. A contemporary tomb at Koumasa on Crete contains a similar figurine, this time grasping a snake.

These goddess figures appear at a time of major economic and social change, as many communities turned to the cultivation of olives and vines. Olive oil and wine became staples of Cretan trade within a few centuries. By 2300 B.C., early Minoan towns appeared along the eastern and southern coasts of the island. No one built large temples, but people lavished great care on their ancestors.

The early Minoans commemorated dead individuals, but also paid close attention to communal rituals. They maintained elaborate cemeteries, like that at Mochlos in eastern Crete, where they carved tombs into rocks facing the sea, as if aimed at the watery "otherworld" where the souls of the dead resided. Some of the larger tombs have decorated facades, and large areas outside the sepulchre where considerable numbers of people could congregate. A few of the empty tombs appear as if they were mortuary chapels. Others, containing disarticulated skeletons and numerous skulls, were obviously ossuaries.

For centuries, local people visited these cemeteries to make offerings of exquisite stone cups and bowls, and gold jewelry. These artifacts formed the first of Evans's uncaptioned pictures (see information in box, p. 229). One libation vessel depicts a goddess clasping

MINOAN CULT OBJECTS

Minoan cult objects are distinctive, and relatively easy to identify in the archaeological record. Here are some of the more common cult objects that appear in Minoan sites:

Horn-shaped objects ("horns of consecration"). These objects, which were first identified by Arthur Evans at Knossos, adorned altars or the walls of buildings. The symbolic meaning of the horns is unknown, but the similarity to bulls' horns cannot be a coincidence. When the horns lay on an altar, there was often a bough of a tree or a double ax (see below) placed between them. The presence of horns of consecration, either in an excavated chamber or in Minoan iconography, is a clear indication of a shrine.

Double ax. A symbol of power in Minoan religion, sometimes thought to be a sacrificial ax. Minoan seals often show them being carried by women.

Rhyton (libation vessel). Many forms of vessels were used for pouring liquid offerings, a central part of Minoan ritual. Rhyta assumed many shapes, and were made of various materials, including clay, stone, metal, and faience (glass). Many were jugs or conical vessels with a hole in the bottom for pouring liquids. Some rhyta were animal shaped, in the likeness of bulls, goats, boars, and other creatures. In these vessels, the liquid emerged from the snout.

Altars. Minoan altars came in many forms, and were occasionally made of stone. They often served as the substructure of a throne or platform where a goddess sat. Sacrificial tables with carved legs appear in Minoan art, often with an animal trussed on top. Libation and offering tables were also used in sanctuaries.

Other ritual paraphernalia included stone maces, which were, perhaps, symbols of power; *kemoi,* which were composite vessels that served as stands for offering bowls; as well as everyday conical cups, which were used to make offerings. All of these distinctive religious artifacts serve as signposts for identifying sanctuaries, shrines, and other buildings in which Minoan rituals unfolded.

her perforated breasts, which flowed with bountiful liquid. Goddesses and bulls are commonplace images in tombs of this period. Occasionally, the artifact pictures give us a more complete narrative. For example, in one scene of the clay dioramas of the Kamilari tomb near Phaistos on the south coast of Crete, four sexless figures are shown sitting against the wall of a portico adorned with the horns of consecration. Designated as a sacred place, the circular sepulchre with rectangular annex chambers remained in use for many centuries. Goddess figures sit at tables; two of the tables have cups on them. Two smaller standing male figures hold jugs and face the seated individuals, who may have been deified ancestors receiving offerings. Two other scenes show domestic activities held in enclosures marked by horns of consecration. In one, two women prepare bread, as a man peers into the bakery. The other depicts four men in peaked caps, their interlocked arms inside a circular enclosure, apparently treading out grapes. Both bread making and wine manufacturing were activities that involved the planting of crops and a seasonal regenerative process of harvesting, fermenting, or composting.

Judging from modern analogies, the early Minoans thought of death as a long process of transformation from corpse to skeleton. They buried or exposed the corpse while the flesh decomposed. Months later, the mourners' descendants collected and stored ancestral skulls and upper leg bones in ossuaries, throwing away the rest of the now-disarticulated skeleton. Similar burial rituals occur among the Dande of Africa, and the Cherokee of southeastern North America. While the corpse still has flesh, the soul is thought to flutter about and be potentially harmful to the living. After decomposition has occurred, the soul reaches its final resting place in the otherworld. Now the living can bury their ancestors, for the dead are no longer to be feared. Cremation was another way of releasing the soul. In Homer's *Iliad*, the soul of Patroclus beseeches Achilles to give him the "soothing rites of fire," for only then can his soul pass the "Gates of Hades." The early Minoans made numerous offerings of clay boats to the reburied dead in their seaside cemeteries, symbolizing the final resting place of the soul in the sea.

Thus, in the Minoan civilization, revered ancestors and a nurturing goddess—who protected the dead—nourished the living, and ensured the regeneration of life.

THE MINOAN PALACES

Five centuries passed. Cretan towns grew larger, island society more complex (table 10.1). Some leaders built larger residences, but these were destroyed by a massive earthquake about 1700 B.C. The Minoans then rebuilt their palaces on a grander scale, especially four great structures: Knossos in the north, Phaistos in the south, and Malia and Zakros in the east. Minoan civilization prospered: Cretan ships carried timber and olive oil to Egypt and the eastern Mediterranean, and exotic goods from many lands flowed to palace storerooms. Expert artisans crafted fine clay vessels, exquisite seals, and many luxury items in bronze, faience, ivory, and semiprecious stones. Cult objects from this period are found everywhere.

Although modest, the palace architecture—along with the abundant cult paraphernalia found in the palaces—reveals important standardized features, as if envisaged according to the dictates of some common ideology their rulers adhered to. All the palaces were oriented to the cardinal points of the compass, despite being built in very different terrain. At Phaistos and Zakros, for example, the practical

TABLE 10.1: GENERAL CHRONOLOGY OF MINOAN CIVILIZATION

1200 B.C.	Postpalatial Period
1311 B.C.	Uluburun shipwreck off southern Turkey
1400 B.C.	Third Palace Period Mycenaean entry into Crete
1500 B.C.	Santoríni explosion
1650 to 1450 B.C.	Second Palace Period Middle to late Minoan
1700 B.C.	Destruction of first palaces
2000 B.C.	First Palace Period Middle Minoan
2900 to 2100 B.C.	Early Minoan Prepalatial Period

demands of the topography would have produced somewhat different buildings, had not the architects adhered to the cardinal orientation. Such orientation is common to sacred buildings in many civilizations, such as the Egyptian pyramids, the ziggurat temples of Mesopotamia, and the pyramids and plazas of the ancient Maya.

All Minoan palaces have rectangular, paved central courts, each oriented slightly east of north. Each courtyard served as an organizing focus of the building. The eminent religious scholar Mircea Eliade has stressed the importance of central courtyards in many early societies as an "*imago mundi.*" He writes: "Just as the universe unfolds from a center and stretches out towards the four cardinal points, the village comes into existence around an intersection" (1959, 46). Everywhere, a series of shrines bounded the western side of the courtyard. At the Zakros Palace in eastern Crete, unplundered shrines yielded large numbers of cult objects (see information in box, following). Greek archaeologist Nanno Marinatos believes that the courtyards and adjacent shrines were the very nuclei of Minoan palaces, serving as symbolic centers for sacred activities.

THE GREAT SHRINE AT ZAKROS

Zakros was discovered as long ago as 1852, and excavated by British archaeologist David Hogarth in 1901. He uncovered a dozen buildings and part of a harbor community before torrential rains and floods prevented further excavation. Local archaeologist Nicholas Platon, then-director of the Heraklion Museum, returned to the site in 1962. He soon uncovered a hitherto unknown palace, undisturbed since the heyday of Minoan civilization. Apart from full storerooms and royal quarters, the palace yielded a central shrine of eleven rooms connected by a maze of narrow corridors. The rooms held a great array of ritual objects, including twelve finely-decorated, four-handled amphorae embellished with plant motifs; three offering tables; six bronze ingots; and three elephant tusks, probably from Syrian animals. To date, the copper ingots have not been sourced, but nearby Cyprus, a major copper supplier of the day, is a strong possibility.

The shrine was small, with a ledge at the back and a low bench with mud-brick support for a single worshiper across from the podium. A dozen clay rhytons, all with pointed bases and painted with interlinked spirals, lay next to the shrine, around the bench, and along the east wall of the room, and were perhaps used for libations. A stairway of eight well-preserved steps next to the shrine led to a basin for purification; a ritual vessel, also possibly used in purification rites, lay in pieces in the fill covering the basin. This masterpiece of Minoan artistry was made of veined marble, with two large, boldly curved, S-shaped handles attached to rim and shoulders. The artist made skillful use of the veins in the marble to produce a marvelous ornamental effect.

A windowless treasury approached by a single narrow passage lay behind the shrine. When the earth fill was removed, the excavators uncovered stone and clay vessels lying in clay partitions, forming eight contiguous cists along the walls, each less than a meter high. The only known intact Minoan treasury, it contained magnificent stone libation rhytons and ritual jugs, sixteen of them in white marble. One superb specimen of elongated shape and pointed base was carved from rock crystal. The artist fashioned the curved handle out of crystal beads, with a curved bronze wire running through them. Other ritual objects included stone mace heads and two large double axes fashioned from a thick bronze sheet. One was decorated with dense vegetation-like patterns of leaves and flowers. Wooden boxes once held small faience, ivory, and crystal objects. A nearby archive room contained thirteen Linear A clay tablets, probably inventories. The remaining temple records were crushed and burned when the palace was destroyed.

The Zakros Shrine reveals the magnificent range of objects associated with major Minoan religious centers. When fully published, the excavations will add much to our knowledge of Minoan ritual and religious beliefs.

Each palace also included a large, paved court adjacent to its western facade. These courts were open to the nearby town, as if they formed a boundary between the unfortified palace and the settlement. They could accommodate large public gatherings, and, thus, serve as places where the general populace could interact with the privileged people who lived in the palaces. Large, circular underground granaries lie in these courts. Raised pathways, perhaps for ceremonial processions, traverse the open precincts at several palaces. Marinatos hypothesizes that the granaries and processional ways were linked, perhaps in the form of an agricultural festival. Well-preserved stone platforms with steps leading up to them may have served as settings for leaders to appear in public. Interestingly, the processional ways link the external platforms with the entrances of Knossos and Phaistos. These elaborate courts on the western sides of the palaces may have been ceremonial meeting places, where the rulers and the ruled met. Obviously important ceremonies unfolded there. The west points in the direction of the setting sun, the "other realm" of death in many cultures, among them the ancient Egyptians. Perhaps ceremonies in these courts linked life and death, storage, planting, and regeneration. These paved courts represent a classic example of how structures in the archaeological record document the religious and ceremonial nature of an architectural complex.

The Minoan palaces were composed of private quarters, storage rooms, courtyards, and small shrines. Designed for a small number of participants, these palaces were cult centers. Most ceremonies took place in open spaces such as the courts outside the western facade. Seals and rings, in themselves symbols of wealth, depict a variety of important ceremonies (figure 10.4). By 1650 B.C., Minoan civilization was in its heyday. The shrines became more specialized. Some were located on palace balustrades, and on raised platforms where people could make formal appearances; whereas others were isolated as private dining chambers, where privileged participants ate ritually prepared food during major festivals. Thus, the Minoan palaces served many secular purposes; their courtyards and shrines must have satisfied a public, as well as a religious, need.

Palace Murals

The standardized architecture of the palaces provided a consistent setting; however, it tells us little about the religious activities that took place within their walls. Gifted muralists, the Minoans

Fig. 10.4. Minoan seal from Knossos showing a goddess with flounced skirt atop a mountaintop flanked by lions. An adorant stands to the right, horns of consecration to the left. Courtesy Oxford University Press.

adorned their public buildings and homes with lively scenes of animals, people, and ceremonial life. The brightly colored painted facades at Knossos—adorned with sacred bulls' horns—depict lively scenes. Arthur Evans selected friezes to be reconstructed under his supervision. His artists completed the paintings with considerable artistic license. Since his day, a larger corpus of Minoan wall paintings have placed Evans's somewhat romanticized and often imaginative depictions into better perspective.

Arthur Evans and his contemporaries thought of the murals as decorative work—art for art's sake. In other western Asian civilizations, palace art had a political or religious message. The Egyptians developed a stylized culture with its own art tradition that glorified divine rulers and deities. No such "ruler iconography" appears on Crete. No kings strut in their glorified roles as great conquerors or divine leaders, as is the case with Assyrian monarchs or Mayan lords of the Mexican rainforest. In dramatic contrast, Minoan palace art clusters around rituals, festivals, and symbolic scenes from nature. For example, a corridor from the west wing of the Knossos Palace bears frescoes depicting a procession of life-sized, long-haired men wearing necklaces and short kilts. They move along one of the walkways, one arm bent, the fist clasped to the chest. Women with raised hands, perhaps carrying vessels, hair resting on their shoulders, walk in formal procession in the foreground. Other figures wear long robes or hide shirts, perhaps denoting different statuses and age groups in the procession. A central female figure appears in each segment of the parade, as if the marchers are bringing offerings to a goddess for the palace storerooms (figure 10.5).

The Knossos Throne Room is reached via the central court through a small anteroom. The stone throne forms the focal point of the room; palm trees and two antithetical griffins flank the seat. The griffin is a symbolic monster with an iconography that originated in western Asia. It combines the attributes of the lion and eagle, and is both a guardian of divinities and an aggressor. In Minoan art, only griffins appear with female deities; indeed they invariably guard the goddesses. Sometimes the goddess is depicted sitting under a sacred palm. Elsewhere in the Knossos Palace, wall painting fragments testify to the important role that women played in the Minoan cult. Men sit on stools facing one another drinking from chalices, a toasting ceremony carried out in the presence of much larger female impersonators of the goddess. A multitude of heads represent the large crowds that attended the festivals. Unfortunately the focal point of the procession is missing. But Nanno Marinatos believes that the women were the protagonists in the ceremony. She identified richly dressed noblewomen of different ages sitting on the stands, supervising the cult activity. The elaborate facades of the surrounding buildings with their horns of consecration appear in the background.

The Grand Staircase of the Knossos Palace has friezes of men in procession carrying offerings. Eight-shaped shield emblems, symbols of ceremonial hunting, adorn the stairwell. Arthur Evans excavated the collapsed upper story, wherein he recovered fragmentary ivory figurines of male acrobats—bull leapers—including a plaque depicting two acrobats in midair and on the neck of a large bull. Just to the east he discovered the famous "bull-leaper" frescoes, a frieze of

Fig. 10.5. Arthur Evans's reconstruction of the Corridor of the Processions at Knossos. Courtesy Oxford University Press.

Fig. 10.6. Arthur Evans's reconstruction of the bull-leaping frieze at Knossos. Courtesy Oxford University Press.

panels once set above dadoes (figure 10.6). Each of the four panels shows acrobats jumping over bulls, landing behind them, or grappling with their horns. Bulls played an important role in the iconography of Knossos's east wing. Evans himself believed that the domestic quarters of the palace were housed in this wing. Nevertheless, ubiquitous bull motifs link this section with ancient ritual associations.

The frescoes at all four of the Knossos Palace entrances present a common theme: that of large bulls' heads and charging beasts. The largest charging bull faces north, and would confront anyone entering the palace. Each animal appears to portray a message of power, of threat.

PEAKS AND CAVERNS

The royal palaces, like Mayan ceremonial centers, were stages for the major ceremonies of the Minoan year. Nevertheless, religious activity spread far beyond the palace walls, high into the surrounding mountains. Every Minoan, whether noble or commoner, had spiritual ties with the natural environment as important as those between villages and palaces. The Minoans revered the deities associated with high peaks, springs, caves, and many other natural features of the dramatic island landscape. They built shrines thousands of meters above sea level, sometimes in caves. One peak sanctuary, high above the village of Kato Syme in southern Crete, is a complex of at least eleven rooms, with stone libation tables, and stands to receive offerings, including dedicated bronze figurines and seals. This shrine of built rooms, set in a remote mountain location, linked religious activities that occurred within a human-made world to the natural world, and makes an argument for a unified Minoan religion. People made pilgrimages to these caves and mountaintop sanctuaries, where they deposited votive offerings—including adorant figurines with clenched hands or crossed arms, expensive jewelry, and day-to-day offerings of food—as a permanent commemoration of their presence. The offerings were displayed on altars and in special chambers and remained there until major festivals and sacrifices took place, at which time they were removed and thrown into fissures as gestures of renewal.

These natural shrines were where elite and commoner met for the commemoration of plentiful harvests, or at times of crisis. Such sanctuary cults, with deep roots in an earlier farming world, unified a scattered and socially ranked populace.

WHO WERE THE DEITIES?

After a century of excavation, we now know more in regard to Minoan shrines, palace architecture, and the images of their religion. We know that their beliefs originated in much earlier farming societies, and that the Minoans honored both gods and goddesses. However, the identities of their deities remain a mystery.

The Minoan goddess first appeared on clay vessels like those from the early Phournou Koryphe shrine. On these vessels, she is depicted as a mother holding a baby, sometimes with perforated breasts, which flow with milk. She sustained life, and symbolized fertility. Although the image of the goddess appeared elsewhere, including several murals and tiny scenes depicted on jewelry and personal seals, the nature and personality of the goddess varied among towns and palaces. On some artifacts, she sits or stands in her shrine or in the countryside surrounded by plants; animals and humans worship her. On others, the goddess supervises women gathering flowers, amidst a landscape bursting with animal and plant life. In an important fresco from the island of Santoríni, 113 kilometers north of Crete, she sits on a platform guarded by a monkey and a griffin, receiving a gift of fragrant saffron from a young girl. Sometimes she rides a griffin or lion, but most often they flank her as animal guardians. Her bodyguards are a symmetry of the most powerful animals in the animal kingdom. On some seals, the goddess wears a crown made of the horns of sacrificed animals, the actual headgear worn by priestesses who *impersonated* the goddess.

The animal kingdom symbolizes the land (lions and monkeys), the sea (dolphins and other fish), and the sky (eagles, doves, and other birds). Snakes had associations with the subterranean realm. Two famous faience figurines from Knossos show bare-breasted priestesses handling snakes without fear, the serpents entwined around their arms (figure 10.7). Terrifying animals like snakes emerged from the underworld in springtime, the time of renewal.

Seals and other scenes show the goddess sitting bare-breasted in a flounced skirt, under a tree, on a rock, or occasionally on a human-made platform—all markers of a sacred place. Only priestesses and animals attend her. Her presence among plants and animals indicates that she was a deity of nature, of fertility, who assumed many identities. Her nurturing essence was the most important perception in Minoan belief.

Male gods appearing at Knossos, and on Minoan artifacts, are strong, vigorous personalities. They are at home in both urban and wilderness settings. Unlike ancient creators in West Asian religions, they are rarely bearded, nor are they martial divines. The Minoan god appears as a Master of Animals, represented by a deity holding two animals in a submissive position. This commonplace iconography appears in the eastern Mediterranean as early as 2500 B.C. The Master of Animals has short, curly hair, and a broad chest. He wears a codpiece and a large belt. Beasts like bulls, wild goats, lions, and

Fig. 10.7. A Minoan priestess with bare breasts and flounced skirt, holding two serpents. Photograph from Hirmir Verlag Munich.

lion-griffins attend him. He holds lions by the tail, as they attack wild goats. The lion may be the king of animals, but the god controls all beasts. On some artifacts, the animals worship the Master of Animals and bring offerings. On others, the god is represented as a hunter with a spear and shield or a bow. Unlike the goddess, who fed and tended wild animals, the god hunted beasts and corralled them. He controlled nature, he did not feed it.

A ring found at Knossos commemorates a male votary greeting the god atop a mountain peak. Another gold ring from an Oxford University collection shows a priestess with the deity hovering next to a tree shrine. When the god and goddess appear together, the Minoan goddess is larger than the male. Her prominence in Minoan religion stresses the continuation and supremacy of women in the chain of life.

EPIPHANIES

Minute scenes on Minoan seals depict numerous divine epiphanies, the manifestations of deities to humans. On one seal, for example, a young god hovers protectively over a town in an epiphanic appearance immediately above his shrine. Epiphanies came to people in remote isolation, in a crowded cult chamber, or in the midst of a wild storm, as moments of deep, and sometimes ecstatic, revelation. These epiphanies were of profound importance to Minoan religion.

Only a few Minoans experienced an epiphany, perhaps, as in other cultures, through fasting, seclusion, or even the ingesting of hallucinogenic drugs. Greek scholar Nanno Marinatos and other archaeologists believe that those who had such visions wore golden rings showing such epiphanies as a mark of distinction.

New studies of architecture and palace murals place Knossos's Throne Room as a formal audience chamber, and as a setting for such epiphanies. A select audience would sit on the benches opposite the throne. The elaborately costumed priestess, the goddess impersonator, would appear bejeweled through the door flanked by painted griffins. She would sit in state on the throne as the goddess, surrounded by her mural of powerful guardians. The Throne Room reinforced the belief that the goddess dwelled in the palace, the residence of priestly families. Their close relationship with the goddess formed the basis of their political, economic, and spiritual power. Waiting in the courtyard outside, a crowd would participate indirectly in the ritual moment with cymbals, drums, and chants. Today, only the architecture, the murals, and the silent throne remain.

DECODING A SARCOPHAGUS

Decoding the Minoan cult of the dead is an extremely difficult task. The only clues come from artifacts, and from occasional fragments of iconographical evidence, like the painted scenes on a limestone sarcophagus dating to about 1400 B.C., which was found inside a destroyed tomb at Aghia Triadha in southern Crete. The paintings show ancient offering rituals, which probably originated centuries before Minoan civilization. Each of the four panels on the sarcophagus represents two or three ceremonies. One half of one side depicts a dead person—a small man standing in front of an altar and a tree before a tomb (figure 10.8). Three males dressed in hide skirts bring offerings of a boat and two calves. The other half shows a man in a long robe playing a lyre. Two women carry and pour offerings into a bucket, perhaps containing holes, so that the liquid can soak

Fig. 10.8. The Aghia Triadha sarcophagus. A man stands before an altar, as three other men bring two calves and a boat as offerings. A priest with lyre and two priestesses make offerings at another altar to the left. Drawing from the University of North Carolina Press.

into the earth. The opposite side of the sarcophagus exhibits a shrine topped by four sets of sacred horns with a small olive tree between them. A pole with a double ax, an altar, and offerings stands before the sacred place. A woman in an animal-hide skirt stretches her hands, palms downward, above a bowl on the altar. Farther along the same side, the blood of a sacrificial bull which is tied onto a table flows into a vessel on the ground. Two goats await their turn for sacrifice, while a man plays a flute. The officiating priestess, wearing a long robe and plumed headdress, stretches out her hands, palms down. A procession of women appears in the background.

Two major ritual themes of Minoan religion appear on the Aghia Triadha sarcophagus. One side displays rituals of death, with the pouring of libations into the soil. The other displays rituals of renewal, with a young tree on the shrine. On one end, two males carrying offerings and two females in a chariot pulled by wild goats head toward the dead man. The regeneration ritual by the tree shrine invokes the goddess, who rides with a priestess in a chariot pulled by two griffins toward the regeneration scene on the opposite end.

RENEWAL AND HUMAN SACRIFICE

The Aghia Triadha sarcophagus shows formal ceremonies and animal sacrifice, with offerings of flowers and grain—of the public side of Minoan religion. But archaeology has revealed a darker element, practiced in discreet secret, perhaps in times of crisis. Arthur Evans always depicted the Minoans as a carefree, peaceful people living in a colorful island world. The image persists, for they were indeed artists of talent and traders of genius. Nevertheless, archaeology has come a long way since Evans uncovered the central precincts of Knossos.

In 1975, another British archaeologist, Peter Warren of Bristol University, excavated a small building dating to about 1450 B.C., which once faced a processional road 350 meters west of the Palace of Knossos. The largest of its basement rooms contained a diagonally placed bench and a white limestone block surrounded by jewelry and a sealstone against the east wall. A collection of lavishly painted libation vessels once kept in a large storage jar had fallen into this basement from a collapsed upper story. A smaller room across a narrow passage contained the scattered bones of four children between 8 and 12 years old. At least 22 percent of the fragments bore the telltale knife marks that resulted from the defleshing of bones. A few more children's bones were found in the original storage room, one of them a vertebra bearing a knife cut pathologists associate with slitting of the throat. Warren believes that these remains were those of children who had been killed, then butchered for cooking—the remains of sacrifices aimed at averting a great disaster, such as a series of earthquakes or other natural cataclysms. Perhaps, the act of cooking and eating human flesh from victims offered to the gods was a special act of communion with the divinity.

In 1979, Greek archaeologists Yannis Sakellarakis and Efi Sapouna-Sakellarakis unearthed additional, but controversial evidence for cannibalism in the highlands. Painted Minoan potsherds and a limestone carving resembling bull's horns came to light at the junction of some ancient roadways in an area called Anemospilia ("caves of the wind"), located 32 kilometers inland of Knossos. The two scholars dug into the nearby hillside and unearthed a ruined, rock-walled shrine with three chambers and a front cross corridor looking north over the coastal plain below. The Anemospilia Shrine stood inside a walled, sacred enclosure. Each room of the shrine opened into the cross corridor that extended the width of the building. Rows of vessels that once contained sacrificial offerings, including charred fruit seeds, stood in the passage and dated the shrine to about 1700 B.C., just before the heyday of Minoan palace civilization.

A violent earthquake had overthrown the shrine, collapsing the roof and walls into the interior without warning. The skeleton of a woman lay in the cross corridor, apparently a victim of the sudden disaster. When the excavators dug into the central chamber, they found burned wood fragments and a pair of clay feet that had originally stood on a raised platform, the remains of a life-size wooden statue once adorned with gold ornaments and ceremonial raiment. A large fragment of hillside rock lay at its feet, conceivably a symbol of the earth, one of the three eternal elements of Minoan life, along with sea and sky. The earth was regularly nourished with blood.

The room to the east contained offering vessels arranged on, and before, a stepped altar that was just like one depicted on a rhyton (cup) from the palace at Zakros. Exact duplicates of sacrificial vessels at Anemospilia also appear in one scene on the Aghia Triadha sarcophagus.

The west room yielded little at first, until the excavators reached the original floor level. There they uncovered two human skeletons lying on the floor, clearly victims of the earthquake who were killed by falling debris. A third human skeleton lay on a platform altar like that depicted by Minoan artists. A pillar with a trough at its base once stood close by, used to collect blood as it dripped from the sacrificial altar. An 18-year-old man lay on his right side on the altar, his legs bound together so that the heels almost touched his thigh. Remarkably, the bones on the left side, which faced uppermost, were white, while those on the right side were black. According to Greek biological anthropologist and physician Alexandros Contopoulos, the bones of burned bodies with their blood supply intact will turn black, while the skeletons of corpses drained of their blood will remain white. He believes that the Anemospilia body was at least partially drained of its blood supply before the earthquake caved in the building and fire ensued. Perhaps the young man's left carotid artery was severed by someone (most likely the priest) as he lay bound on the altar. A still-razor-sharp, 41-centimeter-long bronze sacrificial knife was found by the body. The blade bears a depiction of a mythical beast, with the slanted eyes of a fox, ears shaped like butterfly wings, and the snout and tusks of a wild boar.

A woman about 28 years old died with her legs splayed out, some distance from the altar. Her companion, a powerfully built man in his late thirties standing a full 1.8 meters tall, fell on his back with his hands raised to protect his face, as if to ward off falling roof timbers. He was no ordinary individual, for he wore an engraved seal decorated with a depiction of a man poling a boat, perhaps a priest's talisman, tied to his wrist. The little finger of his left hand bore a ring of silver and iron, the latter a very rare commodity at the time. Criminologist Antonios Koutselinis theorizes that the man cut the victim's carotid artery, then laid down the sacrificial knife so he could collect the blood gushing from the wound. The earthquake struck at that crucial moment, preserving the sacrifice under a pile of collapsed debris.

Did the Anemospilia excavations actually reveal a moment of sacrifice interrupted by a sudden earthquake? Some Minoan experts refuse to believe the Cretans engaged in human sacrifice. They challenge the excavators' interpretation of the crouched skeleton on the altar and wonder if he may also have been a victim of the sudden earth tremor. Other scholars question whether or not the unique Anemospilia Shrine was, in fact, a temple, pointing to decades of research that shows much Minoan ritual focused on the palaces and in mountaintop sanctuaries. However, an accurate depiction of such a structure comes down to us on a sacrificial rhyton excavated by archaeologist Nicholas Platon at Zakros. The tripartite layout of the Amenospilia Shrine and its sacred enclosure is identical to the architecture characterized on the rhyton.

The theory that the Anemospilia Shrine was a sacrificial temple is more compelling when viewed alongside the evidence for sacrifice, butchering, and ritual cannibalism excavated by Peter Warren at Knossos itself, again in an earthquake-felled building. We should also remember the bull-dancer friezes on the walls of the great palace. Youths take part in a dangerous bull game, arguably a form of sacrificial ritual somewhat akin to the ceremonial combats to the death which were part of later Etruscan burial rituals. These, in turn, inspired Roman gladiatorial contests. Perhaps the Minoans staged this ultimate of blood sacrifices as a last resort, at moments of crisis when the earth shook repeatedly. A dull sound would resonate from the ground like the muffled bellowing of the bull beneath the earth as he tossed the Minoan world with his horns. The only recourse for mere humans was the gift of human blood to feed the restless, bull-like denizen of the other world.

THE END OF MINOAN CIVILIZATION

The Minoan civilization was at its height in the centuries before 1500 B.C. when the volcano on the nearby island of Santoríni blew up, causing massive destruction throughout the Aegean (see information in box, p. 248). The Santoríni catastrophe rivaled the Krakatoa explosion in southeast Asia in 1883, which killed 38,000 people along the Malayan coast alone. Minoan civilization staggered under the blow, but survived for another half century, only to be overthrown about 1450 B.C., perhaps by Mycenaean invaders from the Greek mainland. Knossos itself endured until 1375 B.C., perhaps as

DATING THE SANTORINI ERUPTION

The Santoríni eruption offers a fascinating example of the difficulties of dating ancient climatic and cultural events. The eruptions devastated much of the Aegean, deposited volcanic ash over thousands of acres of valuable agricultural land, and caused huge tidal waves that may have damaged Minoan harbors. The date of the eruption is controversial. The buried Akrotiri village and its frescoes are dated to about 1500 B.C., from the late Minoan pottery found in the houses, which, in turn, is dated by association with earlier finds of known age discovered in Egyptian buildings along the Nile. Unfortunately, the few radiocarbon dates from the settlement come from imprecise, perhaps contaminated, contexts.

Large volcanic events emit enormous quantities of ash, which circulate in the upper atmosphere, block the sun's rays, and cause a form of "nuclear winter," which should be visible in ancient climatic records such as peat bogs and tree rings. New tree-ring data from California and Ireland have revealed a period of markedly narrower tree rings beginning in 1628 B.C., indicative of a period of cooler climate. Sulphur peaks discovered in Greenland ice cores of the mid-seventeenth century B.C. may also have resulted from a large volcanic eruption. For years, archaeologists have accepted the 1450 to 1500 B.C. date for the Santoríni eruption, arguing that the great cataclysm hastened the fall of Minoan civilization. If the tree-ring evidence is correct, then the event may have occurred some 150 years earlier, which would lengthen the Aegean late Bronze Age by a century and a half. The search for a definitive scientific answer continues. In the meantime, I have used the widely accepted 1500 B.C. date in this chapter.

the headquarters of a Mycenaean overlord of the island, for many Mycenaean artifacts and Mycenaean Linear B script have been found in the palace. Nevertheless, the ancient beliefs in death and regeneration, in a fertility goddess and a young god who was a hunter and Master of Animals, survived for many centuries, until the threshold of classical times. The imagery of these deities was part of a much larger ritual agenda. The Minoans substituted god/goddess impersonation ceremonies for large statues of their divines. Minoan rulers had such an investment in portraying themselves in divine form that the imagery of rulers and deities became fused (figure 10.9).

Cretan kings did not possess absolute power like Egyptian pharaohs or Near Eastern monarchs. They presided over a mountainous land, where ties of kin still held communities together in a fractious world of earth tremors and factional disputes. Instead of ruling by decree, they transformed ancient village beliefs and rituals into a subtle fusion of political power, personal prestige, and compelling religious office.

Fig.10.9. Arthur Evans's reconstruction of a priest-ruler at Knossos. Courtesy Oxford University Press.

chapter eleven

A SHRINE AT PHYLAKOPI

Phylakopi is a humble island sanctuary in an ancient, remote Aegean town. The shrine came into use about 1390 B.C., two centuries after Crete's Minoan civilization collapsed. The town lies on the north coast of Melos, at the southwestern corner of the Cyclades (figure 10.1). I sailed past Melos's cliffs on a fine July evening, after a hectic passage from the mainland and before a boisterous *meltemi,* the northwest wind of summer. The volcanic cliffs shone pink in the setting sun as we sighted the white houses of the old town atop the island. We slipped between Melos and nearby Kimlos and anchored in Apollonia Bay, sheltered from the dying wind, but still several kilometers from the jumbled boulders of the ancient stone town.

A deep sense of history enveloped me as I walked the rugged coastline. Mainlanders had paddled across to Melos in search of obsidian more than 8,000 years ago. Cycladic villages, Minoans, then Mycenaeans had settled here. Melos had sided with Sparta in the Peloponnesian War, and Thucydides tells how the Athenians besieged the island. After a few months, the inhabitants surrendered, and the conquerors slaughtered the women and children.

From 1896 to 1899 British archaeologist Duncan Mackenzie and the British School of Archaeology at Athens first excavated the small town of Phylakopi, which was inhabited by farmers, traders, and fisherfolk over 3,000 years ago. Mackenzie, an excellent field-worker by the standards of his day, published his research in full. He went on to spend more than 30 years as supervisor of Evans's Knossos excavations (see chapter 10). Outside of a brief season in 1911, no one had worked at Phylakopi for 63 years.

In 1974, Colin Renfrew and the British School arrived, armed with Mackenzie's report and notebooks on his earlier work, which proved to be an excellent guide. Renfrew had trained at Cambridge University, and began his career in archaeology during the 1960s, at a time of great intellectual ferment. Both American and British archaeologists at that time were rebelling against the narrow concentration of earlier archaeology, which focused on "culture history"; the chronological ordering of ancient cultures in time and space based on artifacts found in stratified excavations. A "new" archaeology was emerging, one that was consciously scientific, and concerned not with mere description, but with explaining and interpreting the major developments of the past. The "new" archaeologists hypothesized and made conclusions about every area of human experience. They believed all aspects of the past were inherently reconstructible—even religious beliefs.

Renfrew embraced many tenets of this new approach. Using general systems theory to reconstruct Aegean society around 2000 B.C., he developed a theoretical model that placed equal emphasis on all aspects of island life. He also studied early Cycladic farming villages and their remarkable religious figurines (see information in box, p. 253). Renfrew's Phylakopi research was an extension of his earlier, late 1960s, work on Aegean society of 4,000 years ago. His original intention was to obtain a detailed stratigraphic sequence for Phylakopi, especially for the late Bronze Age levels, which had not been a focus of the Mackenzie excavations. He believed that the best-preserved later horizons lay under unexcavated hillslope deposits at the south end of the town. Mackenzie himself had remarked that this area would probably afford exceptional chances of interesting discoveries in the later strata" (as cited in Renfrew 1985, 6). The initial few weeks of work proved that Mackenzie's assessment was correct.

In the 1974 season, two trenches were opened west of Mackenzie's original stratigraphic profiles. These new excavations were anchored to the old ones. Renfrew's diggers uncovered a jumble of boulders overlying undisturbed stone walls. One trench

CYCLADIC FIGURINES: A COMPLEX ETHICAL ISSUE

During his earlier research, Colin Renfrew studied Stone Age (3300 to 3200 B.C.) farming villages of the Aegean Islands and their ritual figurines. These Cycladic figurines offer a classic example of how vital information about religious beliefs is lost when archaeological contexts are destroyed.

A sparse population of farmers and seafarers, the people of the rocky Cycladic Islands clustered in villages near sheltered bays in the heart of the Aegean. They grew wheat, barley, olives, and vines. Their ships sailed between islands and the Greek mainland, carrying wine, olive oil, obsidian, and marble. The harsh environment made living arduous. At a time when the Egyptians experimented with pyramid building, Cycladic artists created masterpieces in clay and marble: cups, vases, and, above all, figurines (figure 11.1).

Cycladic figurines exude a quiet confidence, an artistic tranquillity. They are timeless in their simplicity: The figurines are often tall and slender, with legs that are always together. The arms are folded across the stomach—right arm always under the left, with fingers clasping the sides. The featureless heads, save for a prominent nose and sometimes a hint of eyes and mouth, rest on long necks. Paint traces tell us that some figurines had painted ears and hair. All the figurines stand on tiptoe, as if leaning against a wall. The female figurines have small, bud-like breasts, which are carefully delineated from the simple torso.

What symbolism lay behind this consistent convention? Were the figurines buried with the dead, or used by the living? Were they ancestors or deities? Although deeply compelling, these marble figurines from the Cyclades Islands, with their strict stylistic conventions, remain one of the great mysteries of archaeology, Sadly, Cycladic marbles command high prices on the art market. Rapacious looters have torn dozens of figurines from their context in village and tomb, leaving archaeologists to face an all-too-familiar conundrum of having to place isolated finds into specific cultural context.

The study of the Cycladic figurines raises an interesting question of scientific ethics, to which there are no easy answers. Like magnificently painted pots from Moche graves in coastal Peru, the Cycladic figurines come from private collections and looted sites. Wrenched from their contexts in both time and space, they are unable to tell us anything about the beliefs and rituals of the people—religious traditions that were to culminate in the Bronze Age Minoans and the Mycenaeans of Crete and the Greek mainland. The only cultural information that can be obtained is from the study of dozens of individual specimens owned by private collectors.

Are archaeologists justified in working closely with collectors, who have paid high prices for fine antiquities, thereby fueling the flames of a red-hot marketplace for prized artifacts like figurines or painted pots? No, argue many scholars, who believe that such collaboration gives collecting and private ownership of the past an unintended scientific validity. They push for a code of professional ethics that outlaws all collecting and any form of research involving private collections. Others strongly disagree, including respected scholars like Colin Renfrew, who completed the definitive study of Cycladic figurines, and Andeanist Christopher Donnan, who has spent his career assembling a unique archive of all known Moche-painted funerary pots. It is better to work with the "enemy," they believe, to ensure that some track is kept of as many surviving artifacts as possible. Access to private collections gives them the chance to wrest as much information as possible about the artifacts from specimens that are otherwise inaccessible to the scientific community, as well as to educate collectors about archaeology and science. The results of their research are often deeply frustrating, especially in the case of Cycladic figurines, which have never been found in an undisturbed archaeological context. However, Donnan's archive of Moche ceramics paid triumphant scientific dividends, when Peruvian archaeologist Walter Alva excavated the royal graves at Sipán in the Lambayeque Valley. The friezes recorded on pots and records in the archive enabled Alva and Donnan to identify the Sipán lords as Warrior Priests by their regalia painted on funerary vessels and duplicated in their sepulchres.

Such detective work in collaboration with private collectors will, of necessity, become a major part of twenty-first century archaeology as the archaeological record vanishes in the face of industrial activity and tomb robbing—whether scientists like it or not. In the final analysis, our ultimate responsibility is to preserve knowledge about the past using all means at our disposal.

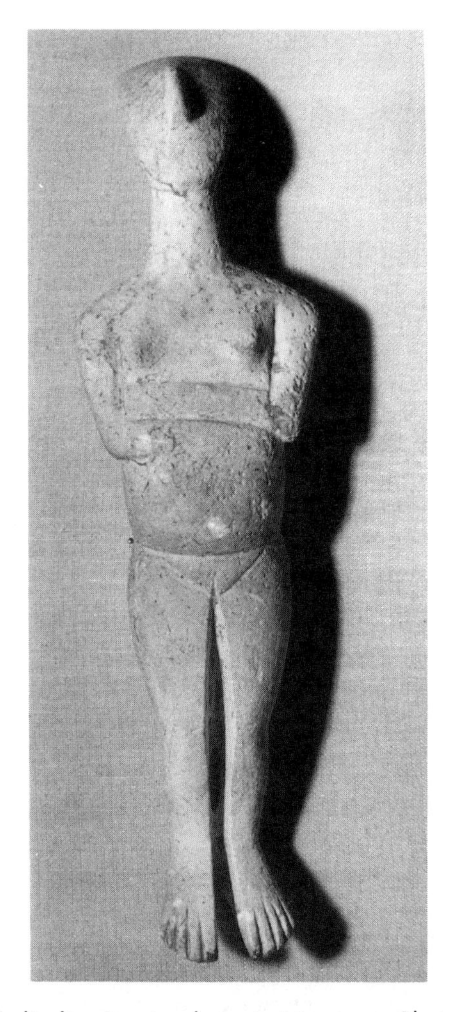

Fig. 11.1. Cycladic figurine in characteristic pose. Photograph from the Metropolitan Museum of Fine Art, New York.

yielded fragments of animal and human terra-cotta figures, some ostrich eggshells, a conch shell, and other items of considerable value, such as sealstones. The latter were not utilitarian objects; more likely, they served as special offerings. Naturally, day-to-day perishable items such as food and drink or sacrificial animals did not survive for the archaeological record. However, on the last day of the season, a wooden human head covered with sheet gold was discovered in one of the trenches. Renfrew realized he was digging into a potential sanctuary.

During his investigation over the next 3 years, Renfrew sensed that he was often treading upon new scientific ground. Most archaeologists at that time believed material remains could never reveal anything substantive about ancient religious beliefs. However, a small but growing faction argued that scientific observation and reasonable deduction could offer valid reconstructions of the "intangible." Renfrew was among this group. Thus, instead of asking whether or not a particular artifact had ritual significance, he asked what *information* a particular ritual structure or object was designed to transmit—what function or purpose it served. The answers could come only from precise "archaeological contexts," that is, from archaeological data, excavated within a consistent, scientifically rigorous framework. Renfrew believed that a portrait of ancient religious beliefs could be developed only from information gleaned from an entire "suite of archaeological data." The cumulative effects and positioning of this data could elaborate, explain, or elucidate a theory (but, of course, could never conclusively say what happened).

CHRONOLOGY OF THE TOWN AND SANCTUARY

The town of Phyakopi was home to an estimated 1,400 to 2,100 people during the late Bronze Age. The settlement was a maze of stone walls, small buildings, and chambers; and the townspeople had rebuilt and expanded their dwellings constantly, creating stratigraphic problems for Renfrew, in every trench he dug. To make sense of the settlement as a whole, Renfrew erected a 10-meter square grid over the entire area, then adopted a checkerboard form of excavation, which gave him horizontal and vertical control over each trench. Grid excavations work extremely well on sites in which buildings are involved. By using a dimensional system of excavated squares separated by narrow "balks" (walls), an excavator can trace buildings from one trench to the next, follow walls over considerable distances with metrical precision, and identify episodes of rebuilding or expansion in both horizontal and vertical planes. Renfrew used 10-meter trenches, with 1-meter thick walls at the north and east sides of each square.

At times, he needed even closer stratigraphic control, so he subdivided these cuttings into 5-meter units. The unfolding excavation became a fiendishly 3-dimensional puzzle of stone walls, rooms, floors, occupation levels, and doorways. With painstaking care and minute stratigraphic control, the excavators reconstructed the chronology and architectural history of the putative sanctuary area by focusing on individual rooms and structures. They built up small-scale stratigraphic sequences, then combined them with others using direct layer links wherever possible, or by using artifacts to cross-date individual layers.

Easily identifiable artifacts such as coins, painted potsherds, or distinctive stone artifacts are the building blocks of archaeological excavation. The archaeologist works on the assumption that different pot decoration styles (or stone artifact types) came into fashion, then passed into oblivion within finite periods of time (see information in box, chapter 9, p. 213). Thus, if the same short-lived painted motif occurs in two occupation layers within a site, even if they are in different buildings or on either side of a wall, it is assumed that the two horizons accumulated at about the same time. This form of artifact cross linking has served archaeology well for generations.

Nearly a century ago, British archaeologist Arthur Evans had used painted Minoan pottery to date the famous Palace of Minos at Knossos on Crete (see chapter 10). Evans was lucky, for both Minoans and Mycenaeans traded with the Egyptians, carrying olive oil, timber, and other raw materials to the Nile. Minoan trading communities flourished at towns like Avaris in lower Egypt. Egyptologists have also found Minoan and Mycenaean potsherds in other well-dated communities farther upstream near Thebes. Such precise chronologies allowed Evans to match different layers of the Palace of Minos with pottery styles in dated contexts along the Nile. As a result, he dated the Minoan civilization to the centuries before 1500 B.C. His chronology has stood the test of time, with later refinements from both radiocarbon dates and tree-ring chronologies from Aegean oak trees.

For many centuries, Mycenaean pots filled with olive oil and wine traveled far and wide in the ancient maritime trade between the Aegean Islands and lands to the east (figure 11.2). The famous

Fig. 11.2. Mycenaean pot of a form traded widely through the Aegean Islands and the mainland. Photograph from ArtWorks.

Uluburun shipwreck off the southern Turkish coast (see information in box, p. 259), dates to about the time of Phylakopi and chronicles the remarkable diversity of eastern Mediterranean late Bronze Age trade. The Uluburun ship carried Mycenaean jars and weapons, linking the doomed vessel to flourishing Aegean olive oil and wine markets. Late Bronze Age trade routes reached even small islands like Melos, so Phylakopi's familiar Mycenaean pottery styles reveal a link to a much wider world. Local modern-day women made their own versions of the same wares, which gave Renfrew the raw materials for dating his trenches.

THE ULUBURUN SHIP

All shipwrecks are sealed time capsules inaccessible to anyone without twentieth-century diving technology. Archaeologists George Bass and Cemal Pulak used sophisticated excavation techniques and a precise recording grid to recover the cargo from the Uluburun shipwreck, located in deep water off the rugged cliffs of southern Turkey. Between 1984 and 1992, 18,648 dives and 6,006 hours of excavation revealed a cargo of dazzling wealth, perhaps a royal payload. Before anything was moved, teams of divers drew cross sections of the wreck, as well as sectional profiles and elevation measurements of the cargo. A specially designed, handheld, acoustic measuring device provided accurate positions for such objects as the 24 stone anchors on board.

The vessel carried mostly raw materials: more than 350 copper ingots and nearly a ton of tin. Noel Gale of Oxford University used lead isotope analysis and neutron activation analyses to source the Uluburun copper (see information in box, chapter 9, p. 195). The lead content closely resembled that from outcrops on nearby Cyprus, a major copper source during the late Bronze Age. Tin was especially valuable in the ancient world and much harder to source, as accurate methods for doing so are still lacking. George Bass believes that the Uluburun tin came from central Turkey or possibly Afghanistan.

The investigations also attempted to determine which way the ship was traveling. The most conclusive artifacts were some faience cylinder seals, which were widely used in northern Mesopotamia and Iran between 1450 and 1350 B.C. Seal expert Dominique Collon of the British Museum believes that they may have originated in a workshop near the city of Ugarit on the Syrian coast. Based on these artifacts, Bass and Pulak think that the ship was voyaging from east to west, taking advantage of well-traveled routes that benefited from the prevailing winds. Bass and Pulak made a seemingly prosaic, but fortunate, discovery—a piece of firewood with tree rings dating to 1316 B.C., plus or minus two years—which provided a remarkably accurate date for the wreck.

The Uluburun ship, with its numerous weapons and king's ransom of metal ingots and other treasures may well have been carrying a royal cargo. Sourcing methods and other high-technology scientific approaches have identified trade materials from all over the eastern Mediterranean world.

A century of excavation has given Aegean specialists a chance to develop dated master sequences of changing Mycenaean pottery styles. Using multivariate statistical analyses and sophisticated computer programs, researchers have compiled an enormous database of painted and plain island and mainland pottery forms, which altered significantly over more than 1,000 years. They can now match vessel styles and painted motifs to narrowly defined periods of time, sometimes no more than 2 or 3 centuries long.

The chronology of Mycenaean wares has been determined from a combination of AMS or conventional radiocarbon dates of different occupational levels, all of which were calibrated with the help of the master tree-ring curve for the Aegean area that has been developed in recent years. Occasional rare finds of objects of known historical age, such as Egyptian scarabs or Mesopotamian cuneiform seals, provide added precision to the radiocarbon chronology. With such background data, Renfrew was able to date the levels of the Phylakopi sanctuary within limits as narrow as a few decades, to an overall 300-year period from about 1390 to 1090 B.C.

Most archaeologists would have been content to identify distinctive potsherds as chronological markers. Renfrew's team went further, however, recording the exact position of major broken vessels and clay figurines, while reconstructing the complex architecture of the excavated area down to each individual room. They assumed people had broken clay vessels and figurines. In time, the fragments distributed themselves through different rooms and were trodden into dirt floors. Laboriously, the researchers compared the jagged edges of individual fragments, looking for relationships among different rooms—a process known as retrofitting (see information in box, p. 261). Such research requires great expertise in mending small clay objects, as well as accurate plotting of individual potsherds and figurine fragments in the ground.

Using this method and meticulous stratigraphic observations, Renfrew identified the earliest "West Shrine," built about 1390 B.C. on demolished walls from earlier houses, close to a fortification wall (figure 11.3). The builders had erected a rectangular structure with two rooms, low benches, and a small niche. About a century later, in 1270 B.C., the townspeople extended the original building to form an

RETROFITTING

Few archaeological analyses are more labor intensive than retrofitting fragmented stone tools, animal bones, or potsherds. Retrofitting was first used successfully on stone artifacts. The researchers tried to reconstruct the sequence of events that occurred when an ancient stoneworker took a lump of fine-grained rock and turned it into artifacts. During this process, the ancient stoneworker left waste products on the ground where they fell. The researchers plotted the exact position of each fragment, then laboriously gathered and refitted the pieces together until they resembled the original lump. (A modern analogy would be that of police detectives rummaging through a suspect's household garbage.)

Months of work can yield spectacular results. For example, retrofitting has shown that some 2,000,000-year-old hominids in East Africa were left handed. On the North American Great Plains, the technique has been used to show how artifacts such as broken butchery tools and projectile points moved several meters across the surface of a bison kill. Retrofitting potsherds found at Phylakopi was used to show the destruction of entire buildings at a single moment of time, and was helpful in refining the dating of individual layers with great precision. Sometimes the distributions of breaks were invaluable. One plot revealed at least one incident of breakage in the sanctuary so widespread as to represent the destruction of most of the structure at a single moment in time. In other instances, the excavators recorded the earliest occurrence of an artifact—for example, a group of joined potsherd fragments—then found other fragments from the same vessel in a neighboring room. Thus, they could cross date the object with reference to its earliest, original date in the first room. This approach is very productive when the potsherds or figurines come from strata of known age. Although retrofitting worked effectively at Phylakopi with its well-dated clay vessels, it is not a productive technique when chronologies are less secure and pottery forms are less well known.

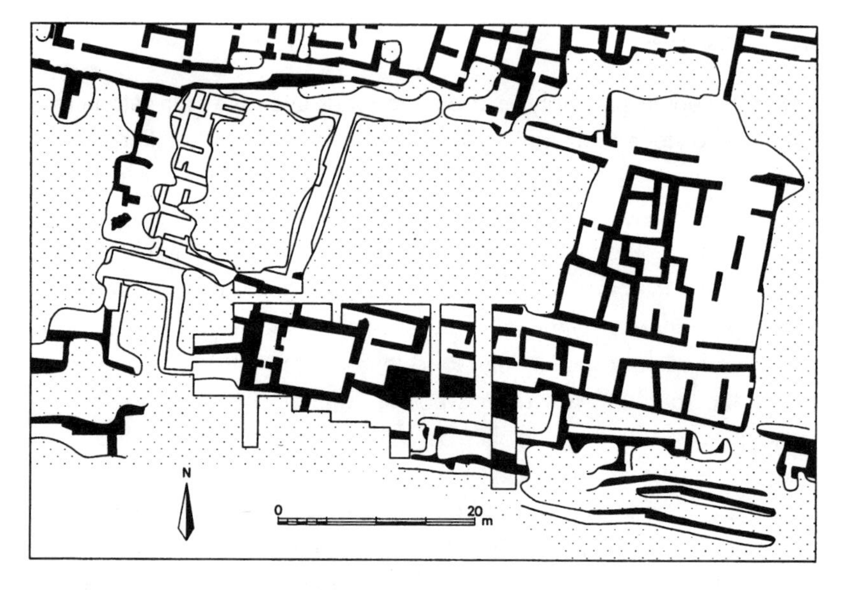

Fig. 11.3. Plan of the Phylakopi Shrine, showing the two chambers. Drawing from the British School of Archaeology in Athens.

"East Shrine," and at the same time, extended the town wall to the south end of the West Shrine. The two shrines flourished until about 1120 B.C., when the entire structure collapsed, perhaps in an earthquake or, less plausibly, during a sudden attack. A remodeled shrine survived a 40 more years before falling into final ruin.

THE CASE FOR A SHRINE

Chronology in hand, Renfrew now confronted the problem of verifying that he had indeed excavated a sanctuary. He had suspected that his 1974 trench, with its broken animal and human figurines, was part of a cult assemblage, for such groupings of figurines occurred elsewhere in Mycenaean sites that were probably shrines. Unfortunately none of the artifacts from these possible sanctuaries came from precise stratigraphic contexts, so Renfrew could not be sure that they were actually part of ritual activity.

In 1975, he removed the uppermost floor in the East Shrine. A stone platform was discovered in the northeast corner, then other exotic objects, including ten sealstones. With no signs of cult activity, he wondered if the room might be a repository for objects of special value. The excavations shifted to the adjacent West Shrine. Here another platform came to light, again in the northeast corner of the room, along with another group of unusual objects, including a female figurine, a double vessel, and several beads. The platform was taller, and perhaps had been used for displaying the objects close by—modest evidence for religious activity.

More excavations revealed new platforms at the western end of the rectangular room that comprised the West Shrine, again associated with female figurines. However, the finds that verified religious activity came from a side chamber in the West Shrine, which contained an accompanying wall niche. A pedestal vase and the fragments of a bovine figure were found in the room itself, while an ugly head without a body; a headless, but naturalistic bovine figure decorated with splotches of red paint and crosses; and several other painted bovine figures lay together within the niche (figure 11.4).

A remarkable female figurine, which Renfrew named the "Lady of Phylakopi," stood upright in the same chamber (figure 11.5). The figurine had a conical stem painted like a long skirt, a bulbous body with small breasts, a painted brown chin, staring eyes, and eyebrows and hair outlined in brown. Another complete female figurine stood to one side. There was no question that these objects had been placed there deliberately.

The case for a sanctuary now seemed stronger: two room complexes (the East and West Shrines) contained a treasury of rare objects that were nonexistent elsewhere on the site. It was apparent that they had been looked after with some care. But questions still remained. Were the chambers used for ritual purposes, or were artifacts merely stored in this area of the site? Back home, Renfrew marshaled the arguments in favor of the two room complexes being a shrine.

The builders had laid out the rooms in such a way that the symmetrically placed platforms were the focus of attention. The objects displayed upon them would have caught the eye at once. The niche containing clay figures in the West Shrine may have had the same

Fig. 11.4. Painted bovine figurine from the Phylakopi Shrine (one-third full size). Drawing from the the British School of Archaeology in Athens.

effect. The original architects had also paid careful attention to illumination, erecting columnar lamps to provide carefully controlled light. Furthermore, the excavations in the "shrines" yielded conch shell trumpets, such as that blown by a priestess or goddess on a seal-stone from the Idaean Cave, high in the mountains of Crete. Bored tortoise shell fragments once formed the bodies of lyres, with the conches the only musical instruments on the site. The priests of

Fig. 11.5. The Lady of Phylakopi (one-third full size). Drawing from the British School of Archaeology in Athens.

Phylakopi were well aware how light and sound played important roles in ritual performances.

The same buildings also yielded fine pedestal vases and double vessels, which were found nowhere else at Phylakopi, but had been discovered in ritual contexts elsewhere in the Aegean. Additionally, the figurines and figures from the two buildings suggest the presence of the translucent. The Lady of Phylakopi gazes upward. She stood close to a number of bovine figures that may have served as cult images.

Renfrew believes that this evidence makes a case for a town shrine or sanctuary, a place for "the performance of expressive actions of worship and propitiation by human celebrants towards a transcendent being or beings" (ibid., 126). He himself was the first to admit that his case was rather weak, even though no plausible alternative hypothesis was apparent. Unlike earlier workers, he was reluctant to ascribe a religious function to any object or symbol he did not understand. Therefore, to further substantiate his theory, Renfrew looked outside the island of Melos. He asked himself, "Were there religious buildings—comprised of simple rooms containing internal benches or platforms with human figurines set on them—in other areas of the Aegean during Mycenaean times?" In his attempt to answer this question, Renfrew studied four other well-known Minoan and Mycenaean finds from Crete, the Aegean Islands, and the mainland. All of these discoveries, taken by themselves, had presented weak, but not conclusive, evidence of religious activity.

At Minoan Knossos, Arthur Evans had unearthed the "Shrine of the Double Axes," where he found a bell-shaped female figure with upraised arms, and a series of distinctive articles, distinguished as known cult objects from previous sites (see information in box, chapter 10, p. 229).

Renfrew also studied the famous sarcophagus from Aghia Triadha, also on Crete, where similar sacred artifacts had been found. The decorated coffin, described in chapter 10, bears painted scenes of four women and a man playing pipes (see figure 10.8). One woman is placing her hands on a trussed bull or calf, lying on a four-legged table. Small platforms carrying a tree flanked with horns of consecration and a pole surmounted by a double ax stand nearby. The scene may represent a funerary ritual rather than a religious rite, but no one can doubt the association of the horns of consecration and the double ax with religious activity.

Renfrew discovered other examples of sanctuaries containing religious artifacts. At the small Minoan town of Gournia, archaeologist Harriett Hawes, the first woman excavator on Crete, unearthed a small building with stone-rubble benches at the end of a narrow street. The building contained a bell-shaped figure of a woman with arms upraised, and a snake coiling around her body. At the Gazi site, also on Crete, Greek archaeologist Spyridos Marinatos found a small room with a flat-topped table and five bell-shaped figurines of women with upraised arms. On their heads were birds, poppies, and horns of consecration.

These isolated finds repeat the same basic symbols, and often show the female figurine with uplifted arms. In three cases, Gournia, Gazi, and Knossos, the female figure with upraised arms came from a special location. Thus, the archaeological case for religious activity at Phylakopi comes not from one context, but from the repetition of symbolic elements at several sites, that, taken together, show that a general set of religious practices was shared over a wide area and reflected in small village shrines.

DIVINES AND DUALITY

The Phylakopi Shrine was an imposing structure in its heyday. On entering the West Shrine, you came into a rectangular room 6 meters long and 5.8 meters wide (figure 11.6). The roof was probably flat with a reed covering; possibly it had cross beams supported by pillars. The walls were apparently plastered. Most light came from the doorway, which faced another door leading into two smaller chambers at the rear. Narrow platforms or benches on either side held the cult objects, with niches above them, which allowed for intercommunication between the rooms. Sometimes a cult or bovine figure stood in one of the niches. The East Shrine was much smaller and was approached from the south, its floor about a third of a meter above the nearby street. Here, the cult equipment stood on a low platform at the northeast corner of the room. Outside, a street from the east opened into a small courtyard. Amphorae, two jugs, a dipper, and three *kylikes* (shallow cups with small stems) were found there. Since only a few clay vessels came from other parts of the shrine, Renfrew believes that this was an area for libations, perhaps poured in an unpaved quarter of the courtyard, where liquids would drain away.

Fig. 11.6. Reconstruction by Alex Daykin of the West Shrine at Phylakopi during its heyday. Drawing from the British School of Archaeology in Athens.

The excavators recorded the locations of figurines in the West Shrine. They discovered that male figurines were restricted to the northwest platform and female depictions to the southwest one. Bovine figures lay on the northwest platform, accompanying the female figurines. Two of the bovines had bowls on their backs, presumably to hold liquid offerings. Two animal figurines accompanied the human males on the northwest platform. Renfrew believes that the distribution of finds hints at a strong duality, between male and female. No human figurines came from the East Shrine, where bovine figures were well represented, as if there was another duality between animals and humans. However, the gold-sheathed head found in the first season may mean a wooden human figure covered with gold sheet once stood in the shrine.

The Lady is the only plausible cult image at Phylakopi, for no signs of a male deity were found in the shrines. Nevertheless, the clear dichotomy between male and female suggests that both deities were worshiped here. The goddess may have sat in a wall niche on occasion, perhaps accompanied by the female figure found with her, conceivably an attendant or a votive. A scatter of beads found near the platform in front of the room where she was found may represent offerings to the cult figure. Since no such finds came from the chamber itself, Renfrew believes that the chamber was a storage area, and not a holy shrine, where the figure could be worshiped privately.

Phylakopi offers a glimpse at a village shrine; a modest shrine, capable of holding a tiny proportion of the town's population, with its humble offerings and understated cult objects. Phylakopi's sanctuary has challenged twentieth-century archaeology. However, science has provided sufficient evidence to determine that the Phylakopi Shrine was a place where torches flickered, trumpets and lyres sounded, and the god and goddess stood in imposing state surrounded by the huddled dwellings and streets of the town. Artifacts and other finds suggest that people came to make their modest offerings, and to seek inspiration, protection, or solace. The ancient religion, with its rituals rooted deep in history, reveals a relationship with a spiritual world honed over many centuries in villages, towns, and great palaces.

chapter twelve

DIVINE KINGS ALONG THE NILE

I climbed the Pyramid of Kephren at Giza many years ago, before such a feat was forbidden. The suburbs of modern Cairo lay at the foot of the pyramids, symbols of Egypt's renowned and ancient past. Kephren's pyramid is almost a meter shorter and more compact than Khufu's Great Pyramid nearby. The climb is a stiff one. Its sides angle up toward heaven at a precipitous 52 degrees. I climbed the eastern side before dawn, when heavy white mist hung loose over the waking city and the Nile Valley.

I could barely see my guide's gray figure in the mist. The first four or five stone courses vanished into the swirling gloom overhead. I shivered in the chill, but not for long. The guide hitched up his robe and climbed briskly above me, his feet pointing the way. I puffed and panted along after him. Each friable weathered block made an uncomfortably long step up, a bit like taking average stair steps three at a time. My calves ached with the unaccustomed near-vertical strides.

Finally I stumbled onto the ledge at the end of the limestone casing that once covered Kephren's tomb. Through the lightening mist I saw a patch of blue sky over the rosy apex high above. With the sun, now hot on our backs, we clawed our way farther up, our fingers and feet clinging to fissures in the slabs. Suddenly, the summit was underfoot—a crumbling, flat platform floating above the dispersing mist. I sat looking west beyond the suburbs of Cairo to the undulating, light-brown desert, once the land of the dead. The green Nile

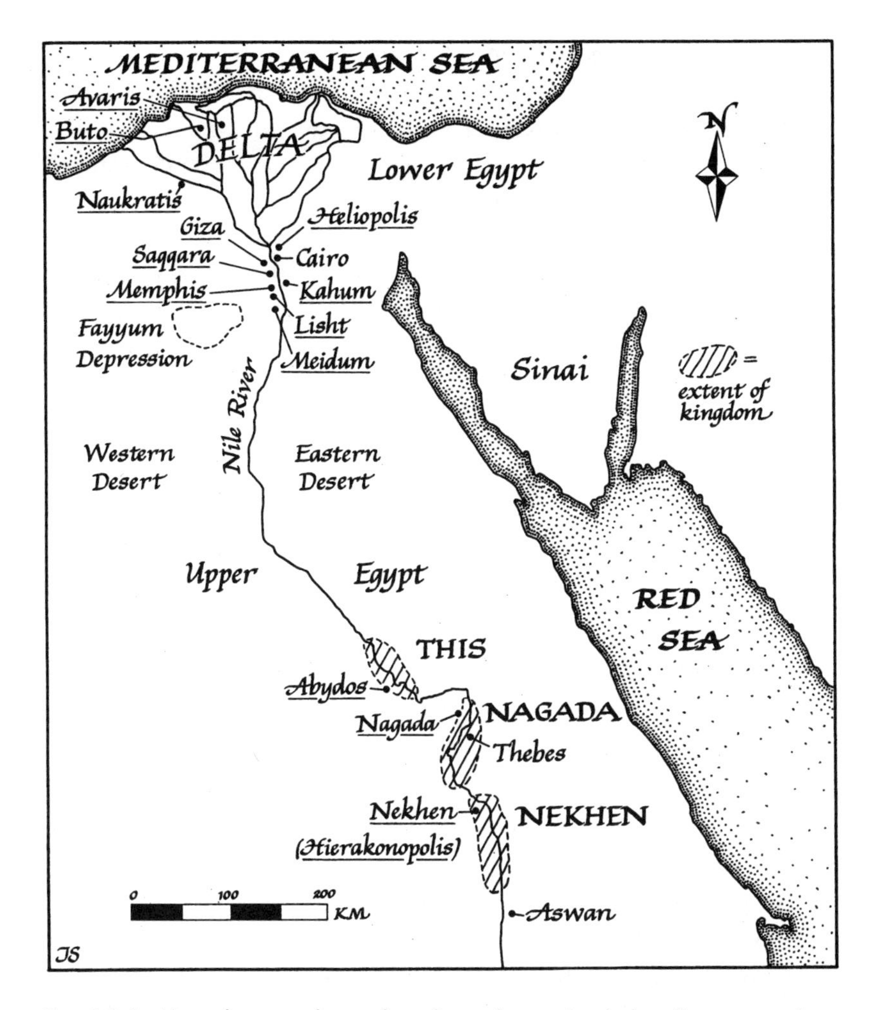

Fig. 12.1. Map showing the archaeological sites (underlined) mentioned in chapter 12.

Valley, the land of the living, snaked away into the haze off to the far south. Sitting up so high, a person can't help but feel the past. I felt as if ancient Egypt lay in a time capsule at my feet (figure 12.1).

The Great Pyramid of Khufu at Giza and its companions overshadow the Nile Valley. These silent colossi—made of 2,000,000 limestone blocks—are mountains of masonry that cover over 5 hectares, an area comparable to 2 square city blocks. Every time I witness the dark walls of the pyramids obscuring the moon or see the dark shadows they cast at sunset, I wonder about the megamonarchy that built them. Among the first monarchs in the world, the pharaohs gave form and meaning to their reign by erecting these particular tombs.*

The pyramids have fascinated people for centuries, not only for their size, but for the secrets that they hold. Many questions have been raised over the years: Was the function of the structures only to preserve 3,000 years of ancient Egyptian civilization for eternity? Can we look at these simple structures as being more than amazing engineering feats? Are they more than mausoleums? And finally, how did the Egyptian state and its divine rulers come into being? The answers to these questions lie along the Nile River where the disciplined formalities of civilization began. Here, gifted rulers and their officials developed a long-enduring vision of divine kingship within a cosmology that celebrated the land.

THE BLACK LAND

The River Nile slashes through the arid landscape of extreme northeast Africa like a green arrow. Its course runs more than 4,800 kilometers, from high in the Ethiopian highlands and Lake Victoria in Uganda north to the Mediterranean Sea. For most of Egypt's 1,300 kilometers, the Nile cuts a deep gorge through some of the driest landscape on earth, then fills it with layer upon layer of deep, fertile river silt.† The Nile's source rivers, the Atbara and the Blue Nile, rise in the mountainous Ethiopian plateau. Lifeblood of ancient Egyptian civilization, these waterways govern the unchanging cycles of farming life. They have risen, flooded, and flowed far downstream since time immemorial. Each year, they carry the runoff from summer tropical rains into the heart of Egypt. During *Akhet*, the season of inundation, the water surges downstream, swelling the Nile above its banks, turning the countryside into a vast, shallow lake. Towns and villages

*The term "pharoah" is a biblical corruption of the Egyptian word *per-aa,* which means "the great house" or "palace."

†This chapter refers to the Nile Valley *before* the construction of the Aswan Dam in the 1960s controlled water flow through Egypt.

become islands on low mounds above the floodwaters. As the current slows, the river drops its silt on the flooded lands, then recedes as the farmers plant their crops on the muddy ground. Then comes *Peret*, the season of growing, when crops of wheat and barley ripen slowly in the temperate late fall and winter sunlight without the need for watering. After the harvest in March or April, the early summer sun hardens and cracks the ground, aerating the soil and preventing the accumulation of harmful salts in the earth. *Shemu*, the season of drought, ends with the onset of the next inundation as the year's cycle begins again.

The Greek historian Herodotus called Egypt "the gift of the Nile," and the ancient Egyptians themselves named their homeland *kmt*, "the black land." The river fertilized and watered their dark-soiled fields, and supported lush marshlands and grazing grass for domestic and wild animals. Fish and fowl teemed along the river banks. In good flood years, food was plentiful. Yet with all this bounty, life was a form of agricultural Russian roulette. A fast-moving Nile could sweep villages, fields, and animals away in a torrent of muddy rapids. The wise farmer stored a reserve against sudden floods or poor inundations, or for when famine spread across the dusty fields. Despite this uncertainty, the Egyptians considered the great river the source of life. The divine Nile was part of the Egyptians' cosmic order, an integral part of their lives from birth to death. Their notions of the environment, life, religion, kingship, and government flowed from their dependence on the organized oasis that was the Nile.

I have traveled by felucca (a lateen-rigged sailing vessel) upstream of Thebes during a Nile winter. The slow-moving river flows through a patchwork of lush green fields, irrigation canals, groves of date palms, and sycamore trees. For the six winter months, the sun rises in the Eastern Desert, soaking up the heavy dew. At dawn, low mist stirs over mirror-still canals; only waterfowl fly overhead. As the day progresses, the sun passes over the valley, and shadows vanish. Lucent sunlight fills the wide, cloudless sky. The white desert hills shimmer in the heat. Come late afternoon, they turn sandy-gray, mauve, or pink. The afternoon shadows lengthen, and the landscape turns turquoise and gold as the sun sinks behind the Libyan hills to the west. The sun is changeless, the soil and the land persistent. The passing days and months repeat themselves with an inevitable rhythm.

My slow voyage helped me to understand why the ancient Egyptians considered the ideal society on earth to be a reflection of a primordial divine order, embodied by the Nile and the passage of the sun across the heavens. Since the very beginnings of cultivation along the river, small farming communities have embedded themselves symbolically in the fertile floodplain. These well-watered lands were the domain of the living. The ancestors lay in cemeteries on the fringes of the valley, in arid desert. Close ties with the land endure in Egyptian villages today. Pharaohs have come and gone; Assyrian and Roman armies, that conquered cities, have moved on. Christian and Islamic beliefs have left their marks on Nile society. But the humble farming community survives with little change, still tied to the cycles of the river and the sun.

PREDYNASTIC SOCIETIES

People anatomically identical to us settled along the Nile's floodplain at least 100,000 years ago. At the end of the Ice Age some 15,000 years ago, the verdant Nile Valley was a paradise for Stone Age foragers. The river supported large bands of hunters and fisherfolk along its marshy banks. With higher rainfall than today, they spent much of the year ranging over the nearby desert, which then supported arid grasslands. From pollen diagrams of the old Nile Valley flood basins, western Asian lakes, and dried-up Saharan lakes, as well as data from Syrian tree rings and Greenland ice-core data, it is known that after 6000 B.C., severe drought conditions developed over much of western Asia. The people of the Nile moved closer to the river, concentrating in better-watered areas. Faced with serious food shortages, some Egyptian communities domesticated plants and animals as "insurance" against a harsh and unpredictable environment.

The origins of farming along the Nile remain a mystery, because the earliest villages are inaccessible under many meters of floodplain alluvium. Most Egyptologists believe simple farming based on cattle herding and cereal agriculture took hold along the Nile as far south as the Sudan by 5000 B.C., and perhaps much earlier. With population densities relatively low, the average Egyptian subsisted much as Upper Nile villagers do today, by growing wheat and barley and herding animals along the river. Dozens of small communities, each with its own patchwork of farming land tied to surrounding ancestral territory, competed and traded with their neighbors.

The compelling forces that nurtured Egyptian kingship originated in somewhat later riverside societies discovered a century ago by British Egyptologist, Flinders Petrie. A pioneer of scientific excavation along the Nile, Petrie started his career by surveying the Pyramids of Giza in 1880. He set up camp in an abandoned tomb and scandalized Victorian tourists by working in his red underwear. Petrie went on to become a champion of scientific excavation, in an era when nothing had any meaning to an Egyptologist other than an inscription or a sculptural find. Above all, he paid close attention to small artifacts such as painted potsherds.

While digging Naukratis, a late Greek trading city in the Delta, he discovered the value of early Greek pottery as a way of dating Egyptian temples and other structures on-site. Coins and inscribed ornaments in the tightly packed soil of temple and building foundations allowed Petrie to date individual structures and occupational layers, an innovation for the day. Petrie employed 107 men organized into teams, with only 2 trained supervisors to watch over them. Though rough and fast moving by today's standards, he achieved miracles, working with minimal funds and few facilities. He even stored mummies under his camp bed.

In 1894, Petrie heard rumors of a vast cemetery of pottery-rich graves in the desert near Nagada, 24 kilometers south of Thebes in Upper Egypt. He descended on the necropolis in force, and with characteristic thoroughness, cleared nearly 2,000 graves. First, he would send boys out to search for softer spots in the sand. They cleared the edge of the grave, then moved on. Petrie's workmen would then clean out the burial pit until they located undisturbed clay vessels. Next "first class men" would move in to expose the pottery and skeleton, after which a skilled Egyptian would clean up the bones and grave goods for photography and recording. The Nagada dead were humble folk. They lay in shallow graves wrapped in mats or linen shrouds, sometimes accompanied by a few strings of beads or clay pots. At times, plain clay human figurines were also found with the dead.

Who were these simple farmers buried in enormous cemetery complexes? Their graves displayed no signs of wealth and no distinctive kingly regalia. Were they the ancestors of the pharaohs? Petrie soon discovered that the Nagada "predynastic" graves were different from later dynastic tombs. Locating an early royal grave near Nagada itself, he obtained a chronological and cultural link between the earlier sepulchres and the culture of the earliest dynastic Egyptians (see information in box, p. 277).

Since Petrie's research, no one has doubted that ancient Egyptian civilization originated among predynastic farming societies along the Nile. A century later, we know that Nagada and Nekhen, upstream of Thebes, were flourishing centers of predynastic culture by 4000 B.C. Neither town was anything more than a small, dense settlement of flimsy, oval huts and animal enclosures with an adjacent enormous cemetery complex. However, they both lay in vital strategic locations, opposite desert valleys leading to remote gold outcrops and to the Red Sea—gateway to more distant lands. The villages were located on exceptionally fertile soils. Demographic archaeologist Fekri Hassan has calculated that the Nagada settlements grew enough grain on land at the edge of the floodplain to support about 76 to 114 persons per square kilometer. By clearing trees, removing dense grass growth, building dikes, and digging drainage canals to clear still-inundated acreage, the farmers opened up much larger tracts of agricultural land. By the time the villagers had cultivated four, or even eight, times more ground, they could support as many as 760 to 1,520 people per square kilometer, including officials, traders, artisans, and other nonfarmers. This intensification of agriculture required supervision, careful storage of grain surpluses, and leaders to organize communal projects such as canal digging.

Nagada and Nekhen (Predynastic Societies)

A mythic world rising from the primordial waters could already have its living equivalent in the exposed mounds of farmland in the midst of receding river water. The beliefs and mythology of the ancient world of the predynastic village and the Nile endured to

SEQUENCE DATING PREDYNASTIC EGYPT

As previously mentioned in chapter 9 (see information in box, p. 213), seriation (formal artifact ordering) is one of archaeology's fundamental analytical methods, and is especially effective with vessels of the same form. Flinders Petrie, a pioneer of artifact ordering, established relative chronologies through a process he called "Sequence Dating." He used this process to develop a chronology of predynastic Egyptian cemeteries. Petrie worked with hundreds of sealed grave lots of funerary jars from Nagada and other predynastic cemeteries. He found gradual changes in the vessel shapes, especially in the handle designs on one jar style. At first the handles were highly functional, intended for daily use. Then they became more decorative, finally degenerating into nothing more than a series of painted lines, as if function had become an art form. Petrie unearthed similar jars in other cemeteries that were associated with changing, but characteristic, artifact groups. He assumed that the functional handle was the earliest; and the painted lines, the ultimate form of the appendage. Within a few seasons, he had developed a series of "sequence dates" based on the gradual handle changes on predynastic pots. He started with "ST 30," arguing correctly that he had not yet found the earliest predynastic societies. "ST 80" coincided with the beginning of dynastic history. Pots in an early royal grave at Nagada enabled him to link dynastic and predynastic. He wrote: "This system enables us to deal with material which is entirely undated otherwise; and the larger the quantity of it, the more accurate the results" (Petrie 1889, 299).

Sequence dating placed predynastic graves in some kind of chronological order, but did not provide an actual chronology for the societies that preceded the dynastic state. Petrie himself identified several predynastic cultures by their distinctive pottery styles. Subsequently, several generations of archaeologists have applied radiocarbon dating and increasingly sophisticated, and even computerized, multidimensional scaling to the classification and ordering of predynastic wares. They believe that their pottery classifications reflect regional differences in farming cultures along the river, radiocarbon dated to between 4000 and 3100 B.C.

become a foundation of pharaonic rule. With their unusual spiritual and close ties to the land, the leaders of Nagada and Nekhen gradually assumed the trappings of divinity on earth.

Archaeologists have worked at Nekhen since 1897–98. They discovered that Nekhen was a walled settlement by 3500 B.C., as if fierce rivalries pitted each small kingdom against its neighbors. More recent investigations focus as much on environmental conditions, mapping, and survey work, as on excavation, in attempts to recover detail ignored by earlier, cruder research. The surveys reveal that in 3800 B.C. Nekhen was a village with a few hundred inhabitants. It lay on a protected area of the Nile floodplain, within easy reach of fertile soils and wooded savanna. Over the next 3 centuries, the population mushroomed to as many as 10,500 townsfolk.

By 3500 B.C., a serious drought had gripped the Nile Valley. The people living at Nagada and Nehken had moved from their dispersed villages and into densely populated towns of rectangular mud-brick buildings. Now the realities of political and economic authority changed. The villagers became townspeople, many of them non-farmers. Nevertheless, we can be sure that their closest spiritual and emotional ties were to their lands.

The early Egyptians' view of the cosmos and creation still revolved around the cycles of the river and the farming year; the months of harsh aridity and the benign sun of winter ripening growing crops. The first town leaders were still influential kin leaders. They had strong ties to ancient village lineages, to riverside lands, and to the cosmos that linked the realms of the living and the dead. Unquestionably, this spiritual world of village and early town life revolved around fertility, Nile floods, and the predictable journeys of the sun.

The people of Nekhen lived in mud-brick houses clustered together. Archaeologist Michael Hoffman traced these houses over almost 40 hectares. Most Egyptians lived in small villages, but Nekhen, like Nagada, was a place where economic and social differences cut across society for the first time. Thus, a few of the more important people owned residences that were set in fenced compounds.

Why did some people acquire more wealth than others? Hoffman believes that Nekhen's leaders went from being farmers to artisans, traders, and pot manufacturers. More than 50,000,000 potsherds litter the Nekhen area. He identified and mapped at least 15 large pottery kiln areas covering an area of more than 950 square meters. Most produced the coarse, ordinary domestic wares used in every local household. A much finer "Plum Red Ware," often found in graves of wealthy individuals, came from kiln sites located in natural wind tunnels in the cliffs overlooking the town cemetery. The potters used the prevailing winds to create the high temperatures needed to fire fine vessels. Each wind-tunnel kiln specialized in different types and quantities of pots, producing many more than could be used locally. Hoffman hypothesizes that these clay products passed up and down the Nile to neighboring communities. The high-grade vessels also served as offerings that were buried with the dead, part of the wealth that accompanied the deceased into the afterlife.

Nekhen's leaders grew wealthy on the Nile trade, despite ecological disaster. A combination of drought conditions, rapacious goat and sheep herds, and deforestation by potters degraded the surrounding grassland. Hoffman believes that some of the local rulers wisely invested their wealth in irrigation agriculture instead of dry farming, thereby insulating the community from famine. They developed water-control systems, raised agricultural productivity, and controlled the stores of grain. These same men developed such secular power from their mastery of the inundation, that they may have become symbolic of the authority of the sun-god—they became rulers on earth of a world created by the sun. These leaders, thus, were the ancestors of the first pharaohs. Their notion of kingship formed the foundation for dynastic Egyptian ideology. In later times, the king was creator on earth, a perfect reflection of the sun in the sky, the two forming a partnership against chaos.

A SUCCESSION OF KINGS

Order and stability: these two words epitomize the ideological foundations of Egyptian civilization, which are reflected in the pharaohs' official records. In the third century B.C., an Egyptian priest named Manetho wrote a work entitled *Aegyptica*. Acquiring information from fragmentary ancient king lists, he set down a list of Egypt's kings

TABLE 12.1: A GENERAL CHRONOLOGY OF ANCIENT EGYPT*

30 B.C.	Roman occupation
322 to 30 B.C.	Ptolemies bring Greek influence to Egypt
1070 to 322 B.C.	Late Period Gradual decline in royal authority Foreigners invade Egypt repeatedly
1550 to 1070 B.C.	New Kingdom Great imperial period of Egyptian civilization
1640 to 1550 B.C.	Second Intermediate Period Hyksos rulers from western Asia rule Egypt
2040 to 1640 B.C.	Middle Kingdom Thebes rises to prominence
2134 to 2040 B.C.	First Intermediate Period Political instability and local rule
2580 to 2134 B.C.	Old Kingdom Despotic pharaohs build pyramids Great culture of ancient Egypt finally established
ca., 2920 to 2580 B.C.	Archaic Period Unification and consolidation of the state
5000 to 3000 B.C.	Predynastic cultures
Before 6000 B.C.	First farmers along the Nile

*Based on archaeology, king lists, and written records.

and divided them into 30 dynasties. Egyptologists have used Manetho's chronology and other synthesized lists to reconstruct a succession of monarchs extending from more than 3,000 years ago up to the Roman conquest of 30 B.C. (table 12.1).

The lists contain significant historical information, and reveal political statements about the continuity of kings. The New Kingdom pharaoh Seti I inscribed a list on the walls of his temple at Abydos around 1290 B.C. He recorded no less than seventy-five royal ancestors, each represented by an individual cartouche. There were serious oversights, however, by Seti's scribes, who omitted periods of political instability or disputed succession. Nevertheless, the inscription sent a message: that the past was a model of good order, a continuous succession of kings, who handed down their thrones to their successors in a single, linear sequence. This is exactly what happened during the Old, Middle, and New Kingdoms, the great eras of Egyptian civilization.

Though fragmentary, the so-called Turin Canon, compiled in the reign of Ramesses II (1290–1224 B.C.) appears to contain the list of every pharaoh from Ramesses back to Narmer (or Menes), the first king of unified Egypt. The anonymous author, however, did not stop at Narmer. He ventured even further back to a time when the cosmos and its divine beings met. The Turin scribe added several lines, which satisfied pharaonic ideology. He summarized the reigns of several "spirits"; gods who reigned over Egypt for thousands of years showing how the royal succession descended to a time when the gods themselves were kings.

Thus, the ancient pharaohs and their pyramids, temples, and sepulchres form a linear history that begins with the gods who created the world, projecting a continuous kingship. The Egyptian past is a history of authority and comfort. In fact, pharaohs would sometimes consult their archives for precedents set, in order to adjudicate legal matters or to ensure correct ritual procedures.

A thousand years before the Turin Canon scribe set down his version of Egyptian history, other officials indulged their passion for lists. They recorded kings and the principal events in each year of their reigns down to the Fifth Dynasty. The Palermo stone, carved about 2350 B.C., shows us what Egyptians of the day considered to be important events: religious festivals, creation of statues of the gods, wars, regular taxation, and the height of that year's Nile flood. Administration and piety toward ancestors also formed a part of their history. Official archives journeyed back to the catalytic historical moment when the first pharaoh, King Narmer, unified Upper and Lower Egypt and formed the unified Egyptian state—whose beliefs reflected a divine order that required constant care through pageantry and ritual.

EGYPTIAN MYTHOLOGY

About 3000 B.C., the Egyptian vision of civilization—of human existence—stressed the unity of Upper and Lower Egypt, the role of the king as an intermediary to the gods, and the notion of equilibrium as a way of life. The state used myths and mythological scenes to reinforce these ideas in the popular mind, much as we use political propaganda today. Practical knowledge, for example, on how to construct a pyramid or irrigate the fields, was useful, but theoretical research extended spiritual associations drawn from known theological ideas. The scribes taught through rote learning: naming the names of things.

Amenemope, a scribe of the late New Kingdom (ca., 1000 B.C.), wrote a pedagogical text entitled: *Beginning of the Teaching for Clearing the Mind, for the Instruction of the Ignorant, and for Learning All Things That Exist.* His book is a list of names for everything from towns to gods, elements of the universe, parts of animals, and all manner of basic knowledge. Egyptian knowledge came from the classification of a known world. It became a convenient method for defining gods, names of places, seasons of the year, and any knowledge absorbed by a well-educated Egyptian.

The names of gods and places allowed the Egyptians to indulge in symbolic geography, an idealized image of the state, in a mythworld created from common experience. The Egyptian world began with primeval waters of nothingness. The god Atum, "the completed one," was the creator, preeminent over the cosmos. He caused the "first moment," raising a mound of solid ground above the primordial waters. The life-giving force of the Sun arose over the land to cause the rest of creation. The Egyptians expressed the moment of creation in a familiar metaphor of the annual Nile inundation, when the brown land emerged from the receding waters. The primordial mound was a compelling image in the Egyptian world. The *benben*, a sacred stone shaped like a miniature, angular mound, stood in the Temple of Atum at Heliopolis. The shining, gilded capstones of the Pyramids of Giza, *benebet*, were also symbolic tumuli.

The myth of Osiris reinforced the notion of cosmic unity. It was the basis for one of ancient Egypt's most enduring cults, and was told in the Pyramid Texts, the oldest religious literature in Egypt, which was inscribed on the walls of Fifth and Sixth Dynasty pyramids (2465–2150 B.C.). Osiris began as Asari, a man-headed deity of agriculture, but in his later form was slain by his brother Seth. The goddesses Isis and Nephthys found his corpse, stopped its decomposition, and twice brought Osiris back to life when Seth attacked him again. Dead pharaohs became identified with Osiris, for they were equated with Horus, his son, while on the throne. Osiris embodied cosmic harmony. His insignia was the rising Nile, while the moon's constant waxing and waning symbolized his bestowal of eternal happiness.

Nekhen's local god, Horus, and other divines combined human and animal forms, and could even change guises with different roles. Horus, depicted as a falcon-headed human figure, derived his name from the word *hry*: "the one who is above/far from." The name associated him with the sky. Even the cosmos and the environment eventually assumed concrete form. The earth, Geb, was a woman; the sky, Nut, also a woman. Once a deity took form, a myth or legend developed to explain its origin and associations. The god or goddess received a name during this process.

The gods played a mythical role in the unification of Egypt. The twin deities Horus and Seth, gods of Upper and Lower Egypt, became a cosmic balancing act reflecting the new circumstances of a united land. Their representative on earth resided in one living person: the king, who interceded between the living and divine worlds, just as his village predecessors did for centuries. The king was "the living embodiment" of Horus, the solar deity and protector of kings. The antagonist was Seth, murderer of Osiris and instigator of confusion. Seth brought storms, drought, and even foreigners, into the harmonious Egyptian world. But Egyptian texts speak of Seth as reconciled to his divinely judged role as an opposing force in the ideal balance of harmony.

An Egyptian myth that forms part of the "Memphite Theology," written down during the Late Period, tells how Geb, the deity of Earth and Lord of the Gods, mediated between Horus and Seth, ending their quarrel. "He made Seth king of Upper Egypt in the land of Upper Egypt, up to the place where he was born. And Geb made Horus king of Lower Egypt in the land of Lower Egypt, up to the place where his father [Osiris] was drowned, which is the

'Division of the Two Lands.' " Geb then passed his inheritance to Horus, the son of his firstborn son. " 'I have appointed Horus, the firstborn, he is Horus who arose as King of Upper and Lower Egypt, who united the Two Lands in the Nome of the Wall [province of Memphis], the place in which the Two Lands were united.' " Horus and Seth were reconciled, and united in the "house of Ptah, the 'Balance of the Two Lands' " (Lichtheim 1973, 52–3).

Ancient Egyptian myths shape an intellectual aesthetic, composed within the minds of their creators as an inner world, quite different from the ancient world unearthed by the archaeologist's spade. The origins of kingship along the Nile stem in considerable part from such pervasive mind-sets as order, serenity, and unity, told through these ancient myths.

ORDER, SERENITY, AND UNITY

People have always feared the forces of the natural and spiritual worlds. But early Egyptian art transcends the spiritual and reflects a desire for order in the living world. In 3500 B.C., dozens of small towns and some large kingdoms flourished along the Nile, despite enemies and special rivals. The harsh environment, surrounded by encroaching desert and the risk from fluctuating Nile inundations formed the background of daily life—a life that included military and political rivalries between ambitious local leaders, who were equipped with their own gods and spiritual authorities. A constant undercurrent of violence, war, and siege from neighbors gave life an uncomfortable sense of disorder. Equilibrium gave way to conflict and political confusion. Within a few centuries, a new intellectual order embraced the fundamental ideas that obsessed Egyptian civilization for nearly 3,000 years: order and stability, and their counterpoint—disorder. The rule that kings maintain order in the presence of a supreme divine force (the power of the sun) became the political reality and the intellectual view that brought the universe together in a long-lived relationship with human life and the natural world.

Every pharaoh was king of Upper and Lower Egypt, the Nile Valley, and the fertile delta downstream. The cosmology stressed three ethical values: piety, justice, and order, a concept the Egyptians called *ma'at. Ma'at* embraces the universe, a cosmic order transformed into a goddess, the daughter of the sun-god Re. Re established *ma'at* at the time of creation; thus, the structure of the state was not a product of the human brain, but an image of the cosmic order. The goddess Ma'at personified the orderly nature of the sun-god himself. She also manifested herself in the domains of the lord Osiris: earth, vegetation, and water.

Egyptian deities "lived on *ma'at*," a phrase often used by the pharaohs to describe their own behavior. Ancient Egyptian religion stressed a close relationship between the living and the gods. Only if humans lived life in harmony with *ma'at* could they conquer death. Because divine life rose again, ancient Egyptians lived with the unshakable faith that *ma'at*, the divine order, was absolute and eternal.

Nekhen art reveals a changing ideological world, one in which animals and humans become tangible forces in the cosmos. It is as if Nekhen's rulers began to devise a set of anthropomorphized concepts as a way of organizing ancient religious beliefs, and their own authority, into a form that made them comprehensible to the general population. The forces of nature became humanized, adorning artifacts, temples, and tombs. In time, they became an integral part of writing. Deities took human form.

Egyptologist Barry Kemp believes that the myth of the Egyptian state and its harmony originated in much earlier beliefs in an ideal world from Upper Egyptian villages. Age-old beliefs of harmony and balanced opposites came together as a single superior force embodied in the figure of the pharaoh, who co-opted ancient Horus and Seth mythology to create a national framework of myth and symbolic geography that was to influence a state more than 1,400 kilometers long. This flexible framework could accommodate the diverse beliefs and religious cults among the 23 *nomes* (provinces) that made up the kingdom.

The earliest form of Egyptian kingship survives only in occasional prestigious art objects, buildings, and material objects, and in royal sepulchres. Archaeologist Michael Hoffman and Egyptologists Barbara Adams and Carter Lupton excavated curving rows of the sand-filled burial places of Nekhen's elite on the banks of a dry stream named Abu Suffian, located outside the town. The Abu Suffian tombs were humble by the standards of later pharaohs' burial places, but impressive for their day. Though looters had ravaged the sepulchres in ancient times, they left a jumble of rich artifacts behind, including finely made, black-topped clay jars, some bearing complex graffiti scratched into the clay. Flint arrowheads, basket fragments, and pieces of furniture testify to the wealthy leaders. A beautiful disk-shaped mace head of polished green and white porphyry was found under a pile of broken artifacts. It is one of the earliest symbols of leadership known from the Nile Valley. Its owner may have been one of the mythical "Divine Souls of Nekhen," the primordial rulers of legendary Egypt.

With meticulous care, Hoffman and his colleagues excavated each tomb in the Abu Suffian cemetery. They undertook a complex salvage operation, using brushes, trowels, and sophisticated recording devices instead of the crude picks and shovels of their predecessors. To their surprise, they discovered that the cemetery was a symbolic map of a unified Upper and Lower Egypt, created at one of the centers of early Egyptian kingship.

The graves straddled the dry Wadi Suffian, which represented the Nile River. The towns at the downstream end of the wadi were in the ancient Lower Egypt style, dug out of the subsoil and lined with mud-brick. Hoffman found groups of three to five graves at this end of the cemetery, as if close members of the family, or the royal court, lay together. Almost nothing remained in the looted sepulchres, except some fine pottery and the remains of a sycamore funerary bed with carved bull's legs. One large tomb excavated many years ago had walls painted with motifs denoting spiritual power, including a motif found on knife blades and other artifacts of a man holding two lions apart (figure 12.2). When the excavators scraped the soil around the pits with trowels, they located circular, brown soil discolorations, the rotted stubs of stout wooden posts. These uprights once supported small painted palaces and shrines that covered the tombs, forming a minature city of fenced enclosures and diminutive buildings. The excavators even located the small holes made by the original surveyors' posts used to align the tombs.

Upstream, on the other side of the dry wadi, the dead lay in rock-cut sepulchres, with the burial chamber of each tomb on one side, a design favored in Upper Egypt. The mourners buried

Fig. 12.2. A male figure, perhaps a ruler, holds apart two facing lions on a knife handle from Gebel-el-Arak. Reprinted with permission from Petrie, W. M. F. 1917. "Egypt and Mesopotamia." *Ancient Egypt* 29:64.

assemblages of sometimes mummified wild and domesticated animals around the tombs, potent symbols of a spiritual world where animals and humans now acted as one.

The Nekhen cemetery commemorates anonymous leaders, whose names are lost to history. We glimpse them only from a scene on a decorated artifact buried many centuries later in a pit. The fragmentary "Scorpion" mace head shows a ruler in full ceremonial dress, with a ritual bull's tail, a symbol of kingly authority, hanging from the back of his belt.* He wears the white crown of Upper Egypt and wields a mattock, as if he is digging the foundations for a temple or a town. A scorpion dangles before his face, presumably a depiction of

*Cultural continuity is an essential feature of ancient Egyptian civilization, reflected in formal regalia and art motifs that developed over many centuries. Such is the inherent conservatism and continuity in visual depiction that Egyptologists can be relatively confident of their identification of royal costume and other artifacts shown in tomb paintings and other contexts.

his name. Kinship may already have been synonymous with conquest, for the top of the mace bears a frieze of lapwings hanging by their necks from vertical standards. The hieroglyphic sign (see information in box, following) of a similar small bird, *rekhyt*, later came to mean "common people," as if Scorpion has returned from a victorious campaign with many commoners as prisoners. Since Scorpion wears only the crown of Upper Egypt, we can deduce that either part of the

---

EGYPTIAN HIEROGLYPHS

The Egyptians' love for traditional forms appears in their basic vocabulary. Their reverence for tradition caused them to use the original shapes of hieroglyphic writing for nearly 3,000 years, despite centuries of cultural change. This distinctive Egyptian script appeared, virtually fully developed, in 3100 B.C., just before the unification of Egypt. Predynastic people used many signs, and painted them on pottery and weapons. Although some of these pictograms closely resemble later hieroglyphic signs, the origins of Egyptian script remain a mystery. Hieroglyphs are commonly thought to be a form of picture writing. In fact, however, they are a combination of pictographic (picture) and phonetic (representing vocal sounds) script and were set down on papyrus reed "paper," carved on buildings, or painted on clay or wood. Although vowels were pronounced as *ths smpl xmpl shws,* ancient Egyptian writing was consonantal rather than syllabic. Two cursive scripts developed from hieroglyphs: hieratic, which developed soon after 3000 B.C., was an everyday administrative and business script that eventually developed into a priestly script. Demotic, a more informal hand, appeared after 650 B.C. and was the standard documentary script by the time the Rosetta stone was inscribed in 196 B.C.

European scholars wrestled with the decipherment of hieroglyphs for centuries. When a French soldier found the Rosetta stone in the Nile Delta in 1799, British and French experts at once realized that its trilingual inscription in hieroglyphs, demotic, and Greek held the key. The brilliant French epigraphist Jean-François Champollion finally unlocked the secrets of hieroglyphs in 1822.

---

scene is lost or that he lived before the southern kings conquered the north. Apparently this king ruled before Narmer, the first pharaoh.

The Narmer Palette

A single ceremonial palette from Nekhen commorating Egypt's first pharoah brilliantly expresses the powerful symbolism of unification. His name was Narmer, which means "catfish," but his subjects called him Menes. In 3100 B.C. he became the most powerful person on earth, a leader not of a small chiefdom or a small river valley, but of a unified Egypt.

Two English archaeologists, J. Green and J. Quibell, identified Narmer in 1898. They were digging into ancient Nekhen (Hierakonpolis), located in Upper Egypt, when they unearthed an elaborately decorated stone palette of a type used to hold cosmetic pigment. The 63-centimeter high, green slate palette bears fine carvings of mythical animals with long necks and lionlike heads and commemorates Narmer as the conqueror of the age-old Two Lands of Egypt (figure 12.3). On one side, he wears the white crown of Upper Egypt, and carries a pear-shaped mace head in his right hand. He is about to smite a captive, perhaps from the Delta in Lower Egypt. A falcon head (representing the human-headed falcon god Horus) emerges from papyrus reeds carrying a human head above the victim. A sandal bearer follows the king, who stands on two dead enemies. The other side shows Narmer—wearing the red crown of Lower Egypt—and two high officials inspecting rows of decapitated enemies from the town of Buto in the Delta. A central design of intertwined animals symbolizes harmony, balancing images of conquest in the upper and lower registers. At the bottom, a bull destroys a city wall and tramples on its enemies.

The defining event of Egyptian civilization—the unification of Upper and Lower Egypt, and the founding of pharaonic kingship—balanced the two Nile kingdoms into a satisfying whole unified by the king. The Narmer palette commemorates this momentous occurrence, using inscriptions and the spare, economic style of hieroglyphs to deliver a clear, unmistakable message of symmetry as a symbol of order. The Narmer palette displays animals engaged in human activities like warfare and conquest. The slate surfaces also bear animal-

Fig. 12.3. The Narmer palette (height: 63 centimeters). Drawing from Routledge, London.

topped standards, as if early deities, like later ones, had animal forms but human behavior. Such images envisaged divine forces; gave them multiple images, traits, and descriptions; and communicated the essence of their being. This may have been the time when the ancient relationship with the sun, the river, and the land developed into an implicit duality of the Egyptian world. Unification, the process of creating a single state along the river from the Two Lands of Upper and Lower Egypt, relied on a balanced duality, as did other philosophical couplings, among them cosmic order and chaos, and light and darkness.

Duality and unification make compelling themes for the Narmer palette, turning it into a symbol of kingship. Real and imagined animals become allegories for the forces of life. The Narmer palette shows carnivores such as lions and dogs acting in harmony in pairs. Other early artistic works dictate a deliberate use of harmony achieved in a turbulent world, of balance between opposing forces. Animals as an untamed life force walk in lines or in registers between parallel lines, again implying an aesthetic consciousness for the first time.

On artifacts and inscriptions from later Egyptian times, kings and gods capture flocks of birds with large nets. The accompanying hieroglyphs stress the symbolism of containment. Another palette from a Nekhen sepulchre depicts a conflict between lions, and other pairs of fighting animals. The lions are in precarious balance, while two savage hunting dogs face one another in a display of mutual strength. Control and harmony show the tautness and formality that characterize Nekhen tomb painting and design.

In another famous statement about unification, King Senusret I of the XII Dynasty (1971–1928 B.C.) had the origin myth inscribed on his mortuary temple at el-Lisht. A vertical sign divides the inscription, a stylized picture of a windpipe that signified the verb "to unite." The inscription tells how: "The gods Horus and Seth, once lords of Lower and Upper Egypt, tie the heraldic plants of their two domains around the hieroglyphic sign for 'unification' " (Kemp 1989, 29). Egyptian legends traced the ancestry of the dynastic pharaohs to the "Followers of Horus," the falcon-headed god and solar deity, who was a manifestation of the living king and his protector. Nekhen, "the City of the Falcons," was Horus's town, and perhaps the founding community of Egyptian kingship.

How and why did the estimated half million or so inhabitants of the Nile Valley in 3000 B.C. come together under a single ruler? None of the obvious factors—such as population pressure, food shortages, or long-distance trade—that led to urban civilization in other parts of the world seemed to apply here. Through a combination of archaeological research and evidence from Egyptian documents, we gain a provocative answer to the unification question.

A GAME OF ANCIENT MONOPOLY

Kemp likens the process of unification to a game of Monopoly in which each player maximizes the opportunities thrown out by the dice. In Egypt, both individuals and entire villages took full advantage of favorable locations; access to desirable resources, such as potting clay or gold mines; and chance breaks that came their way. These riverside communities, like Monopoly players, were basically equal opportunists. Inevitably someone, or some hamlet, gained an unforeseen advantage, either from trading expertise, or unusually high crop yields. Equilibrium gave way to the inevitable momentum, where some communities acquired more wealth and more political power than their neighbors—the ancient equivalent of building Monopoly hotels on Park Avenue. Their victory made possible a monopoly over local trade, food surpluses, and other commodities, which overrode any possible threat posed by other political or economic players.

Hundreds of Monopoly "games" unfolded in predynastic times. As generations passed, the number of players decreased and the stakes increased as progressively larger chiefdoms vied for economic power and political dominance. Players changed—some acquired great influence, then lost it as charismatic individuals died or trading opportunities ended. Egypt, with its fertile land and resources, could withstand such changes for many generations. Surplus grain or toolmaking stone formed the foundation of their strength. The Egyptians also had a genius for weaving a distinctive ideology that imbued leadership and authority with elaborate symbols and rituals. These ideologies became an underlying factor in promoting unification of Upper and Lower Egypt, starting with at least three predynastic chiefdoms that flourished in Upper Egypt around 3500 B.C.: Hierakonpolis, Nagada, and This, near Abydos.

Archaeology, when combined with myth, creates a hypothetical scenario for "unification." Through the discovery of Mesopotamian clay seals from Upper Egyptian sites, we know that by 3500 B.C., the kingdoms of Upper Egypt engaged in major trading enterprises. Direct contact with southern Arabia and Mesopotamia enabled them to bypass ancient trade networks in Lower Egypt. Strong evidence shows that trade traveled across the Eastern Desert. One historical scenario describes a ferocious, but historically undocumented, competition between neighbors, and between the kingdoms of Upper Egypt and the Delta. Inevitably, these rivalries led to decades of fighting along the river, which eventually culminated in the politically developed center of Nekhen conquering its Upper Egyptian neighbors.

Nekhen's rulers had embarked on a military campaign, which engulfed all of Egypt between the Mediterranean and the First Cataract. By 3100 B.C., a semblance of political unity joined Upper and Lower Egypt as depicted in the symbolic linking of Horus and Seth.

THE GREAT CULTURE

Narmer and his successors were confronted not only with the major practical problem of creating an intellectual framework for a state ruled by a supreme king; but also with the problem of how to project the concept of a unified kingdom of Upper and Lower Egypt to people living in towns and villages the length of the land. Around 3000 B.C., some gifted Egyptians codified a remarkably homogenous intellectual system, building on the works of earlier priests. They created a "Great Culture," a court-created tradition that became the instrument of royal rule. The new culture embraced hieroglyphic writing, formal commemorative art, architecture, and a basic iconography of kingship based on much earlier ideas. The king stood at the center of the country as a divine force, identified in his first name and title as a manifestation of the power of the god in the heavens, Horus the falcon. Scribes wrote the ruler's name inside a panel depicting his palace, denoting the king as Horus, present, alive, and in residence. In time, the palace became a convenient way to refer to the person of the ruler, as *per-aa*, "the great house." The pharaoh gradually acquired five ceremonial titles, including *nebty*, "He of the Two Ladies," referring to the protective powers of the cobra-goddess of Lower Egypt and the vulture-goddess of Upper Egypt. By 2500 B.C., the king's titles appeared within an oval cartouche signifying the circuit of the sun around the universe. A second cartouche named the pharaoh "Son of Re," identifying an even closer kinship with the sun-god. The palace assumed great importance in Egyptian ceremonies, as did the rare ceremonial "appearances" (*kha*) of the king, which symbolized the rising of the sun at dawn.

The Egyptians elevated the consummate image of what constituted proper form to the pinnacle of aesthetic desirability. The art of the royal court became a self-perpetuating ideal. Carefully selected artists translated these standards into a precise graphical style almost akin to modern commercial art. Dynastic court art combined three major elements: careful use of registers of figures and themes; an intimate connection between figures and hieroglyphs, which became a single channel of communication; and carefully controlled conventions for animal and human figures. The scribes produced informative and truthful pictures, but they were based firmly on elements taken from the mythic and ideal world. Thus, ancient Egyptian art was a carefully constructed style developed as a means of religious communication, which constantly added new components to ancient myths and traditions. Over the centuries, court art and culture expanded to all corners of the kingdom, adding local traditions, and gradually creating a national framework of myth and design.

As unification took hold, ideas of kingship changed considerably. The trappings of kingship assumed ever-greater importance. The king's clothing and regalia became a mantle of divinity—of potency in creation—that symbolized his role as herdsman and protector of the people. In art, inscription, and regalia, the pharaoh became a warrior and a builder. The king passed the goodness of humankind to heaven and received the blessing of the creator and the other gods on earth. Even the earliest types of kingship portray the king alone acting for the common good, surrounded by a chorus of courtiers who praise his every action. His official entourage formed the "followers of Horus," loyal officials who surrounded the king in the palace, and on his royal progresses through the land. They also transmitted his commands to the world outside the palace audience hall. This ideology extended to the Egyptians' classification of themselves as a series of circles around the pharaoh. A solar retinue (*henmemet*) formed his bodyguard; an inner elite circle (*pat*) was comprised of advisers and senior officials; and the rest of the people, (*rekhyt*) made up the outer circle.

The same messages of kingship endured after death, expressed in imposing sepulchres. The earliest dynastic kings lay in a desert cemetery at Abydos, downstream of Nagada. Their brick-lined burial chambers sat in deep pits in the sand, covered by a plain, square enclosure filled with sand and gravel. Pairs of free-standing stone stelae bore the Horus name of the ruler. A separate mortuary temple associated with each sepulchre lay closer to the edge of the flood-plain, just beyond the town. Here priests made offerings to the king as a god during his life and after death.

King Khasekhemui (ca., 2640 B.C.), the fifth king of the Second Dynasty, was a vigorous military campaigner, who contributed much to Egypt's unity. His funerary enclosure measures 54 by 133 meters, and is surrounded by a double mud-brick wall. The inner wall is no less than 5.5 meters thick in places. Doorways intersect both walls. Khasekhemui's enclosure followed the *serekh* style of paneled facades and recessed entrances developed by King Wadji, the third pharaoh of the First Dynasty. Wadji adopted the *serekh* as the heraldic symbol of the king, replicating the facade in a rectangle, with the ruler's name written above, topped by the Horus symbol of a hawk or falcon. The builders created a panel-like effect by making parallel niches in the mud-brick. A single, free-standing funerary palace building stood near the eastern corner of the enclosure, adorned with the same paneled motif.

We know the meaning of Khasekhemui's building because the *serekh* style occurs elsewhere, on the facades of early dynastic tombs at Memphis and Nagada. The buried frontages of these graves preserve elaborate painted decoration, which represents now-vanished methods of adorning the walls. Long strips of brightly colored matting lashed to horizontal poles once draped the narrow spaces between the wall niches. The paneled surfaces gave way at intervals to deeper recesses with red-paneled sides. A broader niche stood at the back of each recess, painted red to give the impression of a wooden door. This pattern of panels, niches, and recesses became a standard way of adorning offering places and sarcophagi in later sepulchres. The nitched and decorated palace facade became a symbol of royal authority, and communicated the sense of "palace," and "court." Art and architecture combined to render an image of a divine pharaoh and his royal court at the pinnacle of human existence along the Nile. Privilege was carefully controlled. The pharaoh permitted a few of his most prominent courtiers to use a variation on the palace facade to adorn their tombs.

The king exercised a royal monopoly over his kingdom, at least in theory. The Great Culture did not transform Egypt overnight, for court patronage was at best fitful outside the major towns.

THE KING AS TERRITORIAL CLAIMANT

The pharaoh was the symbol of a unified land. Inevitably, court architecture began to reflect the ideology of a supreme ruler, a god on earth. The relatively modest architectural statement that King Khasekhemui built pales in comparison to that of King Djoser, first pharaoh of the Third Dynasty (ca., 2695 B.C.). Djoser's brilliant architect and chief minister Imhotep codified the myth of the state and the institution of Egyptian kingship in the first monumental building built entirely in stone—the Step Pyramid at Saqqâra, upstream of modern Cairo (figure 12.4).

Imhotep, a shadowy historical figure, was the son of a famous architect named Kanofer. He achieved fame as a wise court official, architect, scribe, and healer. His Step Pyramid stands in the center of a 278 by 545 meter enclosure. Imhotep fashioned a *mastaba* (tomb) in six levels, diminishing in size to form a pyramid. The base of the lowest level measures 107 meters from north to south and 122 meters from east to west. The layered, limestone-covered pyramid reached a height of 58 meters.

A maze of subterranean passages attempted to foil tomb robbers, disguising a 28-meter shaft adorned with fine reliefs and blue faience tiles that led to the burial chamber. The tiled walls had the same matting effect used on the royal palace at Memphis. A 3-ton rock plug sealed a granite-lined burial chamber with a 4-meter ceiling. Vast quantities of vases, some bearing the names of earlier kings, were found in these passages, perhaps offerings made by Djoser in honor of his predecessors. Only a mummified left foot remains of the king himself. At the north side of the pyramid stands a boxlike limestone cellar containing a life-size, seated figure of Djoser, wearing the white cloak that was worn at jubilee festivals.

The Step Pyramid is part of a much larger landscape. A thick stone wall with external towers surrounds the enclosure. The builders used a simple version of the now-familiar palace facade motif, with

Fig. 12.4. The Step Pyramid at Saqqâra. Photograph from ArtWorks.

211 bastions and 14 entrances. The main gateway at the southeastern corner leads to an entrance hall 53 meters long decorated with columns. The hall in turn opens into a vestibule with four pairs of columns. The so-called South Tomb faces the pyramid on the south side of the enclosure. The king's viscera were buried here, his mummy under the pyramid opposite, fulfilling the need for a northern and southern sepulchre, one for each of the Two Lands (figure 12.5).

The Saqqâra enclosure measures 108 by 187 meters, faced by paneled stone walls. A replica of a royal palace lies at the southwestern corner of the enclosure. A pair of stone, horseshoe-shaped cairns stand at each end of the plaza. They face an elevated throne dias shaded by a canopy at the foot of the Step Pyramid. The steps leading to the dias face the alignment of the cairns. The design of the dias goes back deep into history. A mace head from King Narmer's reign at Nekhen

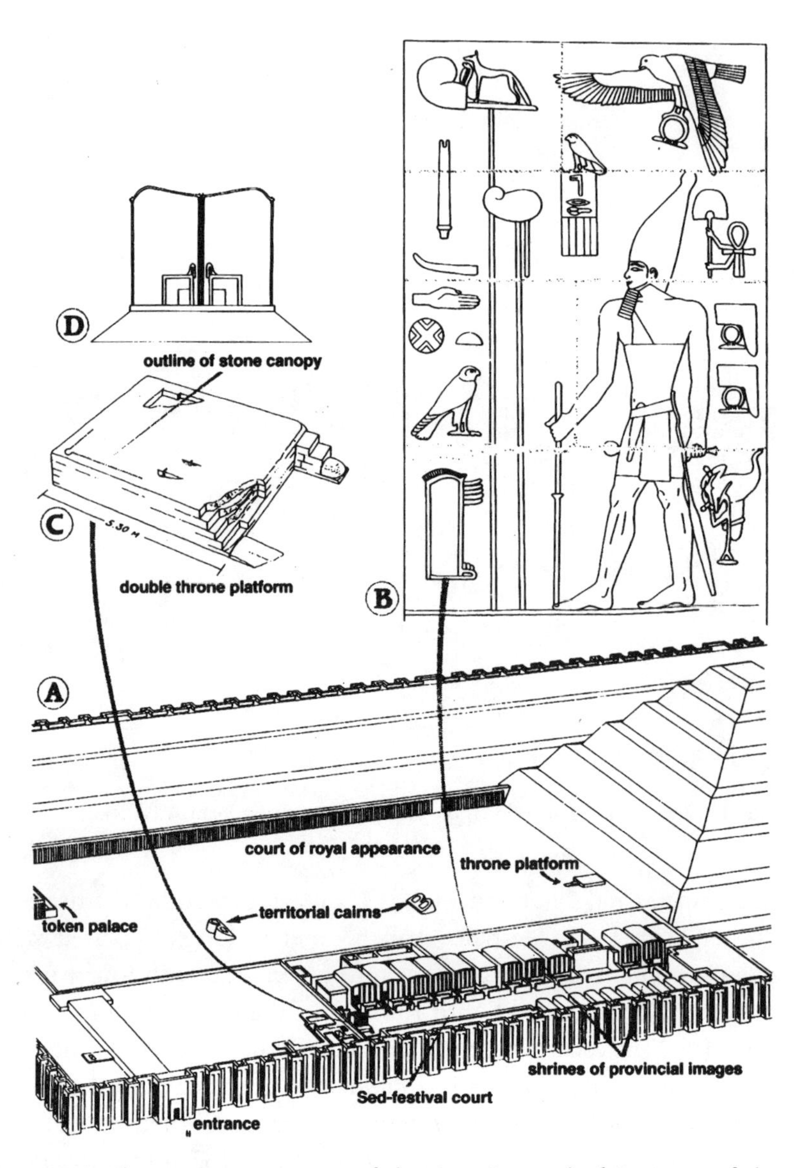

Fig. 12.5. The southern precincts of the Step Pyramid of Saqqâra of the pharaoh Djoser. (A) The eternal plaza of royal display, where the *Heb-sed* festival was enacted. (B) King Djoser before the temporary shrine of Horus of Behdet. The hieroglyphs in front of the ruler read: "Halting [at] the shrine of Horus of Bedhet." (C): Stone platform with double throne at the southern end of the *Heb-sed* court. (D): Ancient representation of the double-throne dais with canopy used at the same festival, shown on a carved lintel commissioned by the pharaoh Senusret III of the Twelfth Dynasty (ca., 1878–1841 B.C.). Drawing from Routledge, London.

depicts the ruler sitting on a similar canopied dias facing prisoners of war and animals captured in battle, arrayed between territorial markers like those at Saqqâra. A god seated in a litter faces the king.

A wooden label from the tomb of King Den of the First Dynasty at Abydos shows a similar scene. Den sits on a canopied dias, a hieroglyphic sign for the year of his reign behind him. The king appears a second time at the foot of the dias, striding between six territorial markers. The same theme appears at Saqqâra in two groups of three carved panels on the backs of imitation doorways in underground passages below the Step Pyramid. We see Djoser performing exactly the same ceremony, striding between the territorial markers. Later inscriptions tell us that the cairns were markers of territorial limits, placed in what was called "the field." The ceremony was called "presenting the field."

Egyptians believed kings received "understanding" from the sun-god while still in the womb. The pharaoh was *netjer*, a god. But deities were of human origin, and, thus, required tombs after death. Unlike the gods, a pharaoh had to observe jubilees of revivification to ensure the continued fertility of the land. The *Heb-sed* festival, one of the great ceremonies of state, was performed exactly 30 years after the pharaoh's accession—and at more frequent intervals later in his reign. *Heb-sed* celebrated the earthly power and vigor of the king and always embraced two elements. First, the ruler would appear in full ceremonial regalia and sit on a special dias provided with two thrones, symbolizing Upper and Lower Egypt. Then he would stride around territorial cairns set in "the field," thereby renewing his claim to the kingdom (see information in box, p. 300). The *Heb-sed* festival had profound significance in ancient Egyptian religious life, as it reemphasized the close relationship between the divine king and a unified land.

At Saqqâra, a courtyard with a complex of dummy shrines lies on the east side of the enclosure. These solid, dummy structures replicate temporary shrines in timber and matting (figure 12.5). Arranged alongside two sides of a rectangular court, these are exactly like buildings built specifically for *Heb-sed* festivals in later times. A once-canopied throne dias with two flights of steps stands at the south end of the court. Scenes of Djoser visiting local shrines appear in the underground galleries.

AMENHOTEP III'S *HEB-SED* FESTIVALS

Such is the fundamental continuity in ancient Egyptian civilization that Egyptologists can use later accounts of major royal ceremonials to interpret earlier structures with considerable confidence. Much of our knowledge of the long-enduring *Heb-sed* festival comes from much later New Kingdom sources. King Amenhotep III (1390–1353 B.C.) celebrated three such occasions at Thebes, the first after 30 years of rule, as prescribed; the next two in the thirty-fourth and thirty-seventh years of his reign. The first two were held at Malkata on the west bank of the Nile south of the royal mortuary temples and the processional perimeter, which joined Karnak, Luxor, and the west bank. (We do not know where the third ceremony was staged.) The setting for Amenhotep's pageantry has been reconstructed from aerial photographs, a site survey, excavations, and contemporary records. At Saqqâra, the king strode around an arena. At Malkata, Amenhotep ordered the digging of a large, artificial basin measuring 1 by 2 kilometers where the floodplain met the desert. The earth from the basin was spread out to make a terrace for the ruler's mortuary temple and adjacent palace, which was shaped into rows of landscaped hills. The climax of the first ceremony saw "the glorious appearance of the king at the great double doors in his palace, 'The House of Rejoicing'; ushering in the officials, the king's friends, the chamberlain . . . and the king's dignitaries" (Sethe 1905, 837). Important officials were honored with gifts and had the privilege of towing the evening and morning barges, which carried statues of the sun-god and reenacted the sun's daily journey. The excavations and site survey revealed the ruins of the transitory palace, demolished soon after its one day of prominence, to make way for an enlarged lake for the next ceremony.

The Step Pyramid complex was an elaborate formal setting for the display of kingship—and of the ruler himself—either to his courtiers, or to the populace at large. The setting always comprised a combination of key elements: open space, an elevated place where the king could be seen by large numbers of people, and a token palace where the ruler could don his ceremonial regalia or rest between appearances. In later centuries, the "appearance of the king" was an important occasion, requiring elaborate devices, but the roots of these later ceremonies were already in place at the very beginnings of Egyptian civilization, when kings were both divine and human.

King Djoser's Step Pyramid brought Egyptian royal tombs to a new architectural height. The pyramid and enclosure formed an arena for the eternal pageantry of the king on earth. As supreme territorial claimant and the focus of adoration and worship, the king could appear within the sanctity of his own enclosure on earth.

During the Fourth Dynasty, however, Djoser's successors changed the stepped pyramid into a true pyramid. Around 2575 B.C., King Huni built the pyramid of Meidum. It lay outside the floodplain in the desert. Soaring toward heaven, the pyramid symbolized the sun's rays descending through the clouds to earth, which absorbed the pharaoh into the sun itself. A small mortuary temple and place of offering for the king's spirit stood on the east side, dwarfed by the smooth walls of the pyramid. A walled causeway led to a temple in the valley, that was a display of linear architecture quite stark, planal, and powerful. In coming centuries, the architectural effect softened at Giza and elsewhere. The pharaoh was now "Son of Re," worshiped as a god while still alive. By this time, the official theology that was mirrored in formal architectural styles codified a "Great Culture" of Egyptian civilization, which evolved slowly over nearly 2,500 years of pharaonic rule, in what Kemp calls "a cauldron of tradition." Ideology legitimized the rule of a king for over 500 years.

Ancient Egyptian society achieved long-term stability because centuries of pharaohs and their officials wove a tapestry of myth and reality—a "Great Culture"—to maintain pervasive doctrines of stability and unity. They used architecture and art to assault the senses with mythic statements expressed in their pyramids, shrines, temples, and tombs—all stunning examples of how art forms develop out of ritualistic roots and the psychology of its inhabitants. As a result of this powerful message, ancient Egyptian civilization endured for nearly 3,000 years.

chapter thirteen

XUNANTUNICH: "THE MAIDEN OF THE ROCK"

The ancient Mayan city of Naranjo in Guatemala's Petén spread out before me, more than 60 hectares of platforms, plazas, pyramids, and abandoned buildings mantled in dense rain forest. Archaeologists must travel with chain saws, and use rough logging roads to reach ancient Mayan cities (figure 13.1). Only a handful of scholars have studied the site, among them an Austrian engineer named Teobart Maler, who made a sketch map of the major architectural features a century ago. More recently, Harvard University archaeologist Ian Graham made a quick foot-and-compass survey to amplify the Maler plan. As I walked among the palm trees, dodging patches of undergrowth, I sensed, but did not see, the high steep-sided pyramids on either side. The trees reached toward patches of blue sky visible through the green canopy. Fallen palm trunks green with damp lichen littered the once spotlessly clean, plastered plazas where Mayan lords of long ago walked in procession. I stopped to look at Maler's rough map and was reminded of the lyrical description written by nineteenth-century archaeologist, John Lloyd Stephens, about another Maya city—Copán in Honduras: "The only sounds that disturbed the quiet of this buried city were the noises of monkeys moving along the tops of the trees and the cracking of dry branches broken by their weight. They moved over our heads in long and swift processions, forty or fifty at a time. . . . Amidst these strange monuments, they seemed like

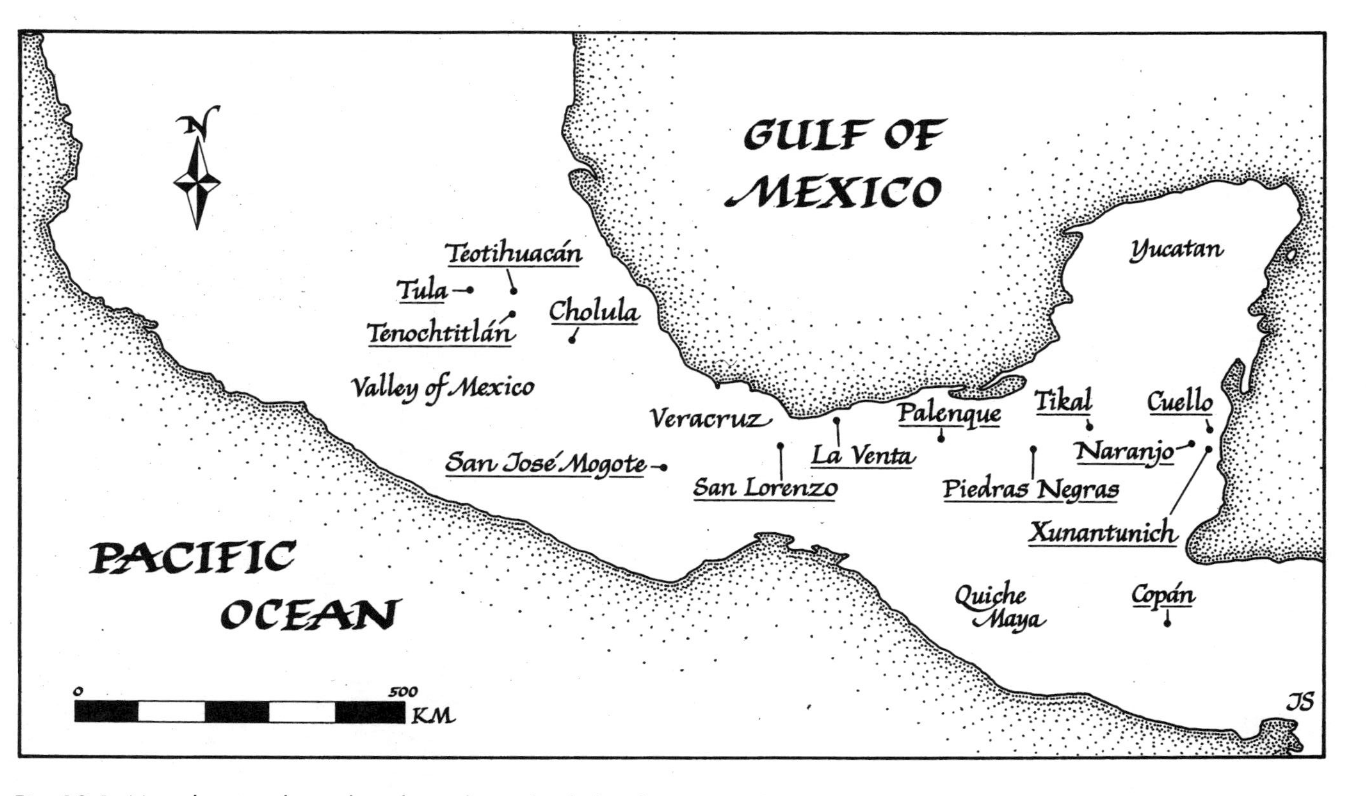

Fig. 13.1. Map showing the archaeological sites (underlined) mentioned in chapter 13.

wandering spirits of the departed race guarding the ruins of their former habitations" (1841, 175).

Today, no monkeys call Naranjo home. The wandering spirits have moved on. I later learned, back at camp, that Naranjo was an important classic Mayan city, founded between A.D. 550 and 600. Deciphered glyphs revealed that a powerful lord, Smoking Squirrel, had come to power a century and a half later. His mother, born into a noble family, came from the city of Tikal, 65 kilometers away. Although he was a scion of one of the greatest Mayan states, Smoking Squirrel's kingdom collapsed by A.D. 900 and the powerful city was abandoned, perhaps a victim of the crosscurrents of Mayan diplomacy and political rivalry. A millennium later, overgrown pyramids, tree-covered plazas, and a handful of inscribed stelae testify to this ruler's magnificence. Four other stelae stand in the town square at nearby Melchor de Mencos, upon which I could discern long-forgotten Naranjo lords standing in triumph on the bodies of captives taken in battle.

Naranjo's pyramids look inward, their steep staircases heading upward from plazas to temples high above. Each plaza is enclosed on four sides, sometimes with barrier walls or platforms restricting access. I explored the small plazas that were once the heart of residential quarters for Mayan nobility; they appeared to me like small valleys. I walked across public plazas where large straight tree trunks formed narrowing corridors. Everything that I glimpsed and explored, whether earth-covered pyramid, weathered stelae, or overgrown natural cave, all pointed to a grand design.

On this memorable day, I climbed several steep-sided pyramids, clinging to tree trunks, my feet slipping on the loose earth and stones that mantled the ancient masonry. I peered into vaulted buildings on their summits, the stonework still intact after 10 centuries. Only a few pyramids penetrated the forest canopy. From the summit of Pyramid C-9 I gazed out over densely forested, rolling hills of grays and greens splashed with brilliant red flamboyant flowers. The pyramids rose in stone-like mountains in the forest, reaching toward the wide arc of puffy trade-wind clouds that marched in procession high overhead. I scrambled back down to the plaza below, skidding and sliding from tree to tree.

Intensely symbolic, the Naranjo landscape in stone, stucco, and clay casts a powerful spell; its stelae and pyramids replicate the Mayan world of trees and sacred mountains. Scattered across a heavily farmed, deforested landscape, the city of Naranjo shone and glistened in A.D. 700, her pyramids and temples plastered and painted in brilliant hues. Ancestral friezes blazoned with vivid color commemorated deities and ancestors. It is at these sacred places that Mayan lords appeared before their people, standing on depictions of their ancestors wrought in plaster and stone. When Smoking Squirrel ruled, more than 200,000 people lived close to Naranjo in the Belize Valley. Now the same number inhabit the entire country of Belize. The Maya have long dispersed, exotic diseases brought by the Spanish conquistadors have decimated the Maya people.

Atop one pyramid, archaeologist Julie Miller and I stopped to study the stone work. She traced the outline of a great serpent's head, which once adorned the facade of the temple on the summit. At first, I couldn't discern the figure until my finger traced the staring eye. We studied how two rows of stones formed the upper jaw from a seemingly jumbled set of boulders. Only a few experts are able to decipher such a portrait. Miller has had years of experience interpreting Mayan friezes; she has worked with the rocks themselves, and with photographs taken in different lights.

In another spot, a noble family's house stood atop a low pyramid. The blackened and weathered facing stones still supported a plastered cornice with red paint clinging to it. Miller showed me where one wall had lost its casing stones; rubble fill cascaded down the steep slope. She pointed out the intricacies of the architecture. I saw how complex the archaeological undertaking of this highly symbolic world can be, and wondered if archaeologists will ever fully unlock its mysteries.

XUNANTUNICH

Just over the modern frontier in Belize, approximately 16 kilometers away as the crow flies, stands the great Castillo pyramid of the Mayan ceremonial center called Xunantunich—"The Maiden of the Rock" (figure 13.2). Ten centuries ago, Xunantunich was the hub of a small valley kingdom, a sacred place that, like Naranjo, replicated the Mayan universe. The first time I saw it was at dawn, when the pyramid seemed to float on billows of morning fog that ebbed and flowed at its foot. As I took the ramshackle ferry across the Mopán river, fine cascades of mist rose from the calm water. By the time I reached the site, dense fog shrouded the summit high above. Despite the gloom, my eyes were drawn irresistibly upward, toward a dark orifice in a building high atop the Castillo—just as the architects intended. From this vantage point, the effect has full impact, for the center of the site has been cleared. Without trees and forest canopy, the site draws you in and upward, to the seat of spiritual and regal power high above.

In the still dawn, dry leaves rustled under the excavation workers' feet. Rust-colored lichen clung to the weathered, great limestone walls of the Castillo, which dominate the rolling landscape and the Belize River Valley for kilometers around. But a thousand years ago, the brightly painted white mountain would have stood out starkly against the blue sky. This smaller ceremonial center has all the authority and mystery of much larger Mayan cities, but the advantage of being more compact, and thus easier to investigate. Richard

Fig. 13.2 The Castillo at Xunantunich, Belize. Photograph by Brian Fagan.

Leventhal has been working at Xunantunich for 7 years with a changing team of expert subordinates, using local laborers who have worked on archaeological sites for years. His project is part basic archaeological research, part conservation and stabilization, and part tourist development: Xunantunich lies only 1-1/2 kilometers from a road leading from Belize City to the Guatemalan border and receives almost 30,000 visitors a year.

Mayan farmers have lived at Xunantunich for a long time. A small Mayan village flourished on the site between 500 and 400 B.C., but was abandoned 200 years later. Classic Mayan civilization was in full swing 800 years after that, when a lordly family of unknown origin built a new ceremonial center on the site, between A.D. 650 and 700. Leventhal has built a small museum to display the surviving stelae. Unfortunately, the few glyphs found at Xunantunich are so badly weathered that they defy decipherment. Despite lengthy examination with oblique and enhanced lighting, no one can decipher the eroded glyphs. "Stela 3" shows a ruler figure wearing a quetzal-feathered ornamental backrack (ceremonial regalia worn on the back) and carrying a scepter in the form of a serpent-footed deity associated with ancestors. "Stela 6" commemorates a somewhat similar ruler wearing a beaded hip cloth, with a carved greenstone head hanging from his belt. Below his feet is a tantalizing historical clue—a barely decipherable outline of the distinctive glyph associated with the city of Naranjo. Leventhal believes that this may indicate that Xunantunich was once threatened, or even controlled by her larger neighbor, Naranjo. Although the latter city had collapsed by A.D. 900, Xunantunich continued to prosper for another century, reaching its apogee in A.D. 1000. Some of the most active building took place after the collapse of Naranjo, when the lords of Xunantunich may have presided over a small, but populous, kingdom of small towns and villages in the nearby river valley. A short time after A.D. 1000, the entire kingdom collapsed and Xunantunich's 400-year history came to an end.

SURVEYING THE KINGDOM

Excavating and surveying even a medium-sized Mayan ceremonial center is an enormous undertaking, for it is impossible to understand such a sacred place without placing it in a wider historical context. From the beginning, Leventhal combined his investigations at Xunantunich itself with a survey, run by Wendy Ashmore of the University of Pennsylvania, of the surrounding countryside.

The Xunantunich survey is modeled on a highly successful project conducted on the hinterland of the great Mayan city of Copán in Honduras (see information in box, p. 309). Ashmore and her team surveyed four transects across the surrounding countryside. Three of them were 400 meters wide and of variable length, up to 5 kilometers. They extended in different directions, and ended at known major sites. Their purpose was to enable investigators to study—through various samples—how an ancient settlement related to different topography, soil types, river drainage, and distance from Xunantunich. The fourth transect was 2 kilometers square and surveyed some of the area between the center and the Guatemalan border, considered to be of vital importance, given the proximity of Naranjo to the west. The survey team also examined important areas lying beyond the transects for any obvious traces of Mayan settlement.

The survey crews worked in teams. They used a Brunton compass mounted on a tripod to lay down the central baseline for a transect, then used machetes to cut a 1- to 2-meter strip along this line, moving the compass frequently to ensure an accurate line. A wider swathe might have been more convenient, but much of the land was under cultivation and local landowners did not want wide cuts through the vegetation to become footpaths across their land. About every 18 meters, the crew cut meter-wide branches at right angles, which extended out for about 200 meters. They cleared any sites that they found during this process. As the crews cut the transects, the archaeologists followed them, recording the topography, vegetation and land ownership, as well as archaeological sites. They documented the information in field books, including the direction of slopes, and a standardized-scale rating on the vegetation, so that they could assess its effect on visibility across the land. Meanwhile, a Belizean member of the team walked the areas between the cuts, looking for important natural features and sites.

When the crews encountered house mounds, they cleared the brush from them, sketched the dimensions, and later, mapped them, recording all the details on a standard form. Then they excavated

SETTLEMENT SURVEY AND OBSIDIAN HYDRATION AT COPÁN, HONDURAS

Large-scale settlement surveys are an essential part of Mayan archaeology. At Copán, successive research teams have located and mapped more than 135 square kilometers around the urban core, focusing on the areas most suitable for human settlement. A combination of aerial photographs and foot surveys has recorded more than 1,425 archaeological sites in the Copán Valley. The researchers mapped and surface collected each one of them. They developed a hierarchy of different site types, from complex villages to isolated hamlets and dwellings as a way of studying changing settlement patterns on the landscape. The survey revealed an urban core, a densely occupied area surrounding the city, and a rural region with a much lower settlement density. The researchers combined selective test pits with their field surveys, digging "keyhole" trenches into 142 sites, chosen on the basis of statistical sampling to ensure unbiased results. The changing pottery styles and radiocarbon samples from the trenches were sufficient to link individual sites to the master chronology for Copán, developed from stratified trenches in the city.

The sample excavations also yielded large quantities of stone artifacts fashioned from obsidian, a highly prized volcanic glass traded all over Central America. A freshly formed surface of obsidian absorbs water from its surroundings. This process forms a measurable hydration layer that is invisible to the naked eye. The hydration layer contains about 3.5 percent water, which increases its density and allows it to be measured accurately under polarized light using a microscope. The depth of the hydration layer represents the time since the object was manufactured or used. However, local environmental variables, such as variation in soil acidity must be taken into account. The Copán Obsidian Dating Project has yielded over 2,300 dates from 202 sites—over 14 percent of the sites in the Copán Valley. A comprehensive sampling strategy ensured that dates came from a variety of obsidian artifacts representing the entire span of the deposits. Copán's obsidian artifacts came from several easily identifiable sources, which allowed researchers to use the distinctive chemical characterizations from the dated artifacts to provide information on changes in obsidian trade through time.

The Copán surveys show that, between A.D. 550 and 700, the Copán state expanded rapidly, with most of the population concentrated in the urban core and immediate periphery zones. There was only a small, scattered rural population. Between A.D. 700 and 850, the Copán Valley saw a rapid population increase to between 18,000 and 20,000 people. These figures, calculated from site size, suggest that the local population was doubling every 80 to 100 years. People were farming less fertile foothill areas, as the population density of the urban core reached over 8,000 people per square kilometer, with the periphery areas housing about 500 people per square kilometer. Eighty-two percent of the population lived in relatively humble dwellings, an indication of the extreme stratification of Copán society. The ruling dynasty ended in A.D. 810. Forty years later, rapid depopulation began. The urban core and periphery zones lost about half their population after A.D. 850, while the rural population increased by almost 20 percent. Small regional settlements replaced the scattered villages of earlier times, in response to cumulative deforestation, overexploitation of even marginal agricultural soils, and sheet erosion near the capital. By A.D. 1150, the Copán Valley population had fallen to between 5,000 and 8,000 people. Years of remote sensing, field survey, and test excavation hint that environmental degradation was a major factor in the Mayan collapse.

---

shovel tests to find diagnostic potsherds, which would enable them to date the settlement. Finally, the surveyors entered the site coordinates into a software program called Generic CADD, which produced a computerized and easily manipulated map of the survey area.

The survey collected evidence that documented changing settlement patterns over many centuries, beginning about 1000 B.C. About the time of Christ, local political and administrative centers appeared. During the next 10 centuries, the fortunes of these centers waxed and waned, with the densest farming population and largest centers occurring between A.D. 700 and 900, as Xunantunich began to reach the height of its powers. Ashmore believed most farmers lived in village-sized clusters of houses, corresponding to kin-based residential groups. These farming communities were dispersed widely over the landscape, unlike those of Caracol, Copán, and Tikal, which exercised close control over the villages in their immediate hinterlands.

Leventhal's researchers excavated homesteads of different sizes to better understand this settlement pattern. Some hamlets were made up of well-defined houses, while others were virtually undecipherable arrangements of boulders and stones. During my visit to one settlement site, archaeologist Cynthia Robin uncovered the stone foundations of a small house and a storage pit lying on a slope above a small stream. Her workers sank shovel pits—small cuttings set out in straight lines through the surrounding woodland—to establish the borders of the settlement. Almost nothing of the ancient village could be seen on the surface, except for some discolored stone and a few potsherds, but she managed to estimate the size of the tiny settlement on the basis of surface survey and shovel trenches.

The villages surrounding Xunantunich were economically self-sufficient. This was advantageous for Xunantunich's rulers, who presided over their small kingdom during a prolonged period of political and economic uncertainty in the Mayan homeland, during which many larger cities collapsed. Xunantunich survived for more than a century, and its rulers enjoyed considerable spiritual and political authority. They used their perceived ability to mediate with the ancestors and the gods to control surrounding communities, which supported them and the great center with tribute in the form of food, exotic goods, and labor.

EXCAVATING A MAYAN SACRED PLACE

At Xunantunich, excavators had to balance preservation and tourism concerns with basic research. Fundamental questions also needed to be addressed: What was the history and chronology of Xunantunich? When were the various buildings and plazas erected, and how long were they in use? What were the relationships between the different major building groups and outlying structures? How were they linked? To answer these questions required a distinctive excavation strategy that used carefully placed and highly selective trenching.

A large site can be approached in two ways. A common approach is to excavate one large stratigraphic trench that cuts through the underlying strata, thus providing a detailed history of the site from one location. Boston University archaeologist Norman Hammond used this approach at a preclassic Mayan site at Cuello in northern Belize. Cuello is a small Mayan ceremonial center, today consisting of an acre-square, 3.6-meter-high platform with a low pyramid. Hammond knew that the Maya habitually rebuilt shrines and major centers on the same sites, in the belief that supernatural power accumulated over the generations. Therefore, he excavated layer after layer of occupation under the main plaza, until he exposed a pole-framed, palm-thatched house set on a lime-plaster platform dating to about 1000 B.C. The Cuello people maintained the same basic plaza layout for many centuries, making it larger and larger until about 400 B.C. At this point, the villagers converted their ceremonial precinct with its wood and thatch temples into a large public arena. They burned their existing shrines, tore down their facades and desanctified them, and then filled the square with rubble to create a raised platform covering more than an acre.

Hammond's assumption that he could obtain a cultural sequence from one place paid off. The same cutting yielded the fragmentary skeletons of more than thirty sacrificial victims in the rubble of earlier shrines, some with hacked-off skulls and limbs, others sitting in a circle around two young men. Human sacrifice, especially of children or prisoners of war, was commonplace in Mayan civilization. The victim was either beheaded or his or her still-beating heart ripped out with a knife, a customary ritual in later Aztec times (see chapter 14). Six carved bone tubes buried with the Cuello victims bore the interlacing, woven mat motif symbolic of later Mayan kingship. The appearance of the motif here may document the appearance of a Mayan nobility by 400 B.C., since such mats were actually royal thrones.

Hammond's meticulous excavations at Cuello revealed some of the earliest evidence for the development of Mayan civilization within a limited area. Using this same method at Xunatunich, however, was not possible, for the major buildings are scattered over an area of about 81 hectares. Therefore, Leventhal adopted a highly effective, but different procedure, which was used at many larger sites than Cuello. With this method, each structure is approached in a specific way; the trenches are carefully placed to solve different archaeological problems.

Hammond's Cuello site showed that the stratified layers, which were built one upon the other, were accumulated for spiritual reasons. Although Xunantunich has a shorter history, sacred structures were also rebuilt one on top of another similar to Cuello. However, other buildings and plaza and pyramid groups were also erected at some distance from the center. Thus, effective excavation of the site required samples of diagnostic potsherds from different stratified levels for dating purposes; large numbers of radiocarbon dates; and, when possible, the use of obsidian hydration samples to confirm the carbon 14 readings. The site also demanded a detailed architectural understanding of pyramids, plazas, and individual buildings. Leventhal spent his days walking from one trench to another, talking to his staff on a portable radio. He balanced recording with discussion, and gave advice when inspecting site notebooks. His excavation was a "hands-on" seminar—a discussion of architecture, artifacts, and layers that continued without a break from one day to the next. He juggled detail with an ever-changing larger picture of a Mayan sacred landscape wrought in plaster, stone, and stucco.

Forty meters high, the Castillo dominates Xunantunich. Standing high on the roof of the uppermost temple, you can see far across rolling forest into Guatemala to the west, and scan an enormous area of the Belize Valley to the east. But when and how the pyramid was built, and what the configuration of the Castillo was in both its final and earlier manifestations remain a mystery. We know that the Maya invariably built their important shrines at the same location, as successive lords erected increasingly larger pyramids on the sacred sites established by their predecessors. Theoretically, then, the earliest structure lies under the center of the pyramid. Archaeologists must resort to tunneling to test this theory, as open excavation would be prohibitively expensive and highly destructive.

Julie Miller, an expert at tunneling into Mayan pyramids, learned her craft at Copán, where archaeologists' tunnels twist and turn under the major buildings of the Acropolis (see information in box, p. 314). At the time of my visit, she had dug two tunnels into the Castillo. One penetrated a few meters into the upper part of the

TUNNELING AT COPÁN

Tunneling, a well-established specialized excavation technique in Mayan archaeology, has reached its greatest refinement at Copán. Tunneling is expensive and slow moving, and, therefore, used only when all other methods are impracticable. Such was the case with Copán's Acropolis, the main political and religious precinct at the site between A.D. 420 and 820. The nearby Rio Copán has eroded away part of the eastern edge of the Acropolis, revealing a 400-year sequence of ancient buildings. The eroded slope allows the archaeologists to tunnel into the core of the Acropolis, following plaza surfaces, walls, and other features buried under meters of compacted earth and rock fill. During the excavations, more than 3 kilometers of tunnels that expose the architectural history of the eastern part of the Acropolis have been dug. Tunnel excavation is noninstrusive, in the sense that it does not disturb standing buildings. It also enables the excavator to identify plaza floors and other levels in the walls and ceiling of the tunnel, giving insights into construction methods and now-invisible minor building remodels. Using computer-based surveying stations, the diggers have created three-dimensional representations of changing building plans.

Mayan rulers had a passion for commemorating their architectural achievements and the rituals that accompanied them. The tunnelers have a valuable reference in an inscription on "Altar Q," a critical textual guide to the Copán dynasty. Sponsored by the sixteenth ruler, Yax Pac, the glyphs record the arrival of the founder Yax K'u'k Mo' in A.D. 426 and portray the subsequent rulers who embellished and expanded the great city. Unlike other cities like Tikal or Palenque, where ceremonial and residential quarters were separated, the Copán Acropolis was a compact royal precinct, and, therefore, easier to study through intensive research at one location. The tunneling project linked individual structures with the recorded history of Copán's sixteen identified rulers. The tunnels uncovered seven texts and their associated architecture—the earliest dating to the reign of Copán's second ruler—allowing researchers to reconstruct a highly detailed history of the entire ceremonial complex.

pyramid at the level of a small plaza halfway up the north side of the structure. The second major tunnel cut into the foundations of the Castillo at ground level, following the natural bedrock more than 20 meters into the heart of the pyramid.

The Castillo tunnel could accommodate a worker with a wheelbarrow (figure 13.3). Electric fans set at strategic places provided adequate ventilation and kept the air moving so that the temperature in the tunnel was bearable, and often cooler than that on the surface. A Honda generator powered a string of fluorescent lights. One skilled worker dug away at the pyramid fill, following the bedrock. Another man passed all the deposit through a fine screen, searching for artifacts that would date the various levels.

"We start off not knowing what we're doing, then we learn a lot," stated Julie, who showed me where the clay bedrock rose inexorably toward the center of the pyramid and the ceiling, making the defile smaller and smaller.

"Why not dig upward," I asked. Julie pointed to a small hole in the ceiling. I poked my finger through and encountered loose earth and stones—soft earth fill. Soon, it was necessary for the diggers to retreat some distance, turn left, and attempt to dig around the higher bedrock, then excavate in toward the center of the pyramid again.

I had never dug a tunnel—on any of my excavations. With some trepidation, I glanced at the sides of the shaft, where the layers of loose fill, sterile soil, and dumped earth from pyramid construction could be clearly seen. Miller drew my attention to the roof over my head, where the stratigraphy continued in three dimensions. It took some time to become accustomed to thinking three dimensionally, and to identify the salient characteristics of different kinds of strata—loosely packed fill and stones, thin layers of clay, piles of earth, and stones from rough "containers," or small enclosures, which the builders had filled with soil. Once one becomes aware of the subtle nuances, however, it is possible to read the strata like a map. By recording all the layers and major features electronically, Miller can produce a three-dimensional CAD map of the pyramid's layers, which reconstructs the strata history, combining computer mapping with dating evidence obtained from late classic Mayan sherds found in the different layers. The known development of late classic pottery dates each of the major layers

Fig. 13.3 A tunnel penetrates the Castillo at Xunantunich. Photograph by Brian Fagan.

within a century or so. Leventhal feels confident that he can date these layers more accurately than he could with radiocarbon dating, which has considerable statistical errors, in the order of 150 to 250 years.

The excavators cleared about a meter of tunnel a day, working with meticulous care so that they could recover each artifact in precise association with the surrounding layers. I must confess that I found the cramped, hot work claustrophobic, but Miller and other tunnel excavators enjoyed it.

At the time of my visit, in March 1996, the trench extended 19 meters into the pyramid, under the small plaza on the north side, with no signs of anything earlier than late classic fill. In an interesting reflection of the goals of modern archaeology, Leventhal's tunnels had been deliberately offset, to avoid any possible burial chambers under the pyramid: the excavations were concerned with architecture, not spectacular finds.

Early the next morning, I stood on the summit of Structure A-11, at the northern end of the site, reputed to be the ruler's residence. Once again, thick fog lay close to the ground, leaving only a shadow of the two pyramids to the north. Gray mist mantled details of staircases and terraces, but the soaring sides of the two structures formed arrow-like lines that pointed toward heaven, and the building high on the Castillo where the ruler had once appeared to the people. The staircases of the three pyramids, large, smaller, and still smaller, are slightly offset, whether by accident or design, we do not know, but the edges lead one irresistibly toward the summit, toward heaven. This effect of looking skyward becomes even more intense inside Plaza 2, which lies at the foot of the Castillo. Here you are more closed in and your eyes look at once toward the wide, steep staircase that ascends the southern side of the pyramid.

The partially cleared staircase rises to the first level, where a sloping, rubble-strewn bank leads to another, steeper staircase, which, in turn, leads toward the buildings on the summit. Leventhal believes that a building with open doorways stood at the top of the first steps. He theorized that the lord of Xunantunich vanished from sight through these doors, after climbing in solemn ceremony from the plaza, only to appear dramatically, as if by magic, some time later at the very summit of the Castillo.

At the time, the main pyramid excavations were on the south side of the Castillo, where another staircase led to what appeared to be a small, walled plaza in the shadow of the upper slopes of the structure. Here student archaeologist David Wilson was cutting test pits in various areas of the plaza, tracing the extent of the open space, and searching for stairways, buildings, and other structures. Unfortunately, such modern technologies as ground-penetrating radar or magnetometers do not work at Xunantunich, where the architectural fills and soils are very porous. Therefore, excavation is the best approach. Wilson's trenches were only a few meters long, each carefully placed to find the margins of the plaza or to clarify puzzling fragments of masonry wall. At the higher level, he traced the outer wall of the plaza, while on a lower step he puzzled over a melange of collapsed boulders and plaster in a small trench.

"You can feel the soil texture just above a plastered bench with your trowel," he stated as he scraped the soil carefully. Leventhal, Miller, and Wilson watched closely as a worker exposed what appeared to be a plaster floor abutting a masonry wall. After a long discussion, they agreed that they had located a bench of a building virtually identical to one on Structure A-13 on the far side of Plaza 1, a long rectangular building that has at least eight rooms on either side. Leventhal believed that this was the same kind of building as the one that had once stood on the north side and planned excavations to check his hypothesis.

On the northern and western sides of the Castillo, I saw the entire history of the pyramid exposed in the consolidated walls of the artificial mountain. A picture of Xunantunich taken as recently as 1973 showed the Castillo heavily overgrown with clinging vegetation. Only the upper structures had been consolidated by combining modern cement with the ancient stones. Unfortunately, the original building stone is softer than the surrounding concrete and is literally beginning to melt in the rain and humidity. Leventhal's consolidation work has involved long experiments with a more indigenous form of cement, which includes lime, coloring, and other local constituents that are both durable and more sympathetic to the soft limestone boulders of the pyramid. After 6 years of steady work, the pyramid was largely cleared of vegetation; the rubble slopes slowly giving way to consolidated walls. He was also conserving a newly exposed frieze on the west side of the pyramid, which shows the ruler reenacting the Creation and emphasizing his vital role as intermediary between the different elements of the Mayan world (figure 13.4). The frieze helps incorporate the making of the universe into the historical charter of Xunantunich, linking the local lords with the very act of Creation—and with the symbolic world that they visited in trance atop the Castillo in the presence of the people.

A Guatemalan artist was making a precise copy of the frieze in local clay, which would become the basis for a mold used to create a near-perfect fiberglass casting of the figures. Leventhal planned to cover up the original frieze with the same ancient deposits that formerly covered it, then mount the fiberglass replica over the covering, thereby meeting the dual goals of conservation and tourism.

The conservation of the Castillo friezes is an important element in deciphering the architecture at the summit. As I walked along the northern facade, Leventhal pointed out how the architects had filled in nearly all the rooms of the first summit temple with rubble. They

Fig. 13.4 Frieze at Xunantunich, Belize. Photograph by Brian Fagan.

then used the roof to erect another, identical building above it, finishing off the 40-meter pyramid with a triple comb, which represented the rulers of the sky. All of this architecture had a vital symbolism: the ruler would emerge standing above the ancestors on the frieze and below the sky rulers, on a stage that symbolized his role as a link between the ancestors, the people, and the gods.

The Xunanuntich project encompassed the entire 81 hectares of the site, not merely the Castillo. Three plazas extend northward, each one plastered and brightly painted. As you walk north, Plaza 2 at the foot of the Castillo is small and relatively closed in by surrounding pyramids. A *sacbe*, or raised ceremonial road, joins the plaza at its southeastern corner, linking the ceremonial precincts with a residential complex about 400 meters away. This relatively small plaza may have been the ceremonial focus of the site, especially during its final years. Leventhal has excavated a low wall, which cut off a defile between Plaza 1 and Plaza 2 about A.D. 1000. Trenching at the northwest corner of the plaza shows that the boundary of a ball court was set as a barrier from that direction. Leventhal believes that these may be signs of an increasingly insecure nobility, which was worried about their eroding authority in troubled times. Downsizing the sacred landscape, they turned the Castillo and Plaza 2 into a smaller sacred world—a last bastion of their power, set high above the dispersed communities of their erstwhile rural subjects.

In its original configuration, Xunantunich's plazas, pyramids, and residential complexes extended out from the Castillo in every direction. Only the northern portion of the site, with its three plazas, is cleared of vegetation. Plaza 1 is larger and is thought to have been the site of a market. Smaller pyramids surround this open space. Structure 13, a long, rectangular building with four rooms on each side tops a low pyramid on the north side of Plaza 1, with a pass-through to Plaza 3, which Leventhal believes to have been the place where the elite lived. Another student archaeologist, Ellie Harrison, was digging into the sides of this structure, attempting to establish the layout of the rooms on either side, and the relationship between the pyramid and the plaza below. Four small trenches sliced into the side of the ruined building exposing the fillings of an archaeological layer cake. I was able to identify the collapsed wall filling in the upper part of the trenches, but the lower levels baffled me. Harrison showed me the entrance to a room, complete with lintel, and the plastered bench that once surrounded Plaza 3 at the foot of the structure. She was just starting to cut a new trench into a pyramid at the western side of the plaza and had already found the same bench running along the west side. Leventhal was plotting all of the floors and benches that Harrison had excavated electronically, turning his measurements automatically into another part of his three-dimensional map of the site.

An elaborate building with several small chambers stands on top of the pyramid at the north end of Plaza 3. The building faces inward. A steep slope falls away on the north side, with only a discreet back stairway leading around to the plaza. This residence, if it is indeed such a structure, appears small and humble, until you look closely at the finely plastered walls and imagine it brightly painted and colored, with jaguar pelts and other lavish furnishings on bench and stoop. From the front door of this palace you look southward to the rising mountain of the Castillo.

Field archaeology, excavation, and survey provide a vital framework for understanding the history of Xunantunich. But a greater understanding comes from combining archaeology with the Maya's own writings.

A MAYAN BIBLE

To understand Naranjo and Xunantunich, you must begin with the Mayan world itself. Every Maya lived in a forest world where mythic geography was all-important; where the great cities and even small ceremonial centers like Xunantunich became symbolic replicas of both a living and spiritual world. This Mayan world still survives, in their intricate glyphs, and in a few remarkable indigenous books set down in Spanish centuries ago and that still serve as an inspiration to modern Mayan communities. One such classic of Mayan literature is the *Popol Vuh,* of the Quiché Maya.*

Few books have had such a colorful history. The European invasion of the Mayan world began in the sixteenth century. In the

*The Quiché Maya live northwest of Guatemala City on the highlands.

wake of Hernán Cortés and his conquistadors came an army of administrators and missionaries, who established a monopoly on art, architecture, painting, and all forms of public expression. Catholic friars burned hundreds of indigenous hieroglyphic books, and even tried to ban Mayan clothing, on the grounds that their textiles carried symbolic messages. The Maya responded by turning in on themselves and shunning colonial society as much as possible. European missionaries taught the Indians the Roman alphabet, so that their converts would set down Christian texts in their native tongue. But the Maya soon used alphabetic writing for their own political and religious purposes, and left behind a priceless legacy of literary works that are a passageway into the ancient Maya's world. The *Popol Vuh* is part of that legacy, something its original authors called "the first eloquence."

That the manuscript survived until the twentieth century is a minor miracle and typical of the checkered history of many early native American writings. Three Quiché Maya aristocrats wrote the *Popol Vuh* in the sixteenth century. They sought to preserve the history of their kingdom, which survived on some fading murals on ancient Quiché ruins. In 1703 a friar named Francisco Ximénez copied and translated their manuscript. Like so many other clerical writings, the only surviving version of the *Popol Vuh* languished in Dominican archives until the early nineteenth century. When French priest Brasseur de Bourbourg saw the manuscript in the University of San Carlos library in Guatemala City in 1861, he realized its importance and somehow smuggled it out of the university to Paris, where he translated it into French. Eventually, the *Popol Vuh* crossed the Atlantic to the Newberry Library in Chicago, one of the great repositories of native American literature, where it now resides. In 1985, anthropologist Dennis Tedlock published a definitive translation of the *Popol Vuh* in English, which is rapidly becoming a classic of its kind.

When I explored the *Popol Vuh*, I found myself in a "Book of Council"; what Dennis Tedlock calls "a complex navigation system for those who wished to see and move beyond the present" (1996, 59). The authors set down the *Ojer Tzij* (the "ancient word")—a re-creation of a precious hieroglyphic version of the Book of Council—based on their memories, and perhaps on a surviving, and carefully hidden, copy of an original codex. The *Popol Vuh* as a whole is a cosmic vision of what the Quiché called *kajulew*—"sky-earth."

The narrative begins in a world with an empty sky and still, primordial waters; a world of darkness, a universe inhabited only by supernatural beings. "Whatever there is that might be is simply not there; only murmurs, ripples, in the dark, in the night. Only the Maker, Modeler alone, Sovereign Plumed Serpent, the Bearers, the Begetters are in the water, a glittering light. They are there, they are enclosed in quetzal feathers in blue-green. . . ." (ibid., 64). The gods converse and conceive of an earth that emerges from the water and sprouts both people and plants, as well as the heavenly bodies. After four tries, they create humans from ground corn and water.

The *Popol Vuh* encompasses a vast sweep of time, from the creation of the world to the time of the authors. I read of the dualities of Mayan life, where the affairs of humans always have a divine dimension, and vice versa. The Quiché lords, like the rest of the Mayan nobility, sought to identify themselves with the gods, through pilgrimages, fasts, and shamanistic trances. They nourished the gods with offerings, but also required their subjects to give them tribute—food and drink, and turquoise, jade, and bright blue-green quetzal bird feathers. Tedlock tells us that such precious objects were the "ultimate fruits of the blue-green world of earth and sky" (ibid., 55). The *Popol Vuh* tells how the Lords Cotuha and Plumed Serpent came to the Quiché homeland and stayed: "And they built many houses there. And they also built houses for the gods, putting these in the center of the highest part of the citadel. . . . After that their domain grew larger; they were more numerous and more crowded" (ibid., 183–4). The same story might apply to the small kingdom of Xunantunich.

I once attended a conference at the University of Texas with Barbara and Dennis Tedlock, and was humbled by their extraordinary knowledge and perceptions of Mayan culture, which extended far beyond the confines of archaeology, anthropology, and the *Popol Vuh*. Dennis Tedlock had translated and analyzed this Mayan "Bible"—a journey into a world where the mythic and the historical became a single unity. He started with the manuscript itself, trying to eliminate imprecise spellings and usages, as well as "slips of the pen" accumulated during copying, which left some parts of the manuscript open to multiple interpretations. He spent weeks talking to Quiché lineage leaders, calendar keepers, and ritual experts, who checked his translation. Then he and his wife Barbara "traveled through the Quiché text" with Mayan diviner and lineage leader Andrés Xiloj Peruch, who taught them how to read "dreams, omens, and the rhythms of the Mayan calendar" (ibid., 18).

The patterns of darkness and light in the *Popol Vuh* impressed Peruch. They gave him, as a diviner, many rich insights into the deep meanings of this remarkable work. Next, the Tedlocks consulted Quiché scholar Enrique Sam Colop, who had recently completed his doctoral dissertation on Mayan poetics. Tedlock drew on new perceptions of Mayan art, on interpretations of painted pots and stucco friezes of such characters as the Heavenly Twins, Hunahpu and Xbalanque, whose deeds feature prominently in the *Popol Vuh* and extend back far into ancient Mayan history (figure 13.5). Then he took into account the latest advances in the decipherment of Mayan glyphs, and fresh observations regarding Mayan astronomy and history. Trained by a Mayan master, Tedlock turned himself into what one reviewer called a "poet of performance," restoring a masterpiece of religious writing to its rightful place in the world.

MAYAN GLYPHS

Centuries of Mayan literature and indigenous writing lay behind the *Popol Vuh*. The Maya were unique among native American societies in that they had a sophisticated written script, that was recorded on temples, specially erected commemorative stelae, painted clay vessels, and codices written on bark paper or deer skin. Unfortunately, zealous Franciscan friars burned hundreds of these codices. The most fanatical of all the missionaries, Fray Diego de Landa wrote: "As they contained nothing in which there was not to be seen superstition and lies of the devil, we burned them all, which they regretted to an amazing degree." Ironically, de Landa wrote a description of Mayan script, which has been of considerable assistance to modern scholars in deciphering the glyphs.

The intricate Mayan glyphs have puzzled epigraphists for more than a century. Until the 1960s, most experts, notably British Maya

Fig. 13.5 A classic Mayan bowl depicts Hunahpu (*left*) and Xbalanque (*right*) resurrecting their father. He assumes the form of a maize god, emerging from a cleft in the back of a turtle. By Karl Taube. Courtesy Regents of the University of California.

specialist J. Eric Thompson, believed that they were little more than calendrical and astronomical observations. Thompson wrote: "The dates on stelae surely narrate the stages of the journey of time with a reverence befitting such a solemn theme" (1971, 222). He considered the Maya a civilization of peaceful astronomers and priest/rulers totally preoccupied with the "endless progress of time." In the late 1950s, Harvard Mayanist Tatiana Proskouriakoff took a much more systematic approach than Thompson and his contemporaries. An artist herself, she inventoried thousands of motifs and elements from stelae and artifacts both large and small. So thorough was her research that she was able to identify the "name glyphs" of individual rulers and their dynasties. She applied a minutely developed classification of glyphs to seven groups of rulers' monuments at a major Mayan city named Piedras Negras in the western Petén. Proskouriakoff was even able to establish that each Piedras Negras monument had been erected on the fifth anniversary of the ruler's accession. She believed that the glyphs were not merely calendrical markers, but rather potential evidence for dynastic history.

As Proskouriakoff worked on stelae, a Russian epigraphist named Yuri Knosorov argued that Mayan writing was phonetic. Knosorov published his theory in 1952, but it was not until 1973 that a small group of Mayan epigraphists combined his phonetic approach with Proskouriakoff's historical analyses. After months of often passionate debate and argument, they established that Mayan writing represented a spoken language, with a specific word order for verbs, nouns, and so on. Soon they were able to paraphrase inscriptions and treat them as texts. In one of the great triumphs of twentieth-century science, this group finally deciphered Mayan writing, revealing a startlingly intricate historical landscape, in which the living and supernatural worlds mingled constantly, just as they did in the architecture of Mayan cities.

We now know that the glyphs taught history, gave supernatural meaning to the events of daily life, and provided justification for kingly accession and daily deeds. There are ritual almanacs, sacred texts, narratives, and dedicatory offerings. Collectively, they provide a window into the ancient Mayan world. But, like so many ancient writings, the glyphs are often political propaganda: texts used to glorify individual deeds, to justify genealogies, and to commemorate actual or perceived victories over hated rivals. Therefore, the scientist uses the glyphs with caution. Despite this fact, there is one point on which everyone agrees: the decipherment of Mayan glyphs has redefined our understanding of Mayan kingship—of great lords as powerful shamans.

Linda Schele and David Freidel, epigraphist and archaeologist, respectively, have an almost missionary devotion to the ancient Maya and their writings. At the same conference where I met the Tedlocks,

Schele and Freidel spoke with passion about Mayan astronomy, calendars, and belief systems. During their presentation, they used a computer program to display the heavens over Central America on specific days in the Mayan calendar of more than 1,000 years ago; and talked of excavations where glyphs, translated on site, reveal the names of the builders of pyramids, temples, and staircases, and even the days when the structures were completed. Their passion for the subject is palpable, as if they themselves have gazed on the faces of Mayan gods and found the one true faith.

Schele and Freidel have written two popular books: *A Forest of Kings* and *Maya Cosmos*, which bring together many elements regarding archaeology and glyphs in general. They enlisted the help of a writer to compile these two stirring narratives about Mayan history and the Mayan world, which are as much journeys of personal exploration and revelation, as they are scholarly analyses. Page after page of closely argued notes provides the context for their hypotheses. The books include data to support these arguments, and identify points of controversy and broad agreement. Both books are controversial: most specialists quarrel with the details rather than the general portrait of an ancient society in which shaman-kings acted as vital intermediaries between the living and spiritual worlds.

CALENDARS AND KINGS

J. Eric Thompson and other specialists of a generation ago thought that the Mayans were ruled by peaceful astronomer-priests, obsessed with ritual observances. Recent research paints a very different picture, of a civilization made up of rival kingdoms headed by warlike lords, who lived out their reigns in an atmosphere of constant diplomatic intrigue and internecine warfare. Over many centuries, city-state after city-state rose to prominence, enjoyed a brief period of glory, then sank into political obscurity. Only a few major states, such as Copán, Calakmul, Palenque, and Tikal, enjoyed long periods of political prosperity, and even they were vulnerable to conquest, drought, and other natural and human-caused disasters.

Mayan priests used the movements of planets and stars to mark the passage of time. The calendar was vital to Mayan life, for the complex geography of sacred time helped determine political strategies and political moves. Each Mayan lord developed a relationship to the constantly moving time scale. Some events, such as planting and harvest, were regular events on the calendar. Others, such as dates of rulers' accessions, important victories, and royal deaths and births, became days that assumed great significance in the history of an individual dynasty. Mayan rulers linked their actions to those of the gods and their ancestors, sometimes legitimizing their descent by claiming that it reenacted mythical events. Their society was embedded in a matric of sacred time and space, reproduced in the stucco and mortar of elaborate ceremonial centers.

Glyph inscriptions have revealed how the Mayan institution of kingship was based on the principle that the royal crown passed from father to son, or to brother of a son-less king to son, in a line that led back to a founding ancestor. Other families and clans were carefully ranked by their distance from the central royal descent line. This system of family ranking and allegiance was the basis of political power. The setup worked well, but was dependent on the documentation of genealogies. Social status was highly prized in Mayan society, as shown at the city of Copán, whose pyramids, temples, and stelae are a remarkable record of Mayan kings exercising political and social power. Although cosmology was a feature of Copán's layout, the main function of the civic structures there, as elsewhere, was to commemorate major events in the lives of kings and the political histories of kingdoms. Each building was dedicated by the ruler who built it in the context of a specific event, ranging from a royal accession to an important victory. Even specialized buildings like ball courts had spiritual and political importance. Copán's ball court was a stadium used for an elaborate ceremonial contest in which sacrificial victims and rulers descended through a symbolic "abyss" into the otherworld. The players wore protective padding and used a rubber ball, which they aimed at markers—stone rings or macaw heads set high on the walls. Although we do not know the rules of this ball game, we do know that the contests were associated with human sacrifice and much pomp and circumstance.

The public buildings at Copán and other Mayan centers emphasized the lord's role as the intermediary with the otherworld. Mayan rulers also used the regalia of their office and elaborate rituals to stress their close identity with mythical ancestral gods. The ruler was often depicted as the conduit through which humans communicated with the otherworld.

The Mayan world view created serious and binding obligations among the ruler, his nobility, and the people, which were reflected in his many responsibilities, including gathering and redistributing commodities of all kinds and implementing agricultural schemes that turned swamps into organized, productive landscapes. The lives of Mayan rulers and their subjects were interconnected in dynamic ways. The lord was state shaman, the individual who enriched everyone's life in spiritual and ceremonial ways. His success in organizing trade and agriculture provided all levels of society with access to goods and commodities. The great ceremonial centers built by Mayan leaders created settings for elaborate rituals and public festivals that validated the world order of the ruler and the ruled. The "histories written and pictured by the kings on the tree stones [stelae] standing before human-made mountains gave form to time and space in both the material and spiritual words" (Schele and Freidel 1990, 319).

THE MAYAN COSMOS

We know from both the *Popol Vuh* and from recently deciphered glyphs that the Mayan world ("sky-earth") was set in a universe made up of three layers: the upperworld of the heavens, the middleworld of living humans, and the underworld of dark waters beneath the earth. A World Tree named *Wacah Chan* ("raised up sky") linked these layers, its roots growing in the underworld, its trunk inhabiting the middleworld, and its branches rising through the upperworld. The World Tree was the vital conduit for supernatural beings and the souls of the dead to pass from layer to layer. *Wacah Chan* grew through the center of the middleworld, forming the point from which the four cardinal directions flowed.

These directions were vital points of reference in the Mayan world, each with its own color and god. East, for example, was red, the direction of the rising sun; north was white, the direction of the ancestral dead. *Wacah Chan* was not located at a specific geographical location, but could be fixed ritually at any point on the landscape by a Mayan lord in a trance. All ceremonial centers, however small, were architectural "maps" of the Mayan cosmos centered on the point at which the trunk of *Wacah Chan* joined the three layers of the universe.

Mayan kings embodied the *Wacah Chan*. As they stood in trance atop the Castillo at Xunantunich, they would conjure up the World Tree through the center of the shrine, where a dark doorway opened onto the spiritual world. The king, dressed in white and elaborate regalia, would appear suddenly and dramatically high above the people assembled in a plaza far below. He would gash his tongue or his penis, spilling blood onto a piece of white bark paper. The red blood, which stood out vividly against the pure white, served to nourish and sustain the spiritual world.

Mayan architects built the ceremonial centers as monumental places. Their sheer scale could diminish the individual. At Copán, Naranjo, Tikal, and Xunantunich—all of the Mayan cities—your eyes are drawn upward to the summit of these artificial mountains. The great drama of spiritual transformation, unfolding for all to see, became the supreme moment. Schele and Freidel tell us: "Clouds of incense rose from the temples and coiled like a serpent around the pyramids high above the onlookers. This was the Vision Serpent, the Feathered Serpent, perhaps the most powerful symbol of Maya kingship" (ibid., 269).

Xunantunich has none of the elaborate architecture and plazas of the great city of Tikal, but due to its very compactness, the sacred architectural structures at the site have a powerful impact. On my last morning at the site, I stood in Plaza 2 and imagined Xunantunich a thousand years ago. The dense fog had gone, and warm air blanketed the plaza. The sun sent torches of light blazing down on the stairways, casting the plaza in a bowl of darkness. A light wind blew the heady scent of wood smoke across the silent pyramids. I imagined the ancient lord as he danced on the heads of his ancestors, journeying into the solemn realm of the otherworld.

Flickering light, resonating sound, powerful scents, and dramatic performance; the silent pyramids and buildings at Xunantunich were designed for moments like these, when the great lord appeared before his people not as a human being, but as the intermediary between the living world and the mysterious cosmos that surrounded the Maya people on every side. As I stood on the ancient stage that had been built for the sacred performances that lay at the very center of Mayan life, I sensed the presence of vanished spectators, their sweating bodies pressed together in a silent throng, all tense with anticipation.

chapter fourteen

THE WORLD OF THE FIFTH SUN

"And when we saw all those cities and villages built in the water, and other great towns on dry land, and that straight and level causeway leading to Mexico, we were astounded. These great towns and pyramids and buildings rising from the water, all made of stone, seemed like an enchanted vision. . . . Indeed some of our soldiers asked whether it was not all a dream. . . ." (Diaz 1963, 214). Spanish conquistador Bernal Diaz was a young, impressionable adventurer when he gazed out over the Valley of Mexico in A.D. 1519 and saw the Aztec capital, Tenochtitlán, stretched out before him (figure 14.1). Half a century later he wrote: "I stood looking at it, and thought that no land like it would ever be discovered in the whole world, because at that time Peru was neither known nor thought of" (ibid., 215).

THE LEGEND OF THE FIFTH SUN

Sometime in the 1470s, an Aztec ruler named Axayacatl dedicated a new sacrificial stone in the Temple of Huitzilopochtli at Tenochtitlán. The stone vanished during the Spanish Conquest, but was dug up by accident in 1790. Axayacatl's sacrificial Stone of the Sun (figure 14.2) depicts the origin myth of the Aztec cosmos. The forces of good and evil engaged in endless combat; their battles caused the creation and destruction of four worlds, or suns. The circular carving depicts the face of Tonatiuh, the sun-god, with four square-paneled reliefs

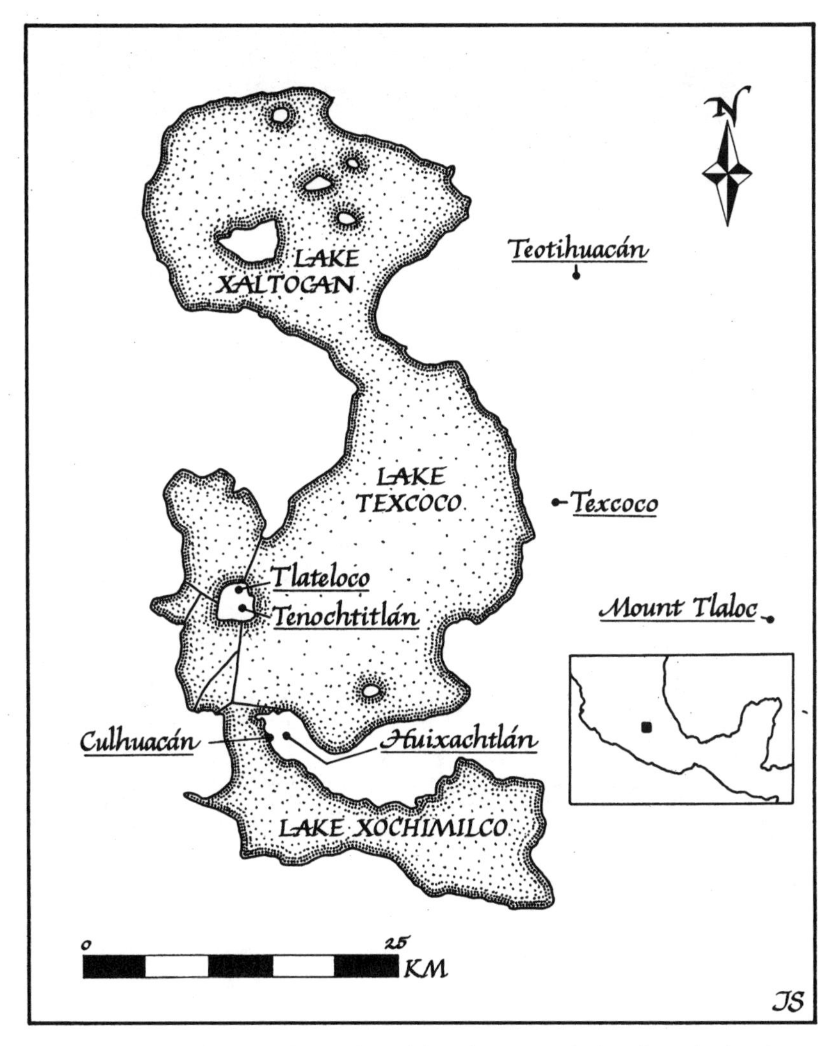

Fig. 14.1. Map showing the archaeological sites (underlined) and other locations mentioned in chapter 14.

recording the dates of the endings of the four previous worlds, which extended over at least 2,000 years.

All was darkness after the destruction of the Fourth Sun. Aztec legends tell the story of the creation of the Fifth Sun. The gods gathered in the ancient ruined city of Teotihuacán—revered deeply by the Aztecs—to take counsel and decide their next course of

Fig. 14.2. The Stone of the Sun. Photograph from the American Museum of Natural History.

action. "Come hither, O Gods! Who will carry the burden? Who will take it upon himself to be the sun, to bring the dawn [of the Fifth Sun]" (Anderson and Dibble 1954, 7). Two gods were then chosen to represent the sun and moon; they fasted for four days. Dressed in the correct regalia, they immolated themselves in a great fire while the others watched. The two deities became the sun and moon, but they were motionless. So the wind-god Ehecatl "arose and exerted himself fiercely and violently as he blew. At once he could move him [the sun] who thereupon went his way. And then when he had already followed his course, only the moon remained there. At the time when the sun came to enter the place where he set, then once again the moon moved" (ibid., 8).

The World of the Fifth Sun was finite, destined for eventual destruction by catastrophic earthquakes. The Aztecs knew that the Fifth Sun would end, however, they believed that they could ensure the continuity of life itself by nourishing the sun with the magic elixir of human hearts. Aztec legends speak of the gods sacrificing themselves in the sacred fire at Teotihuacán to provide nourishment for the sun at the creation of the Fifth Sun. Before the sacrifice, the deities bequeathed their regalia and mantles to their retainers. These sacred bundles and the songs and musical instruments that survived the immolation were those used in subsequent rituals for the gods.

This act of divine sacrifice was insufficient. The sun had an insatiable craving for human blood. Inevitably, humankind had to bear the burden of sacrifice as well. So rulers went to war to satisfy the sun's needs as a sacred mission to obtain sacrificial victims. The hearts of prisoners of war provided *chalchiuhuatl*, the vital sacrificial fluid that fed the sun. The heart symbolized life. The rites of human sacrifice—the feeding of the gods—played a vital role in this ambivalent, tension-filled world.

Like all origin myths, the Legend of the Fifth Sun seeks to explain the relationship between the living and the supernatural. Mexica* legends refer again and again to the ancient city of Teotihuacán, and to a place named Tamoanchan, "the place of descent," a mythic location where civilization began. Aztec beginnings lie in a long period of political confusion that occurred after the collapse of Teotihuacán in A.D. 700 and the rise of Toltec civilization, in A.D. 950. The Toltecs became mythic heroes and revered ancestors of the Aztecs. The highest Aztec nobility claimed descent from one of the Toltec heroes. Militaristic Toltec rulers established precedents of tribute gathering and imperial governance that formed the foundations of the Aztec empire. The Toltec connection was critically important to rulers, who sought to cloak their imperial ambitions in a heroic past.

*In the interests of clarity, I have used the terms "Aztec" and "Mexica" interchangeably in this chapter.

The Aztec place of origin was the mythic Aztlan, an island set in a lake, where the people assembled before setting out on their extraordinary rags-to-riches journey. Legends tell how a heterogeneous group of at least seven Mexica clans traveled together. They entered the basin of Mexico, perhaps in the eleventh century A.D., at a time when the Toltec civilization was at its height. However, the Toltec empire collapsed in A.D. 1150, and left a political vacuum in the Valley of Mexico. Intense military and political conflicts arose between the settled farmers of the valley and nomadic peoples, like the Aztecs.

At first the Aztec were insignificant political players in the overcrowded basin of Mexico, until they were unified under a charismatic (and legendary) leader, who may have borne the name Huitzilopochtli. In later centuries the Aztecs described the early days in the basin as a period of gods and heroic deeds. These myths turned the putative leader Huitzilopochtli into a god. One such myth tells how a priestess named Coatlicue ("Serpent Skirt") was sweeping her earth-shrine on the summit of Coatepec Hill in when she was magically impregnated with Huitzilopochtli—divine patron of the Aztec people—by a ball of feathers that fell from the sky. Coatlicue's daughter Coyolxauhiqui and four hundred other siblings attacked the hill to kill their dishonored mother. Someone warned Huitzilopochtli, who was still in the womb. When the enemy approached the shrine, Huitzilopochtli sprang from his mother as an invincible warrior wielding a *xiuhcoatl*, a "fire serpent" that symbolized a ray of the sun. He conquered his foes and killed Coyolxauhiqui, rolling her dismembered body down the hill.

Huitzilopochtli now became the chosen god of the Aztecs. He was "the humming bird of the left [handed]" (Anderson and Dibble 1954, 111), symbol of rain and human sacrifice. He became Lord of the Daylight Sky, the Rising Sun. His hummingbird symbol came to represent the fallen warriors who accompanied the sun on his daily journey. Among his other attributes, he became the God of War, a fierce deity of fertility who stood for militarism, ambition, and imperial domination. He incited his chosen people to "lift the Mexican nation to the clouds" (ibid.).

The Mexica were unwelcome visitors in the competitive world of the valley. Their neighbors despised their unsophisticated barbarity.

After almost 2 centuries of constant warfare, the Aztecs fled to a reed-covered island near the center of modern Mexico City in A.D. 1325. Their legends tell how Huitzilopochtli appeared before one of the priests, ordering him to search for a cactus where a great eagle perched—Tenochtitlán, "the Place of the Prickly Pear Cactus." The next day, the people found the place. The eagle was the symbol of the sun, of the god himself. The cactus fruit was red, and in the shape of the human hearts that the sun devoured. Immediately, the Mexica piled up a sod platform where they erected a reed temple to Huitzilopochtli. Two centuries later, Tenochtitlán was the largest city in the Americas, and one of the most extensive in the world.

Eventually, the Mexica became allies of nearby Colhuacán, then of the powerful Tepaneca, while under the rulership of Acamapichtli, who died in A.D. 1391. In A.D. 1426, a generation later, Itzcoatl, "Serpent of Obsidian," became Aztec ruler. He reversed a long-standing policy of subservience to neighbors and forged the celebrated Triple Alliance, which linked Tenochtitlán, Tlapocan, and Texcoco in an elaborate tribute-gathering and political arrangement. The Aztec soon dominated the alliance, after generations of quietly consolidating their political position and ruthless military campaigning in the name of Huitzilopochtli. When they conquered a city, they toppled the images of its gods and erected those of their patron god.

Itzcoatl's chief minister Tlalcalel was the principal architect of Aztec imperial ambitions. He prevailed on the ruler to burn all earlier tribal records, on the grounds that they would undermine established authority. He replaced tribal histories with official versions of Aztec history, which promoted Huitzilopochtli to supremacy in the divine pantheon. The sun-god became the great presence, the arbiter, the justification for military conquest and punishment. A succession of talented rulers embarked on ambitious conquests in the god's name.

Itzcoatl's successor Moctezuma Ilhuicamina ("the Angry Lord, the Archer of the Skies") came to the throne in A.D. 1440. After surviving devastating plagues of locusts and catastrophic floods that inundated the capital from A.D. 1450 through 1454, Moctezuma proclaimed war as the principal occupation of the Mexica. His conquests ensured a constant supply of prisoners to supply the insatiable maw of the sun. Moctezuma aimed his campaigns at the well-watered lowlands, so that Tenochtitlán would never be hungry again. A small army of tribute collectors moved in on conquered territories, setting up an efficient tax collection system (figure 14.3). The economy flowed to the center and up to the top, to the ruler. The state catered to the spiritual and material needs of the gods.

Tenochtitlán grew and grew. Moctezuma continued to expand the sacred precincts of the capital. He started work on an imposing temple to Huitzilopochtli, and the rain-god Tlaloc, which was not completed until A.D. 1487, 19 years after his death. At the end of Moctezuma Ilhuicamina's reign, the Aztec society was one of privilege and power; a pyramid in which thousands of merchants and commoners supported a ruler and a small noble class. His successors, Axayacatl and Tizoc, maintained the tradition of conquest and expansion. In the 1480s and 1490s, Ahuitzotl, a gifted general, campaigned as far as the borders of Guatemala. By this time the empire extended from the Gulf of Mexico to the Pacific Ocean, from northern Mexico into Guatemala.

Tenochtitlán's rulers had locked themselves into a vicious circle that forced them to expand their domains simply to obtain more sacrificial captives and tribute. The ever-ravenous Huitzilopochtli craved more human hearts each year, to the point that no one leader could ever hope to change the policy of inexorable conquest. Lavish public ceremonies in the capital's great plaza reinforced the power of the Mexica and their gods. Dominican friar Diego Duran (ca., 1537–1588) heard from informants that "it must have been immense, for it accommodated eight thousand six hundred men, dancing in a circle" (Duran 1964, 75).

In A.D. 1487, Ahuitzotl dedicated the Temple of Huitzilopochtli in front of lesser rulers from all parts of his domains. For 5 days, Ahuitzotl and his priests sacrificed long lines of prisoners until the temple steps ran with blood. Hundreds of captives perished in this orgy of sacrifice as part of a festival when "the streets, squares, market places, and houses were so bursting with people that it looked like an ant hill. And all of this was done with the purpose of lifting up the majesty and greatness of the Mexica" (ibid., 196). Even the Aztecs' contemporaries recognized their genius for elaborate display and public ritual. But nothing survives in the archaeological record of

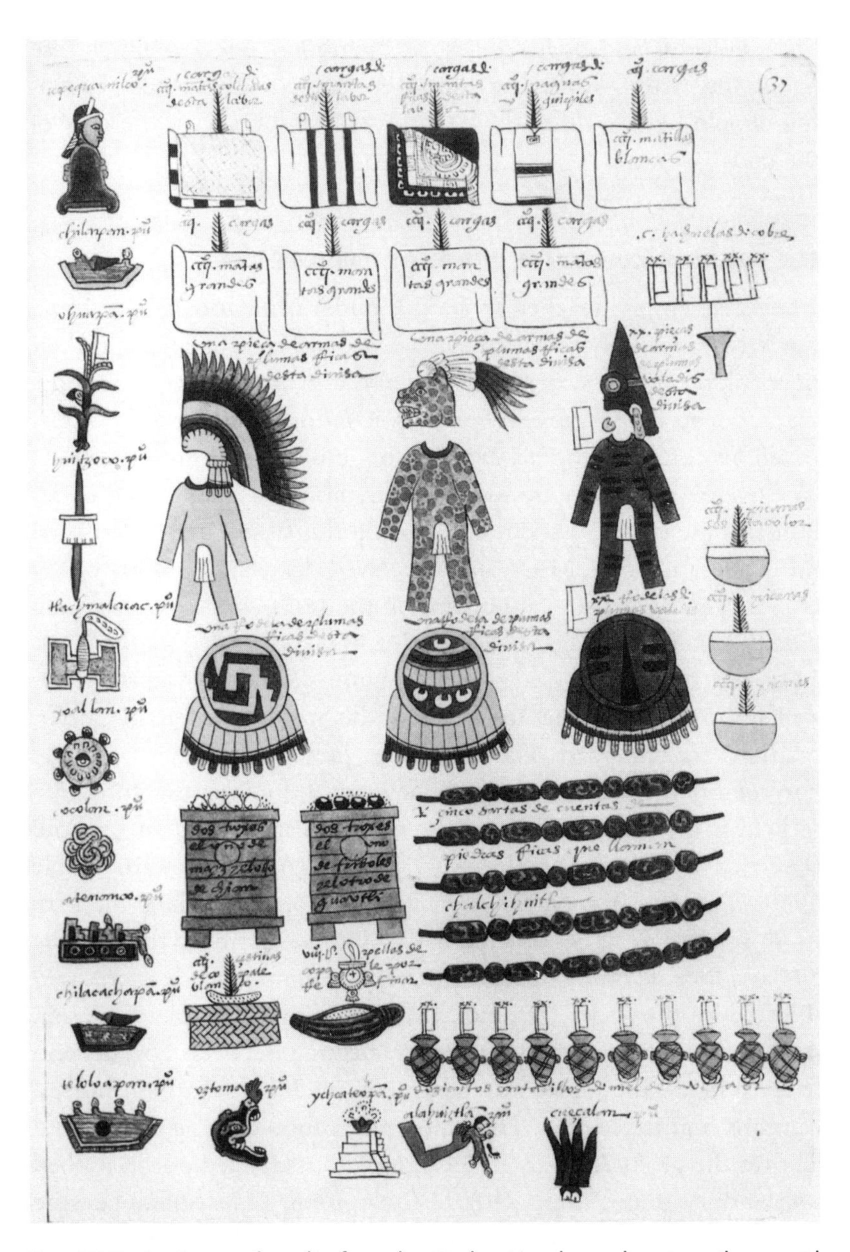

Fig. 14.3. An Aztec tribute list from the *Codex Mendoza*, showing tribute paid by the province of Teochtepec to the Triple Alliance cities. The inventory includes mantles, shirts, tropical bird feathers, and loads of cacao. Photograph from the Bodleian Library, Oxford University. *Codex Mendoza*, Folio 46. (MS. Arch Selden A.1 Folio 46).

these colorful and well-rehearsed ceremonies. Like so much else about the Aztec world, they were ephemeral experiences, just as the people believed that their world was temporary in the eyes of the gods.

THE SPANISH CONQUEST THROUGH MEXICA EYES

The Aztec empire reached its apogee under Ahuitzotl. In A.D. 1502, the Aztec nobles chose Moctezuma Xocoyotzin, his nephew, as their new ruler. A brave warrior, he was also a prudent counselor and a man of deep, conservative religious conviction. Moctezuma concentrated on consolidating his uncle's gains, but at a high military price. The overextended empire with its many nominal allies chafed under burdensome tribute assessments. During this trying period between A.D. 1502 and 1519, Moctezuma received disturbing reports of the arrival of great ships—"mountains on the sea"—from over the Gulf Coast's eastern horizon. White bearded strangers landed on the coast. Lord Moctezuma puzzled over the white-skinned newcomers. He sent ambassadors to greet the leader of the Spaniards, Hernán Cortés, on the Veracruz coast. The emissaries presented Cortés with the correct offerings and regalia. "They put him into the turquoise serpent mask with which went the quetzal feather head fan. . . . And they put him into the sleeveless jacket" (Anderson and Dibble 1969,15). Cortés responded by firing a cannon and placing the men in irons. Moctezuma ordered two victims sacrificed before the ambassadors, "for they had gone to very perilous places, to look into the faces of the gods. . . ." (ibid., 18). The bewildered ruler then sent sorcerers to cast spells and watches, vacillating between hostility and friendship as the Spaniards advanced inland with their horses, cannons, and fierce dogs. They fired off cannon and "smoke massed all over the ground. . . . By its fetid smoke it stupefied one, it robbed one of one's senses" (ibid., 40). The people of Tenochtitlán became fatalistic. "We are bound to die, we are already bound to perish" (ibid., 27).

On November 8, A.D. 1519, Hernán Cortés and his motley band of conquistadors met face-to-face with the Lord Moctezuma Xocoyotzin, the supreme ruler of the Mexica, on an earthen causeway at the gates of Tenochtitlán. Moctezuma arrived in great state, with men who lowered their eyes and swept the earth, and laid down capes so that his feet never touched the ground. The two leaders bowed deeply. Moctezuma gazed on the countenance of Cortés and addressed him with ceremony, as a god: "Thou hast come to arrive on earth. Thou hast come to govern thy city of Mexico; thou hast come to descend on my mat . . . which I have guarded for thee. . . . And now it hath been fulfilled; thou hast come" (Anderson and Dibble 1978, 44).

Moctezuma, with only legends from the past to guide him, believed that centuries before him the Feathered Serpent god Quetzalcoatl and his followers had fled to the Gulf Coast from nearby Tula, the ancient Toltec capital. After a bitter factional quarrel with disciples of the war-god, Tezcatlipoca, Quetzalcoatl constructed a raft of serpents and sailed over the eastern horizon, vowing to return in the year 1-reed. In that very year, Hernán Cortés landed. The Lord Moctezuma, steeped in imperial propaganda and a carefully prescribed mythic and sacred world, assumed that Cortés was the returning god, Quetzalcoatl.

The conquistadors explored Tenochtitlán in bewildered wonderment. The clean and well-organized capital stood in a shallow lake, linked to the mainland by 4 large causeways delineating the quarters of the city. From the market, they approached the central precincts, "a series of courts, bigger I think than the Plaza at Salamanca," Cortés wrote (1962, 66). A double-masonry wall surrounded the entire sacred area and its 78 ceremonial structures. Smooth, white flagstones paved the central, rectangular plaza, some 450 meters square. Cortés wrote that the central area could have held "a town of fifteen thousand inhabitants" (ibid.).

Cortés and Diaz describe how four quarters of Tenochtitlán converged at the foot of the stairway that climbed to the summit of the huge temple of the sun deity Huitzilopochtli and the god of rain, Tlaloc—the temple that archaeologists call the Temple Mayor. The Valley of Mexico's mountains and lakes were settings for elaborate dramas recalling the creation and the birth of humankind; but the center of the Aztec world—the Templo Mayor in the heart of Tenochtitlán—was humanly made (figure 14.4).

Fig. 14.4. Artist's reconstruction of the central precincts of Tenochtitlán, with the Templo Mayor at the left. Photograph from the American Museum of Natural History, New York.

The great temple of Huitzilopochtli and Tlaloc dominated the north side of Tenochtitlán's central plaza. In A.D. 1519, the west-facing temple rose 46 meters above level ground, a stepped pyramid with 2 steep stairways running up the front. Two shrines sat atop the pyramid: Huitzilopochtli's red chapel to the right, and Tlaloc's blue precinct to the left. The stone-built temples were covered with lime stucco and decorated with brightly colored sculptures and murals. Blue bands, symbolic of water, decorated the facade of the rain-god's shrine. Skulls adorned Huitzilopochtli's chapel. Roof ornaments in the form of seashells decorated both shrines.

The conquistadors ascended 114 steep steps to the platform at the summit, where they looked out over a spectacular vista of the city, lake, and neighboring towns. Fresh blood stains smeared the sacrificial stone at the top of the steps. Five human victims had died that day. Moctezuma showed Cortés the twin shrines of Huitzilopochtli and Tlaloc; the great figure of Huitzilopochtli was covered with offerings of freshwater pearls and other items set in paste. Smoking braziers of copal incense burned the hearts of the victims that had been sacrificed that day. Dried human blood caked the walls and floors and smelled "worse than any . . . slaughter-house in Spain" (Diaz 1963, 234). The central precincts included a large rack "full of skulls and large bones arranged in an orderly pattern, and so numerous you could not count them however long you looked" (ibid.).

Although Moctezuma had initially thought that Cortés was the god Quetzalcoatl (and, thus, treated him as such), the doomed leader soon realized that his visitors were not gods, when they forbade human sacrifice, cleaned out the Templo Mayor, and erected a cross on the summit. Cortés returned to the coast to settle a rebellion among his men. Fighting broke out during his absence, and when he returned, Moctezuma was killed. The Spaniards and their allies fled Tenochtitlán with only a quarter of their men unwounded. Cortés returned to besiege a city weakened by a catastrophic smallpox epidemic brought by the foreigners. Street by street, the conquistadors reduced the capital to rubble in 99 bloody days. Aztec priests sacrificed captured conquistadors to Huitzilopochtli. Bernal Diaz remembered: "Then they kicked the bodies down the steps, and the Indian butchers who were waiting below . . . flayed their faces, which they afterwards prepared like leather gloves" (ibid., 387). The Spaniards tortured, raped, and looted in an orgy of destruction and gold fever. They captured Moctezuma's successor Cuauhtemoc, and tortured and executed him in a vain attempt to discover his hidden treasury. When the pillaging was over, Tenochtitlán was a smoking ruin. The Aztecs realized the end of the Fifth Sun was nigh, but took the trouble to record the date: "And when the shields were laid down, when we fell it was the year count Three House; and in the day count it was One Serpent" (Anderson and Dibble 1969, 80).

MISSIONARIES AND THE STUDY OF AZTEC HISTORY

The conquistadors caught a fleeting glance of a civilization with a theology and symbolism as complex as contemporary Catholicism. Aztec traditional beliefs withered, however, in the face of missionary assault. In A.D. 1524, twelve Franciscan friars arrived in Mexico City, after walking barefoot all the way from Veracruz on the Gulf of Mexico. The missionaries were poor, humble men, quite unlike the arrogant conquistadors. They soon confronted their demoralized Aztec counterparts, whose gods had ignored offerings of human blood and allowed the destruction of the ancient order of life. The missionaries learned Nahuatl, the Mexica lingua franca, and embarked on a campaign of mass conversion. At first they succeeded. Thousands of Indians abandoned their loyalty to the old divine order, attracted to the elaborate rituals and colorful ceremonies of the Catholic church. But the friars could not communicate the deeper meaning of Christian doctrine. Many Indians simply added God to their existing pantheon; the saints became minor members of their anthropomorphic lesser divines. The friars found it hard to convince their converts that humans did not have to sustain gods with food and drink. From low-key preaching, the zealots among them turned to forced conversions and a systematic campaign against idolatry.

In A.D. 1531, Bishop Zumarraga of Mexico City boasted that he had destroyed 500 pagan shrines and 20,000 idols. He had also burned hundreds of bark documents, priceless codices of traditional knowledge and history. Fortunately, 2 years before Zumarraga's intemperate boast, a scholarly Franciscan named Bernardino de Sahagun arrived in New Spain. He quickly learned Nahuatl and set out to collect as much information about Aztec traditional culture as he could before it vanished forever. De Sahagun had an omnivorous intellect. He studied records of conversations between a handful of friars and learned Aztecs, who had survived the Spanish Conquest.

"The Words of Wisdom Came Forth"

Less than a generation after the siege, Aztec history was still preserved in the memories of traditionally educated nobles. The Franciscan de Sahagun enlisted the assistance of prominent Aztec elders and former merchants from Tlateloco, now living at Tepepulco near Mexico City. He also chose four youths who spoke both Nahuatl and Spanish and conversed with his informants for nearly 2 years. The Aztecs showed him hidden codices and went through them picture by picture. They used the pictorials as prompts when reciting their oral traditions, taking de Sahagun back to the world of their ancestors, long before the Spaniards came. The written codices were the bold outlines of religious concepts and historical traditions, filled in by reciting spoken narratives of the past handed down from generation to generation. Speaking well in public was a form of courtly protocol, and was fundamental to the education of an Aztec gentleman, just as weaving was to a lady. Trained orators learned to discourse about the past. They greeted diplomats with mellifluous words, and delivered prayers and congratulatory orations to gods and rulers. *Huetlatolli* (*huehue:* "old man"; *tlatolli:* "oration") were special occasions when as the Indians said: "the words of wisdom came forth."

De Sahagun realized that the Mexica perpetuated the fabric of their life, and traditions through oral history. He recorded dozens of orations in his notes; but, not content with formal speeches alone, he also prepared a series of questionnaires. For example, he asked four questions about each Aztec deity: "What were his titles and characteristics?" "What were his powers?" "What ceremonies were performed in his honor?" and "What did he wear?" The answers formed a major part of his research material.

Between A.D. 1547 and 1569, de Sahagun compiled his twelve-volume *General History of the Things of New Spain*, a remarkable compendium of Aztec civilization in Nahuatl and Spanish, the latter text being a paraphrase of the Mexica version. The first four volumes cover the gods worshiped by the Aztecs, religious rituals and sacrifices, and many details of cosmology and divination. The central volumes describe astronomy, omens, values, and theology, "where there are Very Curious Things Concerning the Skills of Their Language" (Anderson and Dibble 1954, 2). Volumes 8 through 10 describe the people themselves, their nobles, the history of their rulers, and "the merchants and artisans of Gold, Precious Stones, and Rich Feathers" (ibid., 3). Volume 11 is a treatise on Aztec natural history and medicine, and Volume 12 describes the Spanish Conquest of Mexico through the eyes of de Sahagun's informants, who lived through the campaign. The *General History* is really a kind of encyclopedia, which reflects the biases of the friar's informants. His ecclesiastical superiors considered the work subversive to the Christian faith and buried it in their archives, where it languished until nineteenth-century scholars discovered it. Today, the *General History* is an essential starting point for anyone studying Aztec civilization.*

Although de Sahagun's *General History* has an immediacy that comes from speaking to people who lived the traditional culture in their youth, he worked a full generation after the Spanish Conquest in a hostile political environment. While de Sahagun considered his informants reliable, he edited their narratives for his book to meet his own Christian objectives. An intellectual chasm separated the informants and de Sahagun's young Nahuatl interpreters, who had been born since the Spanish Conquest into a quite different cultural environment. A similar gully divides modern scholars from the worthy friar. The modern de Sahagun expert wrestles with the problems encountered by earlier scribes copying the prelate's original text and a hitherto unwritten language of highly inflected, compound words for the first time. Nahua discourse is full of allusions and unspoken implications, which escape us 5 centuries later. Historian Inga Clendinnen remarks how she works "with those painfully retrieved words pinned like so many butterflies to the page, remote from their animate existence" (1991, 284).

The Aztecs memorized their history through a combination of structured pictographs on codices and formal orations. This form of "alternative literacy" provided a reasonably standardized account of the Mexica past, which relied heavily on memnonics and formal schooling, where pupils learned by rote, just as ancient Egyptian scribes did. The early friars wrestled with the complexities of a skeletal history transmitted by word of mouth. Thus, any translation is a summary of a once carefully managed past, for the Aztecs inherited from their predecessors a complex legacy of religious beliefs and philosophies, which their leaders refined for their own purposes.

*The first complete translation of the Aztec text was accomplished by Arthur Anderson and Charles Dibble—who worked in English—between the 1950s and 1970s. Their text contains a complete annotation of the Nahuatl original, calling on other texts to amplify the information, including Diaz (1963), Cortés (1962), and other translations. Anderson and Dibble's *Florentine Codex: General History of the Things of New Spain* is worth consulting for its insights into the great complexities of working with de Sahagun's material, collected over nearly 5 centuries ago, by a man with a very different mind-set than modern-day scholars.

Only a handful of fifteenth- and sixteenth-century narratives and books supplement the *General History*. All have serious limitations. Conquistadors Diaz (1963) and Cortés (1962) offer what modern historians consider biased outsiders' accounts, which are imperfect, yet invaluable, written from a Spanish and Christian cultural perspective. Their shock at Aztec religious practices prevented them from writing dispassionate accounts of temples or rituals. Some postconquest codices, commissioned by missionaries or native nobles, copy earlier works, a process that diluted their accuracy and immediacy. About A.D. 1547, Don Antonio de Mendoza, first viceroy of Mexico, commissioned what is called the *Codex Mendoza* from Aztec scribes as a source of information about earlier times. This valuable document combines a pictorial rendering of Aztec conquests and tribute assessments with an ethnographic account of Aztec life from cradle to grave. Most other indigenous works, such as collections of Nahuatl "song-poems" and native accounts set down in the 1560s, have a hybrid quality and sensibility caused by years of Spanish rule.

Early Spanish scholars were usually missionaries with a profound curiosity about Indian life. Dominican friar Diego Duran was one such observer of Aztec civilization. He spoke Nahuatl fluently from childhood. Like de Sahagun, he used informants to collect historical data, including an elderly conquistador turned friar, who had taken part in the siege. Duran compiled an ambitious three-part work, two volumes of which describe Aztec life and beliefs. The third part, *The History of the Indies of New Spain*, is an uncritical essay pieced together from oral and written sources, which covers several centuries of Aztec history.

Tenochtitlán's people dispersed and perished. Considered subversive and heretical by church authorities, the friars' research was never published and gathered dust in ecclesiastical archives for 4 centuries. Their seventeenth- and eighteenth-century successors were mainly casual manuscript collectors with little interest in serious research. Nor did the authorities encourage research into the past. However, Carlos de Siguenza y Gongora (A.D. 1645–1700), priest and university professor, argued that the Mexico of his day was a blend of Indian and Spanish cultural traditions and boldly stated that Mexico should recover the history of early Indian societies from oblivion to achieve a greater understanding of the modern nation. His ideas fell on deaf ears, for the ecclesiastical authorities had no interest in propagating non-Catholic ideas.

Research into Aztec civilization languished out of pure neglect until Boston historian William Prescott published his classic, *The Conquest of Mexico,* in 1843. Prescott, a New Englander with an eye for historical narrative, used an unconventional approach to archival sources. He hired unemployed scholars on both sides of the Atlantic to ferret out and copy manuscripts in the Archives of the Indies in Spain. Unpublished documents from Madrid, Mexico City, and Paris provided rich source materials for *The Conquest of Mexico*. Prescott wove his sources into a gripping, if romanticized, story of Cortés's campaigns. His book was a study in dramatic contrasts between Cortés, the wise instrument of progress, and the weak and superstitious Moctezuma, doomed by his own irresolution. Prescott pictured the Aztecs as ill-disciplined, if brave, warriors, who engaged in human sacrifice and cannibalism. He made no attempt to understand the inner workings of the Aztec mysticism or the society it created and sustained, and compared their civilization to that of "our Saxon ancestors under Alfred" (1843, 54). Ethnocentric, overly romanticized, and dramatically contrived, *The Conquest of Mexico* became a bestseller. Prescott's narrative established the Aztecs stereotype of unsophisticated cannibals feasting on their human victims at banquets "teeming with delicious beverages and delicate viands, prepared with art and attended by both sexes" (ibid., 421).

Aztec civilization bears derogatory epithets in the Western mind since horrified conquistadors watched their friends die on temple altars during the siege of Tenochtitlán. Though de Sahagun wrote how the Mexica cut up sacrificial victims "in order to eat them," Duran remarked: "how cleverly this devilish rite [human sacrifice] imitates that of our Holy Church, which orders us to receive the True Body and Blood of Our Lord Jesus Christ" (1964, 95).

SACRED LANDSCAPE

The Aztec empire encompassed nearly 2,000 years of civilization in the Valley of Mexico. Tenochtitlán's wise men admitted their cultural and religious debt to earlier societies. Abandoned cities like Teotihuacán and Tula became shrines to earlier civilizations (see information in box, p. 350). The silent Pyramids of the Sun and Moon at Teotihuacán witnessed the birth of the Fifth Sun (figure 14.5). The valley's familiar landscape of hills, lakes, and mountains and its powerful spiritual associations had permeated the fabric of human existence long before the Aztecs rose to prominence. However, their rulers and wise men harnessed and expressed this spiritual landscape in masonry and stucco—a symbolic layout of their own within the great capital itself. Tenochtitlán connected the living to the spiritual world, the rulers to the ruled, and the nation and empire to the gods.

As the Aztec empire expanded, its sacred geography developed with its own icons, representing each manifestation of the natural environment. Archaeological discoveries help us link this sacred landscape to annual rituals and actual historical events. For example, the hill Huixachtlan, "thorn-tree place," visible from almost everywhere

Fig. 14.5. The Pyramid of the Sun at Teotihuacán. Courtesy Lesley Newhart.

MAPPING THE CITY OF TEOTIHUACÁN

Teotihuacán was one of the great cities of highland Mexico and a defining catalyst for later Toltec and Aztec civilization. A mere scatter of villages in 200 B.C., Teotihuacán grew rapidly into a town and then a huge city, which reached the height of its prosperity in A.D. 500, and passed into legend 200 years later. Fifteen centuries ago, between 125,000 and 200,000 people lived in her crowded apartment compounds, in a 21-square kilometer metropolis that survived longer than its early contemporary, Imperial Rome.

Investigating such an enormous ruined city presents unusual challenges. The ceremonial precincts alone, with the great Pyramids of the Sun and Moon, cover dozens of hectares. In 1960, the Mexican government commissioned archaeologists José Acosta and Ignatio Bernal to excavate and restore the Pyramid of the Moon and the surroundings of the Pyramid of the Sun. The Acosta and Bernal excavations used potsherds and art objects to date more than 8 centuries of occupation and temple building. Meanwhile American archaeologists René Millon, George Cowgill, and Bruce Drewitt mapped the urban landscape of Teotihuacán, one of Mesoamerica's greatest ancient cities. The Teotihuacán Mapping Project was a staggering undertaking, combining aerial photography with years of meticulous ground survey. Millon and his colleagues walked every accessible precinct of the city. They mapped great plazas and pyramids, and identified about 2,000 apartment complexes. Small teams of researchers collected thousands of figurines, potsherds, and other surface artifacts for dating purposes. The photographs, maps, and small finds came together in a unique archaeological portrait of a Teotihuacán so large it rivaled William Shakespeare's London a thousand years later.

We tend to think of ancient cities, or the past for that matter, in terms of public buildings—palaces, pyramids, and magnificent temples. But the Millon map allows us to look far beyond the facade of fine buildings and lavish display, beyond the imposing Pyramids of the Sun and Moon at Teotihuacán, and into the humblest quarters of the great city. The Millon survey brought a long-dead city alive. Beyond the ceremonial precincts lay the city itself, teeming barrios of rectangular apartment compounds, each a maze of flat-roofed small rooms, courtyards, and passageways. Each covered a large rectangle, separated from its neighbors by a straight street about 4 meters wide. Each was once a self-contained community of between 20 and 100 people, all members of the same kin group. Some housed skilled artisans, families who were expert carvers or artists. Potsherds and house styles tell us traders from distant places like the Valley of Oaxaca and Veracruz lived in barrios, where they used their own architectural styles and traditional pottery designs.

The Millon map provided enough survey data to show that the city did not grow in a haphazard manner. Teotihuacán's architects worked to a master layout devised early in the life of the city. The 5-kilometer-long Street of the Dead formed a north-south axis, lined with fine residences, ancestral shrines, and tombs. An east-west street bisected the north-south thoroughfare at the Ciudadela, where a huge market and the Temple of the Feathered Serpent Quetzalcoatl and the rain-god Tlaloc once stood. The entire city was a symbolic landscape of artificial mountains and foothills separated by open spaces oriented toward conspicuous landmarks on the horizon. The Street of the Dead changes elevation several times, carefully built to give an illusion of visual authority. The Pyramid of the Sun is an artificial mountain with 5 stages. Excavations for the installation of a sound and light show in 1971 unearthed a natural cave under the pyramid. Teotihuacán's sacred cave had 4 chambers, where offerings still lay intact, a natural gateway to the underworld. The Pyramid of the Sun was the most sacred edifice in the city, and marked the passage of the sun from east to west, and the rising of the Pleiades on the days of the equinox. These associations with the forces of the sun, and with life and death, were as important to Teotihuacános as they were to the Aztecs. Unfortunately, their capital, Tenochtitlán, lies under the urban sprawl of modern Mexico City, so a survey like Rene Millon's remarkable map project is impossible. However, the excavations of the Templo Mayor have revealed the sacred core of the city.

---

in the Valley of Mexico, was the place where life began anew every 52 years. The last 5 days of the 52-year cycle had always been days of renewal—which included the New Fire rites. A temple-platform lay on the summit long before Aztec times (see information in box, p. 353).

During the terrible famine years of A.D. 1450 to 1454 an indelible impression was left on the Aztec consciousness. The famine ended in 1454, just as a 52-year cycle of years came to a close. Now the New Fire rites and Huixachtlan assumed even greater importance in the cyclical Aztec life. As the sun went down on the last day of the old cycle, fire-priests proceeded from Tenochtitlán to the volcano's summit. Thousands of people gazed at the dark hilltop, watching the skies for the appearance of the "fire-drill" constellation (probably Orion's belt) at the western horizon. As the "fire-drill" constellation appeared, a young man died on Huixachtlan's sacred platform. (A fire-drill is a wooden twirler used to kindle fire in a log of softer wood.) The high priest kindled flames with a fire-drill placed on the victim's chest. A great bonfire flared. Runners with resin-dipped torches carried the new fire to temple hearths throughout the land. Time flowed anew. The world would continue.

Mount Tlaloc lies east of Texcoco, the highest peak on the eastern side of the Valley of Mexico. The temple on its summit, 4,000 meters above sea level, was a place of offering for many centuries. Richard Townsend and a team of Mexican archaeologists mapped the shrine in 1989. The temple itself is an open quadrangle, surrounded by 3-meter walls, approached from the west by a corridor-like processional way, and enclosed by high, parallel stone walls, which masked the spectacular view. Duran describes how the rulers of several cities, including Tenochtitlán, would make a pilgrimage to the mountaintop in April or May to call forth rain from the depths of the mountain. A simple temple of temporary materials stood in the enclosure, housing an image of Tlaloc and other deities. The rulers entered the precinct first, carrying lavish gifts. They dressed the Tlaloc statue in the correct regalia, then left food offerings, including the blood of a male child.

Townsend believes that the temple's carefully preserved, rock outcrops served as cardinal markers. A rock-cut shaft dug 3 meters into the summit led, symbolically, to the center of the mountain. Townsend hypothesizes that the shrine stood for the womb of the mountain. The rulers entered it as a gesture of fertilization. Here the inside met the outside; the underworld encountered the sky. The rulers made the pilgrimage to the mountain top, gave offerings, and

THE AZTEC CALENDAR

Ancient Mesoamerican calendars are among the most intricate of the ancient world. The Aztecs maintained 2 cyclical calendars, which meshed like 2 gear wheels. *Xiuhpohualli*, "A stem of grass," was a secular calendar that marked off an annual cycle of changing seasons—the times of planting and harvest. This timetable used the number 20, cempohualli, "one full count," based on the 20 fingers and toes of a human being. Each *xiuhpohualli* comprised 18 months of 20 days, known as *metzili* ("moon"), making up a year of 360 days. Thus, the secular calendar was close to our solar year of 365 days. It left 5¼ days to be accounted for, considered days of evil omen and dread significance with no divine patrons. Each month of the *xiuhpohualli* had special religious significance, culminating in a festival honoring its patron god. The new year started by honoring Tlaloc, the rain-god. The *xiuhpohualli* scheduled planting, harvest, and other routine events, including market days at 5-day weekly intervals.

The sacred calendar, *tonalpohualli* (count of day signs) consisted of 260 days, a day count formed by joining 20 signs with the numbers 1 to 13. The *tonalpohualli* originated long before the Aztecs, perhaps even among the Olmec people of 3,000 years ago. The 260-day count may have originated from hallowed ancient astronomical observations that once measured the length of the harvest season. This was a sophisticated divination system, which served as a carrier of tradition and authority. Each day sign combined with any one of 13 numbers, giving 260 possible combinations. The number sequence began with 13 signs associated with the numbers 1 to 13. Then a new number sequence began, linking the second series of numbers 1 through 7 with the remainder of the signs. When all 20 signs combined with all 13 numbers (260 days), the *tonalpohualli* would begin anew. Thus, the sign Alligator first appeared with the number 1, then 8, 2, 9, and so forth. By the time the 260-day cycle was complete, Alligator and all the other signs had linked with every number from 1 to 13. The calendar progressed by revolving a set of digits from 1 to 13, like sprockets of a gear wheel, upon another set of 20 signs, each in a fixed order and named after an animal, another object, or a natural phenomenon. The 260-day *tonalpohualli* comprised 65-day groups of 5 "weeks," of 13 days each carrying a sign oriented toward north, south, east, or west. This orientation kept the sun in motion throughout the year. Each "week" had its own patron deity or deities, as did each of the 20 day-names; the patron giving the day its special characteristics.

The Aztec century, *xiuhmolpilli,* "a bundle of years," was 52 years long, beginning at the point when the sacred and secular calendars coincided. The priests marked the passage of years by putting aside a peeled reed for each one until the cycle was complete. Then they bound them together and ceremonially buried them in a ritual to symbolize the start of the new count. The end of a 52-year cycle was a moment of fear and apprehension, when the rekindling of sacred fire in the chest cavity of a sacrificial victim symbolized the start of a new cycle of finite time. Archaeologists use day glyphs preserved in buildings as a way of dating Aztec structures and the reigns of the rulers who built them.

---

brought life-giving water back with them as a reenactment of the creation myth, where a mythic hero brought rain, the gift of life.

Once the Mount Tlaloc rituals were over, the rulers returned to the eastern shore of Lake Texcoco, traveling in a fleet of canoes to an island called Pantitlan. There they planted a sacred tree by a great spring as a symbol of life. A young maiden perished, her blood mingling with the spring water as an offering to the deity of ground water in support of earthly creation.

THE ARCHAEOLOGY OF THE TEMPLO MAYOR

When the conquistadors prevailed over the Aztecs, they founded their colonial capital atop the Indians' capital, making systematic excavation of Tenochtitlán a virtual impossibility.

In 1790, workers laying water pipes under the Zocalo Plaza at the heart of the city unearthed a statue of the mother goddess Coatlicue and the famous Stone of the Sun of the ruler Axayacatl. The authorities hid the finds away from view on the grounds that they were subversive to the Catholic faith. Invariably, as a result of construction work, occasional excavations have revealed traces of the ancient capital. In 1913, the great Mexican archaeologist Manuel Gamio uncovered the southwest corner of the Temple Mayor when a building was demolished in the Zocalo area. The digging of the Mexico City Metro in the 1960s and 1970s yielded large quantities of Aztec artifacts, including a temple to the wind-god Ehecatl-Quetzalcoatl (figure 14.6). These excavations provided snapshots of the ancient capital.

In February 1978, electrical workers unearthed an enormous monolith of the decapitated and dismembered goddess Coyolxauhqui. The discovery sparked fresh interest in the central precincts of Tenochtitlán, and especially in the Templo Mayor, which lay just east of the cathedral. Under the direction of Eduardo Matos Moctezuma, the National Institute of Anthropology and History undertook a large-scale excavation of the temple.

Moctezuma had many years of experience with excavations in the Valley of Mexico. He had worked on the Tlatelolco excavations of

Fig. 14.6. Temple of Ehecatl-Quetzalcoatl, found during the digging of Mexico City's subway. Courtesy Lesley Newhart.

the 1960s; and had dug at Teotihuacán, Cholula, and Tula. Moctezuma gathered every available piece of information on the Templo Mayor from archaeological finds and historical chroniclers like Cortés and Diaz, who described the temple as a twin-shrined pyramid. The duality commemorated the twin gods that symbolized the pillars of Aztec civilization: war and agriculture. Moctezuma pieced together the limited archaeological data and formulated a 5-year campaign of excavation and conservation work designed to reconstruct the history of the Aztec empire's most illustrious temple.

The research presented extraordinary challenges. Streets, water mains, and sewers crisscrossed the built-over site. Magnificent colonial buildings of great historical significance stood close by. Before excavation began, an Advisory Board on Monuments assessed the importance of any structures threatened by the excavations. The authorities carefully recorded these features before the bulldozers moved in. Then archaeologists opened up an enormous excavation area—between 5,000 and 7,000 square meters—and divided it into numbered and lettered 2-meter squares. They worked with soil engineers, who monitored the water table, some 4 to 5 meters below the modern city streets.

Tenochtitlán lay in the midst of a swampy lake. If the excavators had pumped out all the water from the dig, they would have undermined the temple structure and neighboring colonial buildings. Moctezuma focused his attention on the structures lying above the water table but below the modern streets, leaving the lower tiers of the Templo Mayor for future generations equipped with more advanced water-control technology. The high water table, however, had advantages: Many organic finds such as animal and human bones survived and were processed on-site by professional conservators.

The excavations moved slowly. Architects, engineers, and other experts balanced and supervised the acquisition of historical information with long-term conservation measures. Moctezuma discovered that the latest phase of the temple had been leveled to its foundations by the conquistadors. Underneath lay six earlier temple pyramids, stacked one on top of the other on the same location, as successive rulers built ever-larger shrines (figure 14.7).

Fig. 14.7. General view of excavations in progress on the Templo Mayor, Mexico City. Courtesy Lesley Newhart.

Moctezuma's workers uncovered evidence for seven temples on the same spot. He dated the construction stages by using glyphs carved on excavated objects, working back from the razed temple of A.D. 1519, described by Cortés and Diaz. Fortunately, the Spaniards left earlier stages intact, giving the archaeologists a complete picture of the greatest sacred structure of the ancient Americas.

The earliest temple on the site lies under the water table, and probably dates to the fourteenth century, soon after the founding of the city. Each successive temple had a shrine to Huitzilopochtli and Tlaloc on its summit. The ruler Acamapichtli built a new shrine atop the original temple platform in A.D. 1390, dated by a stone date-glyph with the inscription "2-rabbit." The summit of Acamapichtli's pyramid had a sacrificial stone of volcanic rock set into the floor 2 meters from the front stairway and in front of Huitzilopochtli's chapel. A multicolored *chacmool* figure lay before Tlaloc's shrine. The *chacmool* depicts a man with raised head in a supine position with bent knees and a receptacle on his stomach. He acted as a messenger, bearing gifts from the priests to the gods. Two pillars bearing goggle eyes (eyes projecting like cylinders) of the god fronted Tlaloc's shrine. Black-and-white vertical lines near the base of the pillars may represent rain. The god's image lay on a low platform in the center of the shrine. Huitzilopochtli's image also stood on a low banquette, his chapel adorned with fine wooden door jambs and entrance pillars.

Acamapichtli's successors covered his modest construction with vast amounts of earthen fill, reflecting a surge of building activity as the Aztecs emerged as an independent nation. A new, and larger, pyramid was built on the fill during the reign of Itzcoatl, in the year "4-reed" (A.D. 1431). We know of only the pyramidal base and well-constructed double stairways. Eight life-sized figures of standard-bearers once stood by Huitzilopochtli's shrine. Moctezuma found them lying in a neat row against steps leading to the shrine, as if they had been placed there ceremoniously when the next stage of construction began.

The most spectacular architecture and sculptures come from the fourth temple, constructed by Moctezuma Ilhuilcamina in the year "1-rabbit" (A.D. 1454), with additions as late as A.D. 1469. Moctezuma enlarged the pyramid base, then adorned it with braziers and serpent heads on all four sides. The braziers on Tlaloc's side bear his goggle-eyed face, those on Huitzilopochtli's, the bow-like symbol of the sun-god. The ruler added an imposing stairway over the old one on the main, west facade. Moctezuma's successor, Axayacatl (A.D. 1469–81), placed a new platform in front of the stairway, set with the disk of Coyolxauhqui, carved with her dismembered limbs adorned in all their ritual finery. The sculpture lay at the foot of the stairway commemorating the legend of the newly born Huitzilopochtli. Thereafter every prisoner sacrificed to the all-powerful sun-god walked across this fearsome monument on his way to the sacrificial stone at the top of the stairs. The temple Coatepec commemorates this legend, and symbolizes Tenochtitlán's military supremacy. The north side of the pyramid supported Tlaloc's shrine, identifying the pyramid with

Tlaloc's mountain of life. His effigy stood inside the shrine, adorned with seeds of all the important cultivated plants.

The entire Templo Mayor rested on a platform with enormous, undulating serpents' bodies writhing around the corners, terminating in brightly colored heads. Excavations around the Coyolxauhqui stone between the two central serpent heads marked the point where the stairways to the two shrines joined. Offering rooms with colored marble floors lay at the north and south ends of the platform.

Diaz had heard rumors of rich offerings of gold, silver, and precious stones cast into the foundations of the temple during construction, and "watered with the blood of Indian victims" (ibid., 237). Four and a half centuries later, archaeologists unearthed a staggering array of such offerings in the long-forgotten pyramid, including Tlaloc-masked pots and ceramic vessels bearing the image of the water deity Chalchiuhtlicue. Temple braziers still contained obsidian and silver necklaces, serpentine figures, and decorated sacrificial knives with obsidian blades, along with bundles of coral and gold fragments. Seashells, greenstone canoe models, fish skeletons, and waterfowl commemorated water and rain. A crocodile skeleton symbolized the ancient myth of the earth floating in primordial waters. Numerous trophy objects, including a 2,000-year-old Olmec mask and artifacts in the ancient Teotihuacán style, give credence to the Aztec state's glorious past and its imperial power.

Eighty-six offertory caches yielded more than six thousand artifacts. They came from small chambers with stone walls and floors, from inside stone boxes with stone covers, and from the rubble of earlier construction. Most lay below the front platform of the temple or close to the two shrines. The Aztecs manufactured magnificent sculptures, but most artifacts came from tribute areas of the empire such as lowland Veracruz. Interestingly, most offerings date to between A.D. 1440 to 1469, the period of rapid military expansion, when tribute collections were at their height.

The head of Tlaloc, the god of rain and fertility, appears on once blue-painted stone jars. Eleven such containers made up one offering. Underneath them lay the skeletons of at least forty-three infants, sacrificed to honor the rain-god. Their tears symbolized rainfall. Caches of hundreds of flint and obsidian artifacts included knives, sharp blades, and even minature heads and models of rattlesnake rattles.

Aztec artisans also worked human skulls, turning them into facial masks, made of the front of the cranium and the jaw. They filled the eye sockets with white shell disks. Round hematite fragments served as the irises. Sometimes a flint knife lay between the teeth like a grotesque tongue. In other cases, the knife penetrated the nasal cavity. The skulls were either ritual masks or symbols of death.

Two life-size clay images of Eagle Warriors flanked the entrance to a series of rooms to the north of the Templo Mayor complex. The eagle was the symbol of the sun, and the Eagle Warriors were soldiers of the sun. Fabricated about A.D. 1485, the potter made the figures in four sections, complete with eagle helmet and extended arms symbolizing the eagle's wings (figure 14.8).

The temple offerings form a metaphorical language, revolving around agriculture and war. Food and tribute, water and fire (war), and life and death in sacrifice ensured the survival of the world of the Fifth Sun. The Aztecs regarded conquests as their destiny, a task that was carried over from Huitzilpochtli's victory over Coyolxauhqui. The tribute they accumulated symbolized their takeover of other nations that now belonged to the god.

The Templo Mayor was the true center of the Aztec universe. The Mexica believed that they were the chosen people of the gods, those who maintained order in the universe. The temple was built on the spot where the four cardinal directions of the living world intersected. Each heading had a color, a creator god, and a plant or tree that held up the sky. The pivot of the four bearings was a vertical channel through the Templo Mayor, which joined the three tiers of the heavens, the earth, and the underworld, making it the ultimate sacred place in the Aztec world. The symbolism extended to the capital, Tenochtitlán, which lay in the center of a lake, like Aztlan, another mythic center of the world order.

The Templo Mayor itself was a microcosm of the Aztec world view, where both horizontal and vertical dimensions had cosmic importance. Moctezuma believes that the platform supporting the temple was the horizontal, terrestrial level of the cosmos. The celestial layers were the four slightly tapering tiers of the pyramid that rise to

Fig. 14.8. Clay figure of an Eagle Warrior, Templo Mayor, Mexico City. Photograph from the National Geographic Society.

the summit with its two shrines, the supreme level called *Omeyocan*, or the "Place of Duality." The thirteen levels of the underworld lay below the earthly platform. The temple formed two vertical halves, both sacred mountains. The southern half was Coatepec, where Huitzilopochtli defeated his rivals. The northern half was Tonacatepetl, "the Hill of Sustenance," where Tlaloc ruled. As Townsend remarked: "In the heart of Tenochtitlán the pyramid rose as an architectural fetish charged with the powers of all the offerings and the blood from thousands of sacrificed human beings. The structure was the terrifying center of the Aztec world, and an architectural hieroglyph of the term *atl tepetl*, water mountain" (1992, 154).

Three cosmic regions met at the Templo Mayor, the *axis mundi* (world axis) of Aztec civilization. The temple itself lay in a sacred precinct of 78 structures, each placed in its correct position in the Aztec universe. The entire complex was the hub of a carefully orchestrated setting for the reenactment of public rituals ensuring the continuation of the Fifth Sun.

THE FLOWERY DEATH

The Mexica were indeed tumultuous warriors addicted to human sacrifice. Also sophisticated thinkers, they inherited ancient cosmic myths that originated deep in their past, through poems, songs, and formal orations.

Mexican scholar Miguel Leon-Portilla has pieced together a credible impression of Aztec philosophies and beliefs. He describes how the Aztecs organized their lives within a layered cosmos controlled by a crowded and elaborate group of several hundred deities. Aztec philosopher-priests understood the dilemmas confronting people who inhabited a world controlled by the Creator and lived there to serve him. One wise man pronounced: "We are but toys to Him; He laughs at us. Humans are on earth to serve and nourish the gods" (Leon-Portilla 1963, 66). The Aztecs saw war and sacrifice as an eternal order, as a way of perpetuating a doomed world. Successive Aztec rulers led their people into an existence where human sacrifice became a self-sustaining, integral part of their lives, perhaps even an instrument of political policy. The greatest incidence of human sacrifice occurred when the empire was expanding rapidly. However, history shows that there is a point at which a satiated population turns and rebels against an unbearable burden; where, perhaps the onus of the dead weighs heavily on the minds of priests, ruler, and the nation as a whole. There are tantalizing clues that an increasing intellectual polarity between ferocity and mercy was causing profound tensions in Aztec society at the time of the Spanish Conquest.

In Aztec society, the manner of your death determined your final destiny. When people died, they entered another layer of the cosmos, the world of "the beyond, the region of the dead" (Anderson and Dibble 1954, 6). Most people went to *Mictlan*, the 9-level region beneath the earth. After 4 years of wandering and repeated tests, they came to permanent rest in "the ninth place of the dead" (ibid.). Those favored by Tlaloc, the rain-god, went to *Tlalocan*, where "never is there a lack of green corn, squash, sprig of amaranth" (ibid.). Warriors who died on the battlefield or on the sacrificial stone enjoyed the "Flowery Death." They traveled to *Tonatiuhilhuicac*, the dwelling place of the sun. There the warriors would sing war songs as they accompanied the sun to the zenith of his journey. After 4 years, they turned into humming birds, symbols of the sun-god who sipped from the flowers of heaven and earth.

The offering of human blood was key to the Aztec canon. The custom was acquired from the gods themselves: Human beings were important in the cosmic order as providers of nourishment to the gods. A captive warrior that was sacrificed pleased the gods; the more valorous the warrior sacrificed, the more the gods were nourished. Every prisoner was the property not of his captor, but of the deity. Often the warrior wore the ceremonial body paint and regalia of the god as he climbed the pyramid to his death, thus becoming a symbolic god himself.

Elaborate rites surrounded important sacrifices on occasions like the festival of Toxcatl, which honored the young and virile war-god, Tezcatlipoca. A flawless young man always impersonated Tezcatlipoca. He assumed the role of the god for a full year, walking around in divine regalia, playing the sacred flute. A month before his death, he was married to four young priestesses who sang and danced with him as he walked around the city. On the day of sacrifice, the young man walked willingly and alone to his appointment with death. His decapitated skull held a prominent place on the skull rack in the plaza below the temple. Human sacrifice was not an earthly, but a divine, drama. The young man ruled for a year as an ideal model of the war-god. He displayed the gracious bounty of the deity, then perished as an exemplary captive to show the god's power. His short ceremonial life was glorious, a transitory existence in a transitory world ruled by the deity who watched over human destiny.

Aztec human sacrifice took many forms, including death from arrows, burning, and beheading. Aztecs also sacrificed women and children, to bring rain and to honor fertility deities. Most prisoners died on a sacrificial stone, giving their hearts to the god. The priests painted the sacrificial victims with red and white stripes, then reddened their mouths and drew red circles around them. They glued white feather down on the captive's heads. The victims lined up at the foot of the pyramid steps. Priests escorted them one-by-one to the summit, each symbolic of the young sun rising to his zenith at the moment of sacrifice. The executioners thrust the prisoner backward over the sacrificial stone. Four priests grasped the limbs, and a fifth pressed down on the neck, bending it back under great pressure. Then the high priest smashed open the chest with an obsidian-bladed knife, thrust his hand into the cavity and tore out the still-beating heart, which he held high as an offering to the sun before flinging it into a special bowl (figure 14.9). The priests tipped the still-quivering body over the edge of the pyramid, sending it tumbling down the steps, representing the sun returning to earth.

No one knows how many human victims the Aztecs sacrificed each year. Cortés put the figure at 50 for each temple annually, which would have meant that some 20,000 persons died for the gods throughout Aztec domains each year. The highest numbers died in Tenochtitlán, where as many as 800 prisoners may have perished during a single festival; a far cry from the thousands claimed by early Mexica informants. The skull rack at the base of the Temple of Huitzilopochtli was a lattice-like construction of long poles where sacrificial crania lay in long tiers. Two conquistadors claimed that 136,000 skulls sat on the rack, probably a gross exaggeration. Archaeological researches put these claims into perspective. Excavations at

Fig. 14.9. Aztec human sacrifice. From the *Codex Magliabechiano* Folio CLXIII: 3. Photograph from the Library of Congress, Washington, D.C.

the important temple at nearby Tlatelolco yielded 170 skulls perforated for stringing on poles.

The Aztecs did not invent human sacrifice. The gods had fed off human blood at Teotihuacán during the great city's heyday. The Maya were experts at human sacrifice, too. But the Aztecs developed mass human sacrifice to new and obsessive heights. Offering people to the gods was far more than a political device or a means for a brave warrior to obtain rich rewards. It was a nation's effort to prolong the Fifth Sun, to keep the sun itself in the heavens. The Aztec had no compunctions about sacrificing people. Quite the contrary, it would have been a grave sin not to do so.

As for cannibalism, the Spaniards' lurid tales of stews of human flesh and the butchering of sacrificial victims for feasts labeled the Aztecs as incorrigible cannibals in Western minds. Most anthropologists or historians place little credence on these travelers' tales. Duran describes a sacrifice to Huitzilopochtli where the thigh of the sacrificed became the property of his captors. "It was carried away and eaten, each received his part, the number depending on the number of those who had captured him—never more than four" (1971, 97). The ritualized eating of sanctified human flesh may have been a common practice, but there was certainly no wholesale consumption of fellow humans. To date, no one has excavated any houses in the now-buried capital. When they do, the presence or absence of butchered human bones may settle the cannibalism issue definitively. But most experts believe that the consumption of human flesh was a sacred act. War, sacrifice, and cannibalism required human blood.

The Aztecs lived in a deeply mystical, yet militaristic world that sustained a society that may have been very ambivalent about itself and terrified that their universe would soon end. Two years after Diaz and his colleagues marveled at the sight of Tenochtitlán, the city was a smoking ruin, its inhabitants slaughtered or taken into slavery. Weapons of steel and gunpowder overthrew one of the world's last preindustrial civilizations. The conquistadors razed the great temples, destroyed codices, and threw down idols. When the gods died, the people and their sacred landscape perished as well. Half a century later, Diaz wrote: "Today all that I then saw is overthrown and destroyed; nothing is left standing" (1963, 214). The capital of colonial New Spain, Mexico City, rose on its foundations. Nearly a quarter of a million Aztec people vanished into historical oblivion within a generation, decimated by epidemic disease, inhumane treatment, and slavery.

EPILOGUE

When in April the sweet showers fall
And pierce the drought of March to the root, and all
The veins are bathed in liquor of such power
As brings the engendering of the flower...
Then people long to go on pilgrimages
And palmers long to seek the stranger strands
of far-off saints, hallowed in sundry lands,
And specially, every shire's end
In England down to Canterbury they wend...

(Chaucer, as cited in Coghill 1962, 17)

In A.D. 1170, when the Toltecs ruled the Valley of Mexico, four of King Henry IV's knights hacked Archbishop Thomas Becket to death in the transept of Canterbury Cathedral, mother church of all England. Rumors of miracles spread within hours: invisible hands had lit candles by the archbishop's body, his voice was heard among the cathedral choristers, and the ailing and blind were restored to health as they prayed to the martyr. In A.D. 1174, a disastrous fire burned the church to the ground. A new cathedral rose from the ashes, proclaiming the glory of God and attracting visitors from kilometers around. Two hundred years later, Chaucer's thirty supplicants converged on Canterbury in an April pilgrimage.

Geoffrey Chaucer began his *Canterbury Tales* in A.D. 1386. The great poet never finished his masterwork, but left a concise portrait of ordinary people from England going about their business seven centuries ago: nobles and commoners, townspeople and country folk, clerics and laypeople. The pilgrims had a striking individuality, that is reflected in their tales, all of which embrace an idea, a moral, or a wisdom. They possessed but one common goal: a mission of pilgrimage to an intensely sacred place where heaven met earth. My pilgrimage—a study of ancient sacred places—was born of science. Though known mostly from archaeological research, the sites still link heaven and earth.

I visited Gloucester Cathedral in western England when the research for this book was completed. I had spent the morning in the midst of the sacred landscape at Avebury, studying the relationships among the great stone circle, West Kennet long barrow, and Silbury Hill. My mind was full of archaeological data, of the difficulties of reconstructing long-vanished settings for the secular and the supernatural. Gloucester Cathedral's pinnacled tower beckoned, as a quiet place where I could sort out my thoughts. A Norman edifice built between A.D. 1089 and 1100, the church became an important place of pilgrimage in 1327, when the Gloucester monks enshrined the bones of the murdered King Edward II. The money made from pilgrims enabled them to afford to rebuild in the new Perpendicular style. I sat in the fourteenth-century nave, churning archaeological thoughts in my mind. The overwhelming hush and high piers drew my eyes toward heaven. I felt like a pilgrim myself, drawn to a place where, left alone, my thoughts could flow as if compelled, down through the corridors of the past.

Gloucester's rebuilding happened at a time when church building reached epidemic proportions in Europe. A new Gothic style of architecture introduced flying buttresses and stained glass windows, which gave new cathedrals a crystal-like quality of lightness, opening up their huge spaces, so that air could flow into the central areas. Christianity acquired a new magic and quality with its cathedrals, places where chants and hymns resonated throughout the upper reaches of the vaulted sanctuary. I realized my own archaeological pilgrimage had come full circle bringing me from the remotest past to the frontiers of the modern world.

The medieval cathedrals signified power in time. Their builders oriented them around the Incarnation and Crucifixion of Christ. At the altar, they replicated poetic mythology from the New Testament.

The north side represented the Creation and the Old Testament; the south, the Last Judgement and the New Jerusalem. Each cathedral formed a Latin cross. Their naves were sacred vessels, a Noah's Ark carrying humanity through time. The stained glass windows high above guided worshipers' spiritual courses. Excavations under the naves of Canterbury, York, and other cathedrals show that their builders had built on the same sacred place again and again.

Like Avebury, Stonehenge, Knossos, and Teotihuacán, Gothic cathedrals were places of worship and pilgrimage. Like the Templo Mayor, they were also settings for flamboyant performances, moments of high ritual, and symbolic sacrifices.

As the nave darkened that cloudy afternoon, my thoughts drifted far back in time, to the western Asian farming villages that clung to small fields 10,000 years ago, the lives of the people ruled by cyclical time. I reflected on how archaeology revealed ever-present ancestors—at Jericho, Çatalhöyük, in the Cycladic Islands, at Knossos, high atop Mayan pyramids, and in predynastic Egyptian villages. I recalled the great earthworks and stone circles of Avebury and Stonehenge, and the complex astronomical observatories of the Hopewell Indians. From Hopewell, my thoughts turned to Mircea Eliade's writings on the sacred and the profane, on the *axis mundi*, which linked the centrifugal force of the cosmos. Where I stood at that very moment was metaphorically on an axis, a place where heaven met earth, or as T. S. Eliot called it: "the still point of the turning world." The medieval cathedrals of Canterbury and Gloucester gave me the metaphor of sacredness that I was seeking.

Many years ago, I visited Chartres Cathedral in northern France. Notre Dame de Chartres, France, built in a mere quarter of a century around A.D. 1195, is the sixth church built on the same site; a masterpiece where the infinite becomes a miracle in stone and glass. The cathedral is all windows, the great rose window of the western front symbolizing the Virgin Mary herself. Stained glass windows provide ethereal settings amongst soaring beams and graceful arches. At Chartres they become a form of a new language, and bring together ancient principles of Christianity, many of them derived from even older cosmic beliefs. The rose was a powerful symbol that evoked soul, eternity, wheel, sun, and the cosmos. The rose was sacred to the Egyptian goddess Isis, to Greek Aphrodite, and to Venus—a symbol of human love transcending passion, which signified the Virgin Mary in Christianity. The major rose windows at Chartres depict the Virgin and Child (north), martyrs who spread the Word and the New Testament (south), and the wounded Christ at the center of the Last Judgement (west). Each uses the same vocabulary of color, form, geometry, and symbol. The gem-like transmutation of the light shining through the windows creates transcendental effects that could heal and revivify worshipers crowded in the nave.

Stained glass windows were designed to instruct the nonliterate, those who could not read the scriptures. Like ancient Egyptian artists, the medieval artisans followed a standard vocabulary of forms as far as the deposition of figures and backgrounds were concerned. They made use of unique geometrical compositions to structure the motifs of the windows, often with close ties to astrological and cosmological images, as well as zodiacal symbolism. A stunning window at Lausanne Cathedral in Switzerland, erected about A.D. 1230, depicts the medieval image of the world *(imago mundi)*. This cosmological treatise in glass and stone shows the element water; the signs Capricorn, Aquarius, and Pisces; and the moon being drawn across the sky in a chariot. Like the Templo Mayor, the Egyptian Step Pyramid, and Mayan centers, Chartres's windows are an integral part of a setting that brings heaven to earth and joins the secular and spiritual.

A circular tiled maze lies on Chartres's nave floor, with a rose at its center. Interestingly, if the north rose window high above were hinged at the ground, it would cover the maze exactly. The window represents the Last Judgement, while the labyrinth on the ground depicts the path of the human soul through life on earth. Only one maze path leads to the center, which requires a sequence of movements that symbolically replicate the movements of the planets in the heavens. The confusion of the maze walker reflects the complex journey of life on earth. The pilgrim achieves purification by reaching the center.

Chartres served the same purpose as much older sacred places. Like Cahokia, Knossos, and Saqqâra, the cathedral was a magnet. The permanent population of medieval Chartres may have been no more than 1,500 people. The cathedral regularly attracted 10,000 worshipers, an offering to God as powerful as the human sacrifices of the Aztecs and the propitiated killing of children for Minoan deities. The cathedral provided a way of connecting the divine to the living world, for all things emanated from the Kingdom of God. Requiring enormous expenditures of human labor and sometimes extreme deprivation, Gothic cathedrals were expensive outpourings of love for the Lord, as well as metaphorical sacrifices of stone and material goods offered in the expectation of divine favors in return.

The cathedrals were settings for dazzling spectacles. Sung masses and mystery plays depicting the life of Jesus or episodes in the lives of saints brought on intense emotional reactions among the faithful. The great cathedral bells tolled at times of joy, and at moments of mourning. They sounded warnings, and rang out in exultation and in crisis. Their message was as powerful as the funeral drums of the Ila in Central Africa. Great preachers attracted great crowds; no matter what the event: baptisms, marriages, funerals, and prayers for the dead; ordinations and excommunications; victory celebrations; or public meetings. The cathedrals were the focus of human life.

Like Mesoamerican Indians, medieval Christians worried about the fertility of the land and the continuity of life itself. Every Easter eve, a New Light was kindled, celebrating the Resurrection and the year's start. A thousand tapers were lit and carried from town to village and village to household, as life was renewed. During autumn harvest festivals, churches were decorated with the fruits of the soil, commemorating its bounty, like the Green Corn Ceremonies of eastern North American groups.

Perhaps endowed with a sacred relic, a miraculous image, or a sign of martyrdom, a cathedral such as Chartres or Gloucester was a tangible symbol of eternal, supernatural reality; a witness to the imminent presence of God. Seven centuries ago, the medieval cathedral was the Bible of the poor, provided an image of the cross and of the body of Christ, and represented a corner of God's kingdom.

Other architectural metaphors survive. In 247 B.C., a joyful procession wended its way to Anuradhapura, royal capital of Sri Lanka, along a road sprinkled with white sand and flowers. Priests planted a bough of the sacred bo tree, under which Buddha had attained enlightenment in India. As the tree flourished, so did the royal capital, a walled city ringed with Buddhist monasteries that became a replica of the sacred world. In northwestern Cambodia, more than a dozen Khmer princes built their palaces and temples at Angkor. King Suryavarman II erected Angkor Wat as an observatory, shrine, and mausoleum in the early twelfth century A.D. The temple honors Vishnu, ruler of the western quarter of the compass. The five multi-tiered towers of the temple depict Mount Meru, home of the Hindu gods and center of the universe. Celestial maidens twist and cavort in endless dances on Angkor's walls, depicting the pleasures of paradise.

According to the Koran, Allah created seven heavens and as many layers of earth, each associated with specific gemstones, colors, qualities, and stages of human existence. Islamic architects transformed the mythic universe into an icon to be contemplated in association with the mathematical dimensions of the physical world that established the direction of Mecca. Islamic mosques are models of the cosmos. Their domes replicate the bowl of heaven surmounting the earthly domain. The garden courtyard is paradise, the thrones and wall towers reflect the sacred mountain, and the gates are entrances to heaven.

I followed the intricate tracery high above me in Gloucester's fourteenth-century choir and realized the stupendous power that sacred architecture has in elevating the senses, and achieving the sublime. I remembered the Aztec rulers of Tenochtitlán and their hallowed pyramids, temples, and plazas set at the center of the earthly and symbolic world, and the Egyptian pharaoh Djoser laying claim to his unified kingdom in the shadow of his mortuary pyramid. My mind slipped back to the mysterious rulers of Minoan Knossos, who used colorful rituals set amid richly decorated courtyards to commemorate their influence over the supernatural forces that controlled the eternal realities of cyclical time and human life: birth, life and death. I remembered looking up at the temple atop Xunanuntich's Castillo, where Mayan lords passed in trance into the supernatural world, and recalled how charismatic leaders have always harnessed the supernatural for their own ends, using performance and dramatic settings to authenticate their rule. I realized that archaeology had shown me a direct, if invisible, line that runs from ancient sacred places through the mythic layers of human existence, and connects directly with symbolic architecture.

Science and archaeology have added a new dimension to our understanding of sacred places. Settlement research, environmental reconstructions under earthworks, sophisticated radiocarbon dating methods, and computer-based studies of enormous amounts of archaeological data, allow us to reconstruct entire sacred landscapes. The archaeological record yields evidence of a compelling universal literacy of symbolism that defines archaeological sites as venerated historical masterpieces. These sites survive as hieroglyphs, providing further insights into the intangible world of ancient sacred places.

GUIDE TO FURTHER READING

The archaeological literature surrounding sacred places is enormous. The references that follow were especially helpful in researching and writing this book. Interested readers will find comprehensive biographies of the more specialized scientific literature in any of the syntheses below.

CHAPTER I: THE ARCHAEOLOGY OF THE INTANGIBLE

The Ila people are well described by two early missionaries: W. E. Smith and A. Dale, *The Ila-Speaking Peoples of Northern Rhodesia* (London: Macmillan, 1920).

Barry Lopez, *Crossing Open Ground* (New York: Vintage, 1989) contains thought-provoking essays on people as part of the natural environment.

Landscapes of the mind are generating an enormous amount of literature. Useful titles include: Christopher Tilley, *A Phenomenology of Landscape* (Oxford: Berg, 1994); and Simon Schama, *Landscape and Memory* (London: HarperCollins, 1995). The latter is wonderful on forest environments. See also Barbara Bender, ed., *Landscape: Politics and Perspectives* (Oxford: Berg, 1993).

Douglas Comer, *Ritual Ground* (Berkeley: University of California Press, 1996) is a remarkable monograph on archaeology, world views, and ritual in the context of a frontier fort in the American West.

Travis Hudson and Ernest Underhay, *Crystals in the Sky: An Intellectual Odyssey Involving Chumash Astronomy, Cosmology. and Rock Art* (Socorro, New Mexico: Ballena Press, 1978), is a broad essay on Chumash cosmology and shamans. Two basic surveys of archaeological method and theory—Paul Bahn and Colin Renfrew, *Archaeology,* 2d ed. (London and New York: Thames and Hudson, 1996); and Brian Fagan, *In the Beginning,* 9th ed. (New York: Addison Wesley Longman, 1997)—describe many examples of settlement archaeology, as well as survey recent theoretical advances.

Much has been written lately about ethnographic analogy. Alison Wylie, "The Reaction against Analogy," *Advances in Archaeological Method and Theory* 8 (1985):63–111 offers a good overview.

The archaeology of mind is generating a growing literature base. A good starting point is a series of essays under the title "What is Cognitive Archaeology," in the *Cambridge Archaeological Journal* 3, no. 2 (1993):247–270.

The Panama case study is described by Olga Linares, *Ecology and the Arts in Ancient Panama: On the Development of Social Rank and Symbolism in the Central Provinces* (Washington, D.C.: Dumbarton Oaks, 1977).

A useful summary of the Olmec civilization is Michael D. Coe and Richard Diehl, *In the Land of the Olmecs,* 2 vols., (Austin: University of Texas Press, 1980).

Lawrence Sullivan, *Icanchu's Drum* (New York: Free Press, 1989) is one of those seminal books that has the potential to change entire fields of science. Sullivan combines anthropological and archaeological data with historical sources to produce an extraordinary synthesis of South American indigenous religions before and after European contact, which is a potentially valuable source of interpretative data on ancient religious beliefs for archaeologists.

For Mayan research combining archaeology, astronomy, and glyphs, see Harvey M. Bricker and Victoria R. Bricker, "Astronomical References in the Throne Inscription of the Palace of the Governor at Uxmal," *Cambridge Archaeological Journal* 6, no. 2 (1966):191–229.

CHAPTER 2: DARK CAVES, OBSCURE VISIONS

The literature on late Ice Age art is enormous, but the following general summaries can be recommended: Paul Bahn and Jean Vertut, *Images of the Ice Age* (New York: Viking, 1988), a primary source for this chapter; Paolo Grasiosi, *Palaeolithic Art* (New York: Abrams, 1960); and André Leroi-Gourhan, *Treasures of Palaeolithic Art* (New York: Abrams, 1965). For a recent review of research, see Margaret W. Conkey, "New Approaches in the Search for Symbolic Meaning? A Review of Research in 'Paleolithic Art,' " *Journal of Field Archaeology* 14 (1987):413–430.

Sources on specific sites include the following. The discovery of Grotte de Chauvet: Jean-Marie Chauvet, Eliette Brunel Deschamps, and Christian Hillaire, *Dawn of Art: The Chauvet Cave* (New York: Abrams, 1996). Lascaux: A. Leroi-Gourhan and J. Allain, *Lascaux Inconnu* (Paris: Centre de Researches National Scientifique, 1979). Niaux: Jean Clottes, *Les Cavernes de Niaux: Art Prehistorique en Ariège* (Paris: Seuil, 1995). Jean Clottes, "Paint Analysis from Several Magdalenian Caves in the Ariège Region of France," *Journal of Archaeological Science* 20 (1993):223–235, covers more specialized ground.

Jean Clottes and David Lewis-Williams, *Les Chamanes de la Prehistorie: Transe et Magic Dans les Grottes Ornées* (Paris: Seuil, 1997) describes the role of shamans in Cro-Magnon art, using Lewis-Williams's San research as a guide.

CHAPTER 3: SAN ARTISTS IN SOUTHERN AFRICA

For information on Zimbabwe rock art, see Peter Garlake, *The Hunter's Vision: The Prehistoric Art of Zimbabwe* (Seattle: University of Washington Press, 1995). Richard Lee, *The !Kung San* (Cambridge: Cambridge University Press, 1979) is the authoritative account of these modern-day hunter-gatherers. David Lewis-Williams's research into San rock art is summarized in his *Believing and Seeing: Symbolic Meanings in Southern San Rock Paintings* (New York: Academic Press, 1981), and can also be found in an invaluable summary paper: "Modelling the Production and Consumption of Rock Art," *South African Archaeological Bulletin* 50, no. 162 (1995):143–154, which includes a comprehensive bibliography.

Altered states of consciousness are discussed by J. D. Lewis-Williams, *Reality and Non-Reality in San Rock Art* (Raymond Dart Lecture, Johannesburg, South Africa: University of the Witwatersrand Press, 1988). Wilhelm Bleek and Lucy Lloyd's research into San mythology is largely unpublished, but see Wilhelm Bleek and L. C. Lloyd, *Specimens of Bushman Folklore* (London: George Allen, 1911). Patricia Vinnecombe, *People of the Eland* (Pietermaritzburg, South Africa: Natal University Press, 1976) is a classic monograph on the Drakensberg paintings. Lawrence Sullivan, *Icanchu's Drum*, already cited in the chapter 1 list, is a mine of information on hallucinogens and shamans.

CHAPTER 4: FERTILITY AND DEATH

This chapter is concerned with scientific literature on early farming religion. For this reason, the sources are confined to serious archaeological works that deal with vital issues of context and scientific excavation rather than widely speculative scenarios of great Mother Goddess cults. Lynn Meskell, "Goddesses, Gimbutas, and 'New Age' Archaeology," *Antiquity* 69 (1995):74–86 gives an admirable summary of the literature and controversies. See also Ronald Hutton, "The Neolithic Great Goddess," *Antiquity* 77, no. 271 (1997):91–99.

Sources on specific sites include the following. Jericho: Kathleen Kenyon, *Digging Up Jericho* (London: Michael Joseph, 1957). Çatalhöyük: James Mellaart, *Catal Huyuk: A Neolithic Town in Anatolia* (London: Thames and Hudson, 1967). This book is a widely read summary, but is somewhat outdated. The new, long-term excavations are still at an early stage. The surface survey, however, is covered by Ian Hodder, ed., *On the Surface: Çatalhöyük 1993–95* (Cambridge, England: McDonald Institute for Archaeological Research and the British Institute of Archaeology at Ankara, 1997), which is an exemplary preliminary monograph.

Peter J. Ucko, *Anthropomorphic Figurines* (London: Andrew Szmidla, 1968) is a useful general survey of prehistoric figurines in the Near East. Anthony Bonanno, *Archaeology and Fertility Cult in the Ancient Mediterranean* (Amsterdam: B. R. Gruner, 1985) contains some valuable essays on figurines and ancient Mediterranean religion, with a Maltese slant. Lauren E. Talalay, "Rethinking the Function of Clay Figurine Legs from Neolithic Greece: An Argument by Analogy," *American Journal of Archaeology* 91 (1987):161–169 covers the issue of context very nicely.

CHAPTER 5: POWER AND THE ANCESTORS

The research for this chapter involved an intensive course in anthropological case studies. Three books were especially useful, foremost among them David Lan's classic study of spirit mediums and guerrillas in the Zambezi Valley: *Guns and Rain* (Cambridge: Cambridge University Press, 1985). His chapters on ancestors, rain making, and spirit mediums are the foundation for this chapter. Elizabeth Colson, *The Social Organization of the Gwembe Tonga* (Manchester: Manchester University Press, 1960) is another classic, this time about a north bank group.

Once again, Lawrence Sullivan, *Icanchu's Drum* (New York: Free Press, 1989) is essential reading for anyone interested in non-Western beliefs and cosmology.

CHAPTER 6: AVEBURY: LANDSCAPES OF THE ANCESTORS

Avebury is less well known to the general public than its famous neighbor Stonehenge, although this is changing as bus tours are being diverted from crowded Stonehenge to this less-developed, and in my mind, more spectacular alternative. Caroline Malone, *Avebury* (London: English Heritage, 1989) gives a sound basic description, for much of the technical literature is relatively inaccessible to the layperson. The bibliography is invaluable.

Other useful references include Aubrey Burl, *Prehistoric Avebury* (New Haven, Conn.: Yale University Press, 1979); and I. F. Smith, *Windmill Hill and Avebury* (Oxford: Clarendon Press, 1965).

The technical literature proliferates yearly. A recent summary, which I found invaluable, appears in Alisdair Whittle, "The Neolithic of the Avebury Area: Sequence, Environment, Settlement, and Monuments." *Oxford Journal of Archaeology* 12, no. 1 (1993):29–53.

A great source for information on West Kennet is Stuart Piggott, *The West Kennet Long Barrow: Excavations 1955–6,* Ministry of Works Archaeological Reports, Her Majesty's Stationary Office, London, 4 (1963). A. W. R. Whittle and J. Thomas, "Anatomy of a Tomb—West Kennet Revisited," *Oxford Journal of Archaeology* 5 (1986):27–56 updates the original excavations.

A. Whittle, A. J. Rouse, and J. G. Evans, "A Neolithic Downland Monument in Its Environment: Excavations at the Easton Down Long Barrow, Bishops Canning, North Wiltshire," *Proceedings of the Prehistoric Society* 59 (1993): 197–239 describes these important excavations and demonstrates state-of-the art environmental reconstruction under earthworks.

No one should miss Stuart Piggott, *William Stukeley: An Eighteenth Century Antiquary,* 2d ed., (London: Thames and Hudson, 1985).

CHAPTER 7: STONEHENGE AND THE IDEA OF TIME

Anthony Aveni, *Empires of Time* (New York: Basic Books, 1989) writes persuasively about the idea of time and places Hesiod's work in historical context. Apostolos Athanassakis, *Hesiod: Theogony, Works and Days, Shield* (Baltimore: Johns Hopkins University, 1983) is an authoritative translation and analysis of Hesiod's *Works and Days* for a contemporary audience.

Christopher Chippindale's charming and yet definitive *Stonehenge Complete,* 2d ed., (London: Thames and Hudson, 1994) is the best starting point on Stonehenge.

The ultimate source for specialists is a monumental summary of all excavations in the twentieth century: Rosamund M. J. Cleal, K. E. Walker, and R. Montague, *Stonehenge in Its Landscape: Twentieth Century Excavations* (London: English Heritage, 1995). Quite simply, it is magisterial. Julian Richards, *The Stonehenge Environs Project* (London: English Heritage, 1990) is an authoritative account of the Stonehenge landscape. Both of these monographs contain comprehensive references.

CHAPTER 8: TWO LIVINGS: AGRICULTURE AND RELIGION

Southwestern archaeology has a complex literature, dating back more than a century. Linda Cordell, *Prehistory of the Southwest* (Orlando, Fla.: Academic Press, 1984) is a standard work, but somewhat outdated. Steven Plog, *Ancient Peoples of the American Southwest* (London and New York: Thames and Hudson, 1997) is an admirable introduction incorporating the latest research.

The astroarchaeological literature is often technical and sometimes obscure. Two key references are Michael Zeilik, "The Ethnoastronomy of the Historic Pueblos, I: Calendrical Sun Watching," *Astroarchaeology* 8(1985):S1–24, and

Michael Zeilik, "The Ethnoarchaeology of the Historic Pueblos, I: Moon Watching," *Astroarchaeology* 10 (1986):S1–22. Both are useful on Pueblo Indian ecological chronologies.

Peter Nabokov, *Indian Running* (Santa Barbara, Calif.: Capra Press, 1991) places modern Pueblo Indian running in a mythical context.

Frank Cushing is best visited through his own writings. However, Jesse Green, ed., *Zuni. Selected Writings of Frank Hamilton Cushing* (Lincoln, Neb.: University of Nebraska Press, 1979) is a useful starting point.

For information on Chaco Canyon, a good summary is S. H. Lekson, et al., "The Chaco Canyon Community," *Scientific American* 259, no. 1 (1989):100–109. See also S. H. Lekson, "Great House Architecture of Chaco Canyon," *Archaeology* 40, no. 3 (1987):22–29, and S. H. Lekson, "Rewriting Southwestern Prehistory," *Archaeology* 50, no. 1 (1997):52–55.

CHAPTER 9: THE MOUNDBUILDERS OF EASTERN NORTH AMERICA

For the history of moundbuilder controversies, see Robert Silverberg, *The Myth of the Moundbuilders* (New York: New York Graphic Society, 1968).

The classic source on Adena is William S. Webb and C. E. Snow, *The Adena People* (Lexington, Ky.: Publications of the Department of Anthropology and Archaeology, University of Kentucky, 1945).

David S. Brose and N'omi Greber, eds., *Hopewell Archaeology: The Chillicothe Conference* (Kent, Ohio: Kent State University Press, 1979) is a fundamental source. For information on the Hopewell site itself see N. Greber and K. C. Ruhl, *The Hopewell Site* (Boulder, Colo.: Westview Press, 1989).

A closely argued account of Chachi ceremonial centers is Warren DeBoer, "Ceremonial Centers from the Cayapas (Esmeraldas, Ecuador) to Chillicothe (Ohio, USA)," *Cambridge Archaeological Journal* 7, no. 1 (1997):1–15. A popular account of Chachi ceremonial centers is Warren DeBoer and J. H. Blitz, "Ceremonial Centers of the Chachi," *Expedition* 33 (1991):53–62.

Ray Hively and Robert Horn's two papers on Hopewell geometry, at Newark, Ohio, "Geometry and Astronomy in Prehistoric Ohio," *Archaeoastronomy* 4 (1982):S1–20, and at High Bank, near Chillicothe, Ohio, "Hopewellian Geometry and Astronomy at High Bank," *Astroarchaeology* 7 (1984):S85–100, are useful sources on Moundbuilder astronomy. Another useful reference is J. Pacheco, ed., *A View from the Core: A Synthesis of Ohio Hopewell Archaeology* (Columbus, Ohio: Ohio Archaeological Council, 1986).

For general information on Southeastern Indians see Charles Hudson, *The Southeastern Indians.* (Knoxville, Ky.: University of Tennessee Press, 1976).

Thomas R. Pauketat and Thomas E. Emerson, eds., *Cahokia: Domination and Ideology in the Mississippian World* (Lincoln, Nebr.: University of Nebraska Press, 1997) is an up-to-date and provocative synthesis of Cahokia and the Mississippian in general.

Vincent Steponaitis, *Ceramics, Chronology, and Community Patterns: An Archaeological Study at Moundville* (New York: Academic Press, 1983) and Christopher Peebles, *Excavations at Moundville 1905–1951* (Ann Arbor, Mich.: Museum of Anthropology, University of Michigan, 1979) cover Moundville.

For information on Etowah, see Lewis Larson, "The Etowah Site," *The Southeastern Ceremonial Complex: Artifacts and Analysis*, edited by Patricia Galloway, 58–67 (Lincoln, Nebr.: University of Nebraska Press, 1993). In fact, the entire book is an authoritative collection of essays on the subject.

Lawrence Sullivan's edited *Native American Religions: North America* (New York: Macmillan, 1989) offers a good summary of North American religious beliefs. Patricia Galloway, *Choctaw Genesis 1500–1700* (Lincoln, Nebr.: University of Nebraska Press, 1995) is an exemplary monograph of vital importance to these chapters.

CHAPTER 10: THE BULL BENEATH THE EARTH

Arthur Evans's original monograph, *The Mycenaean Tree and Pillar Cult* (London: Macmillan, 1901) is outdated, but still a comprehensive statement on early Aegean religion. Nanno Marinatos, *Minoan Religion* (Columbia, S.C.: University of South Carolina Press, 1993) is dedicated to Arthur Evans. This volume is a definitive modern statement on the subject, using archaeology and other sources. Peter Warren, *Minoan Religion As Ritual Action* (Gothenburg, Sweden: Paul Astroms Forlag, 1988) is a useful summary.

CHAPTER 11: A SHRINE AT PHYLAKOPI

The Cycladic figurines are controversial as scientific artifacts, since some archaeologists think it is unethical for professional scholars to collaborate with collectors who have acquired looted specimens. Whatever one's views on the controversy, Colin Renfrew's *The Cycladic Spirit* (London: Thames and Hudson, 1991) describes this extraordinary art tradition with fluent authority. Phylakopi is described definitively by Colin Renfrew, *The Archaeology of Cult: The Sanctuary at Phylakopi* (London: British School of Archaeology at Athens and Thames and Hudson, 1985). This exemplary volume includes a comprehensive bibliography.

CHAPTER 12: DIVINE KINGS ALONG THE NILE

Barry Kemp, *Ancient Egypt: Anatomy of a Civilization* (London: Routledge, 1989) is a magisterial analysis of Egyptian civilization based on archaeology, art, architecture, and contemporary literature. This definitive book provided the framework for this chapter. Cyril Aldred, *The Egyptians,* 2d ed., (London: Thames and Hudson, 1986) is a good popular account. Byron E. Shafer, *Religion in Ancient Egypt* (Ithaca, N.Y.: Cornell University Press, 1991) provides valuable insights, as does S. Morenz, *Egyptian Religion* (London: Routledge, 1985).

Bruce Trigger, B. J. Kemp, D. O'Connor, and A. B. Lloyd, *Ancient Egypt: A Social History* (Cambridge: Cambridge University Press, 1983) is a useful summary of the development of civilization along the Nile. Michael Hoffman, *Egypt before the Pharaohs* (New York: Knopf, 1979) and Michael Hoffman and Barbara Adams's monograph *The Predynastic of Hierankonpolis* (Cairo: Egyptian Studies Association, 1982) provide essential background on the predynastic period. Karl Butzer, *Early Hydraulic Civilization in Egypt* (Chicago: University of Chicago Press, 1976) has a wealth of information on the beginnings of irrigation agriculture along the Nile.

I. E. S. Edwards, *The Pyramids* (New York: Viking, 1978) is the classic account, which is updated by Mark Lerner's beautifully illustrated *The Complete Pyramids* (London: Thames and Hudson, 1997). Jamieson Hurry, *Imhotep* (New York: AMS Press, 1978) is a long-established biography.

CHAPTER 13: XUNANTUNICH: "THE MAIDEN OF THE ROCK"

The literature on Mayan civilization is enormous. Michael Coe, *The Maya,* 5th ed. (New York: Thames and Hudson, 1994) is the best general account. Linda Schele and David Freidel, *A Forest of Kings* (New York: William Morrow, 1990) and *Maya Cosmos* (New York: William Morrow, 1993) are flamboyant descriptions of the Mayan world as seen through glyphs and archaeology. These books are written for the general reader but are considered controversial by some critics. The notes alone are worth the price of admission.

Michael Coe, *Breaking the Maya Code* (New York: Thames and Hudson, 1992) tells the story of decipherment, while Dennis Tedlock, *Popol Vuh* (New York: Touchstone Books, 1996) is the definitive and masterly analytical translation of this most important Mayan work.

The archaeology of Xunantunich appears in many specialist papers, but Richard Leventhal, ed., *Xunantunich Archaeological Project. Annual Reports on the 1993 to 1995 Seasons* (Los Angeles: UCLA Institute of Archaeology, 1993–5) is the best.

REFERENCES

Anderson, Arthur J. O., and Charles E. Dibble. 1954, 1969 (12 vols, 1950–1969). *The Florentine Codex: General History of the Things of New Spain.* Salt Lake City, Utah: University of Utah Press.

———. 1978. *The War of Conquest: How It Was Waged Here in Mexico.* Salt Lake City, Utah: University of Utah Press.

Athanassakis, Apostolos. 1983. *Hesiod: Theogony, Works and Days, Shield.* Baltimore: Johns Hopkins University Press.

Aveni, Anthony. 1989. *Empires of Time.* New York: Basic Books.

Bailey, Cyril. 1947. *De Rerum Natura.* Oxford: Clarendon Press.

Bleek, Wilhelm. 1862. *A Comparative Grammar of South African Languages.* London: Trubner.

Bleek, Wilhelm, and Lucy Lloyd. 1911. *Specimens of Bushmen Folklore.* London: Trubner.

Chaucer, Geoffrey, 1962. *The Canterbury Tales.* Edited by J. Coghill. Baltimore: Pelican Books.

Chauvet, Jean-Marie, Eliette Brunel Deschamps, and Christian Hillaire. 1996. *Dawn of Art: The Chauvet Cave.* New York: Thames and Hudson.

Chippindale, Christopher. 1994. *Stonehenge Complete.* 2nd ed. London: Thames and Hudson.

Cleal, R. M. J., K. E. Walker, and R. Montague. 1995. *Stonehenge in Its Landscape: Twentieth Century Excavations.* London: English Heritage.

Clendinnen, Inga. 1991. *Aztecs.* Cambridge: Cambridge University Press.

Cortés, Hernán. 1962. *Five Letters to the King of Spain, 1519–1526.* Translated by J. Bayard Morris. New York: W. W. Norton.

DeBoer, Warren. 1997. "Ceremonial Centers from the Cayapas (Esmeraldas, Ecuador) to Chillicothe (Ohio, USA)." *Cambridge Archaeological Journal* 7(1):1–15.

Diaz, Bernal. 1963. *The Conquest of New Spain.* Translated by J. M. Cohen. Baltimore: Pelican Books.

Duran, Diego. 1964. *The Aztecs: The History of the Indies of New Spain.* Translated by Doris Heyden and Fernando Horcasitas. Norman, Okla.: University of Oklahoma Press.

———. 1971. *Books of the Gods and Rites* and *The Ancient Calendar.* Translated by Doris Heyden and Fernando Horcasitas. Norman, Okla.: University of Oklahoma Press.

Eliade, Mircea. 1959. *The Myth of the Eternal Return.* Chicago: University of Chicago Press.

Evans, Joan. 1943. *Time and Chance.* London: Longmans.

Flannery, Kent V., and Joyce Marcus. 1993. "Cognitive Archaeology." *Cambridge Archaeological Journal* 3 (2):260–267.

Gimbutas, Marija. 1991. *The Civilization of the Goddess: The World of Old Europe.* San Francisco, Calif.: HarperCollins.

Green, Jesse, ed. 1979. *Zuni: Selected Writings of Frank Hamilton Cushing.* Lincoln, Nebr.: University of Nebraska Press.

Hawkins, Gerald. 1965. *Stonehenge Decoded.* Boston: Boston University Press.

Homer. 1990. *Homer: The Iliad.* Translated by Robert Fagles. New York: Viking Penguin.

Jefferson, Thomas. 1794. *Notes on the State of Virginia.* Philadelphia, Pa.: Matthew Carey.

Kemp, Barry. 1989. *Ancient Egypt: Anatomy of a Civilization.* London: Rutledge.

Kenyon, Kathleen. 1957. *Digging Up Jericho.* New York: Frederick Praeger.

Lan, David. 1985. *Guns and Rain.* Berkeley: University of California Press.

Layard, Austen Henry. 1849. *Nineveh and Its Remains.* London: John Murray.

Leon-Portilla, Miguel. 1963. *Aztec Thought and Culture: A Study of the Ancient Nahuatl Mind.* Translated by Jack Emory Davis. Norman, Okla.: University of Oklahoma Press.

Leroi-Gourhan, André. 1964. *Treasures of Prehistoric Art.* New York: Abrams.

Lewis-Williams, David. 1981. *Seeing and Believing.* New York: Academic Press.

Lichtheim, Miriam. 1973. *Ancient Egyptian Literature: A Book of Readings.* 3 vols. Berkeley: University of California Press.

Malone, Caroline. 1989. *Avebury.* London: English Heritage.

Matthews, Cornelius. 1970 (reprint of 1839 publication). *Behemoth: A Legend of the Mound-Builders.* New York: Garrett Press.

Petrie, Flinders. 1889. Sequences in Prehistoric Remains." *Journal of the Royal Anthropological Institute* 29:295–301.

Prescott, William. 1843. *The Conquest of Mexico.* New York: Harpers.

Renault, Mary. 1958. *The King Must Die.* New York: Pantheon Books.

Renfrew, Colin. 1985. *The Archaeology of Cult: The Sanctuary at Phylakopi.* London: British School of Archaeology at Athens and Thames and Hudson.

Schele, Linda, and David Freidel. 1990. *A Forest of Kings.* New York: William Morrow.

———. 1993. *Maya Cosmos.* New York: William Morrow.

Sethe, K. 1905. *Urkunden der 18. Dynastie.* Leipzig: Deitung.

Silverberg, Robert. 1968. *The Myth of the Moundbuilders.* New York: New York Graphic Society.

Squier, E. W., and E. Davis. 1848. *Ancient Monuments of the Mississippi Valley.* Washington, D.C.: Smithsonian Institution.

Stephens, John Lloyd. 1841. *Incidents of Travel in Central America, Chiapas and Yucatan.* New York: Harpers.

Stow, George. 1874. Account of an Interview with a Tribe of Bushman in South Africa." *Journal of the Royal Anthropological Institute* 3:244–247.

Sullivan, Lawrence. 1989. *Icanchu's Drum.* New York: Free Press.

Swan, Brian., ed. 1994. *Coming to Light.* New York: Random House.

Tedlock, Dennis. 1996. *Popol Vuh.* New York: Simon and Schuster.

Thomas, Cyrus. 1894. *Report of the Mound Explorations of the Bureau of Ethnology.* Washington, D.C.: Bureau of American Ethnology.

Thompson, J. E. S. 1971. *Maya Hieroglyphic Writing: An Introduction.* Norman, Okla.: University of Oklahoma Press.

Townsend, Richard. 1992. *The Aztecs.* London and New York: Thames and Hudson.

Zeilik, Michael. 1985. "The Ethnoastronomy of the Historic Pueblos, I: Calendrical Sun Watching." *Archaeoastronomy* 8:S1–24.

INDEX